T0173784

# Selection of Engineering Materials and Adhesives

# MECHANICAL ENGINEERING
## A Series of Textbooks and Reference Books

*Founding Editor*

## L. L. Faulkner

*Columbus Division, Battelle Memorial Institute
and Department of Mechanical Engineering
The Ohio State University
Columbus, Ohio*

# Selection of Engineering Materials and Adhesives

## Lawrence W. Fisher, P.E.

**CRC Press**
Taylor & Francis Group
Boca Raton London New York

CRC Press is an imprint of the
Taylor & Francis Group, an **informa** business
A TAYLOR & FRANCIS BOOK

CRC Press
Taylor & Francis Group
6000 Broken Sound Parkway NW, Suite 300
Boca Raton, FL 33487-2742

First issued in paperback 2019

© 2005 by Taylor & Francis Group, LLC
CRC Press is an imprint of Taylor & Francis Group, an Informa business

No claim to original U.S. Government works

ISBN-13: 978-0-8247-4047-4 (hbk)
ISBN-13: 978-0-367-39300-7 (pbk)

Library of Congress Card Number 2004058288

This book contains information obtained from authentic and highly regarded sources. Reasonable efforts have been made to publish reliable data and information, but the author and publisher cannot assume responsibility for the validity of all materials or the consequences of their use. The authors and publishers have attempted to trace the copyright holders of all material reproduced in this publication and apologize to copyright holders if permission to publish in this form has not been obtained. If any copyright material has not been acknowledged please write and let us know so we may rectify in any future reprint.

Except as permitted under U.S. Copyright Law, no part of this book may be reprinted, reproduced, transmitted, or utilized in any form by any electronic, mechanical, or other means, now known or hereafter invented, including photocopying, microfilming, and recording, or in any information storage or retrieval system, without written permission from the publishers.

For permission to photocopy or use material electronically from this work, please access www.copyright. com (http://www.copyright.com/) or contact the Copyright Clearance Center, Inc. (CCC), 222 Rosewood Drive, Danvers, MA 01923, 978-750-8400. CCC is a not-for-profit organization that provides licenses and registration for a variety of users. For organizations that have been granted a photocopy license by the CCC, a separate system of payment has been arranged.

**Trademark Notice:** Product or corporate names may be trademarks or registered trademarks, and are used only for identification and explanation without intent to infringe.

### Library of Congress Cataloging-in-Publication Data

Fisher, Lawrence W.
    Selection of engineering materials and adhesive / [Lawrence W. Fisher]
          p. cm. -- (Mechanical engineering ; 155)
    Includes bibliographical references and index.
    ISBN 0-8247-4047-5
    1. Strength of materials. [1. Adhesive.] I. Title. II. Mechanical engineering (Marcel Dekker, Inc.) ;
155.

TA405.F4975 2004
620.1'1--dc22                                                        2004058288

**Visit the Taylor & Francis Web site at
http://www.taylorandfrancis.com**

**and the CRC Press Web site at
http://www.crcpress.com**

# Preface

The ability to select and apply the appropriate material or adhesive is as important to the success of a design as the concept itself. The most creative ideas often require equally creative material formulations and application. Electromechanical products that fail prematurely in a mechanical fashion often do so as the result of poor material selection and implementation rather than defective materials. The rapid product development cycles of today's economy necessitate the ability to properly select the correct materials and adhesives quickly, the first time itself, leaving little room for error or time for redesign.

Why are poor material choices made? There is no single answer to this question, but there is little doubt the reason lies somewhere between demanding market requirements and engineering education. Time to market and cost considerations often pressure engineers into making decisions without all the desired facts and verification necessary prior to product release, resulting in unanticipated performance. Many successful designs are often developed with only 60–80% of the design requirements identified for the application. For most engineers this translates into an unacceptably low level of confidence in the design, leaving too much uncertainty over the resulting performance. Unfortunately, most engineers operate in an environment that necessitates this approach because the first to market is rewarded by greater market share and higher profits.

The need to prepare engineers for this environment by providing them with the necessary tools to be successful is paramount. Typical undergraduate mechanical engineering programs provide at least one course in materials and several involving their use in conjunction with engineering analysis–related topics. Upon graduation, design engineers need to be well–informed regarding the range of materials and adhesives available to them and how to select them if they expect to be successful. Unfortunately, this knowledge base is often entirely learned on the job through formal or informal mentoring and without the aid of a reasonably brief, concise reference course book to guide them. The objective of this book is to fill that void.

This book has been arranged by material family and includes a section devoted to adhesives. It is not the intent of this book to be all-encompassing in any of the areas presented, but it is intended to provide the engineer with relevant details regarding the broad base of materials and adhesives available. It is also expected that the information here will provide a foundation suitable to support further investigation into the desired level of detail required by practicing engineers to successfully complete their design work.

Ultimately, material selection is driven and controlled by a design engineer's ability to research and obtain the most appropriate material for an application, after taking into consideration both the engineering and business climate for which it is to be used. Periodic assessment and reselection of the material is not uncommon and is often the best method of learning what is successful. The content of this text is intended to support the design engineer in this activity.

Lawrence W. Fisher, P.E.

June 2004
Torrance, California

# Contents

# 1

# Introduction to Material Selection

Selecting the appropriate material is an integral part of the successful implementation of an engineer's design. To begin with, the functional requirements of the design must be carefully considered if the appropriate material is to be identified. A design engineer's ability to objectively quantify the combined marketing, technical, and manufacturing requirements as they apply to the material selection is critical to the actual as well as the perceived success of the product. Further, most products are composites of numerous materials that must work in harmony when performing their intended function. This complication necessitates balancing the material needs with the product requirements or expectations while avoiding excessive costs.

Consider a product fabricated from a plastic material. The material may have been selected because of its light weight and cosmetic appearance, without taking into consideration the necessary environmental requirements. Although plastic may seem a natural choice for an all-weather application, there are many issues to consider regarding its weatherability. These include its response to ultraviolet (UV) exposure, capacity for water absorption, and ambient temperature characteristics, to name but a few. This type of oversimplification of the selection process is one that has severely affected the use of plastics in engineering applications and continues to have a lasting effect that influences their selection. Engineered plastics have come a long way in providing the desired properties necessary for their expanding role in product design, with many resin and net shape providers offering to formulate special materials to meet a specific application (Figure 1.1).

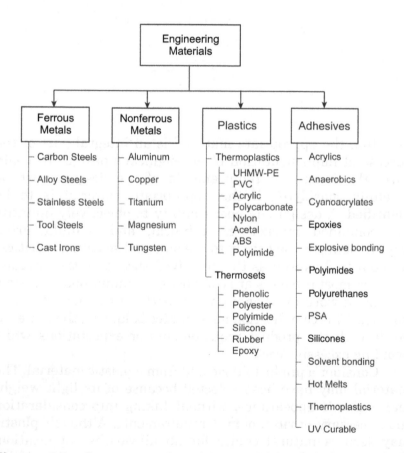

**Figure 1.1**   Common engineering materials and their specific material types.

The material selection process is an integral part of the overall product development process and must be considered in the early phases of the product definition. The materials chosen for the fabrication of a product, whether it be a simple infant's toy or an automobile, can have a profound impact on the success of the design due to the material performance. Casual or improper selection techniques can lead to the use of materials with inadequate properties for the intended application, with the unintended result being a lasting effect on the product's reputation and ultimately its sales and thus company profit. Equally important are the related material, manufacturing and service costs, and lead times. Basing a selection purely on material costs and not considering the effects on these functions could be a costly mistake. Material selection has important ramifications throughout a product's lifecycle, including the fabrication, assembly, and service stages. For example, if a plastic is to replace an existing ferrous metal part, the entire fabrication process could change from bending, welding, and machining to injection molding. It would also likely alter the assembly process by requiring threaded inserts and spacers instead of the tapped holes.

Just as material considerations are an important aspect of any product, so is the process used to develop the product. Each organization has a unique method of developing products that supports its enterprise methodologies. This process is just one part of the composite structure that is its competitive advantage, providing the company with the ability to bring quality products to market quickly.

Throughout the development process the design engineer needs to reflect on the material selection decisions made and how they affect the production-released product. This chapter will investigate the universal considerations that can be applied in support of the product development process with regard to both materials and adhesives. The final selection must be based on knowledge of the application and the relevant parameters that describe the materials' performance. Selecting the parameters and comparing the candidates in a meaningful fashion is the final emphasis of this chapter.

The development of a new product, from concept to full production, involves the comprehensive collaboration by numerous individuals both inside and outside an organization. Although there is no one correct method that must be followed to accomplish this, the general approach is often the same regardless of the specific company protocol—and, of course, the objective is the same: to design and build a quality product that is successful. To put it simply, the product development process is the methodology an organization employs to develop new products.

Generally, the selection of a material by a design engineer will begin with the definition of the specific properties necessary for the design to succeed. At a minimum, these will often include the following:

- Strength
- Hardness
- Toughness
- Ductility
- Fatigue resistance
- Wear resistance
- Corrosion resistance
- Dimensional stability
- Thermal stability
- Optical properties
- Conductivity
- Resistivity
- Dielectric constant
- Magnetic properties
- Formability
- Machinability
- Weldability

Further selection will involve less technical considerations such as determining the scope of available materials, design life, cost and processing issues, and disposal methods.

Disposal methods continue to attract the attention of government agencies and environmentalists because of their desire to protect the environment and gain efficiencies in recycling. Many design requirements now include methods of disposal or reuse as part of a product's characteristics.

Specifically, the automotive and consumer electronics industries are addressing this issue in their product development cycles by creating designs that facilitate rapid disassembly and by using materials that can be reprocessed without excessive effort.

## DESIGN METHODOLOGY

Successful product development methods vary widely among individual organizations and industries, making it difficult to suggest any one method or process that will produce the desired result: an engineered product that performs as intended and is adequately documented to allow for consistent replication. Entrepreneurial organizations, including "start-ups" and many small businesses, rely on an individual or two having an intimate knowledge of the technical aspects of the product and related industry and being capable of driving the product to market, often with few formalities. Conversely, larger organizations must rely on similar individual contributions but in the context of a review methodology that is commensurate with the industry requirements. A large organization must impose controls to avoid chaos and, most importantly, limit its liability in case of product failure. Although product development methodology cannot absolve a company from this liability it can support the company during litigation, by providing the evidence that industry standard methods and processes were employed to produce a product of the highest possible integrity within reasonable limits.

Two generic sets of terminology and related methods have been developed over the years to support product development, regardless of the industry or product. A word of caution is required up front with regard to how rigorously these methods are applied to a development program. The extent to which the methods are applied should be commensurate with the activity and not applied to comply with a one-size-fits-all methodology. The methods and processes must accommodate the size and scope of the development to avoid wasted effort, which requires that they be documented in such a way as to account for the variation. For example, the rigorous methods necessary for the large-scale development of an aircraft would be inappropriate

**Table 1.1**    Common Design Phase Workflow and Related Acronyms Most Often Found When Working with the U.S. Government and Its Agencies

| Design Phase | Description | Objective |
|---|---|---|
| SRR | System Requirements Review | Obtain confirmation that sufficient analysis has been completed to proceed to a system-level design. |
| SDR | System Design Review | Establish that all requirements documents are in place for detailed-level design work. |
| PDR | Preliminary Design Review | Review of concept-level design work and obtaining of approval to proceed with detailed design work and creation of fabrication drawings. |
| CDR | Critical Design Review | Establish that all design work is complete and related documents are of sufficient quality for fabrication. |
| FDR | Final Design Review | Confirm all design documents are completed and validation was successful Release to production. |

for the development of a coffeemaker. This is not to say that the same elements are not required—they are, but the scope of coordination is significantly reduced. The need for formal documents and distribution systems can be simplified: laboratory notebooks and handwritten meeting minutes or actions items can be used, if they are appropriately retained in support of the required domestic and international compliance regulations.

Industries involved in programs sponsored by the U.S. government and its agencies often are compelled to follow their rigorous development process (Table 1.1). Commercial industry has attempted to adopt a similar methodology, which is captured in ANSI/ASQC D1160-1995 and includes terms similar to those found in government contracting (Table 1.2).

**Table 1.2**  Common Design Phase Workflow and Related Acronyms as Defined by ANSI/ASQC D1160-1995 for Commercial Industry

| Design Phase | Description | Objective |
|---|---|---|
| PDR | Preliminary Design Review | Establish that all requirements documents are in place for detailed-level design work. |
| DDR | Design and Development Review | Confirm the design meets the stated requirements and review the analysis, calculations, and test data that support this. |
| FDR | Final Design Review | Verify the completed design meets the stated requirements, including lifecycle costs, and that all documentation is complete. |
| MDR | Manufacturing Design Review | Establish the producibility and environmental impacts of the design by review of the manufacturing, inspection, and disposal methods. |
| IDR | Installation Design Review | Evaluate all aspects of the transfer of the product to the end use point, including shipping, government approvals, installation methods, and all related documentation. |
| UDR | Use Design Review | Establish the success of the design and its acceptance by users while considering how best to incorporate successful features and development process methods into future design cycles. |

The scopes of both these workflows are similar, and they strive for the same result: a successful design, as defined by their respective markets. The progression of development from concept to production-release is obvious but the need for such a workflow may not be: lack of a system to facilitate communication and documentation control results in duplication or missing information, which is almost always detrimental to the timely completion of the development program. Further study of these methods is outside the scope of this text but the reader is encouraged to pursue a greater understanding of them.

The rigor with which these approaches are applied by engineers in their development projects is either mandated, as in government contracting activities, or defined by experienced management for commercial environments. Although following these methods does not guarantee the success of any given program, it does provide a consistent framework allowing the necessary communication to achieve success. It also facilitates conducting a postmortem review of the program to determine the shortcomings and sources of concern so that they can be openly addressed and corrected for future development activities.

The flow of product development can take many forms, but it generally follows a predictable path regardless of the organization (Figure 1.2). The controls placed on activities and their progression are highly dependent on the organizational needs of the development team and are likely related to the size and cost of the development effort. Smaller, entrepreneurial endeavors are not likely to be highly systematic whereas a large military program will have specific approval points and management oversight.

Ultimately, the results will justify the methods employed to achieve them. The important lesson here is that all the methods that are believed to be necessary to complete a successful product development effort should be documented and repeated in future programs. Not every program will require the same detail but, without a doubt, each should follow the same basic processes and procedures. The degree to which they are employed should then be left to the management and the project engineer.

## SELECTION METHODOLOGY

The ability to select the most appropriate material for a given design requirement is the fundamental challenge faced by the design engineer. A senior level engineer has an extensive knowledge base to draw from when making these selections, ranging from personal experience to knowing the right materials engineer to contact within the specific industry of

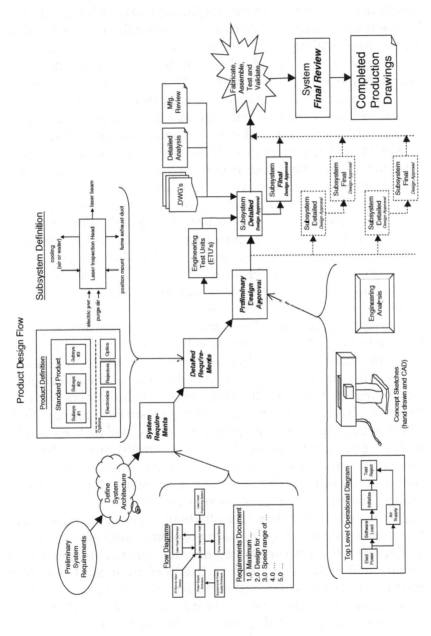

**Figure 1.2** Generic product development flow diagram.

application. Developing this knowledge obviously takes years and certainly includes many decisions that may have been less than optimal at the time, requiring costly redesign efforts. In addition, many industries have specific materials that meet long-standing requirements (local, national, or international codes or standards), immediately reducing the available materials list significantly while making the use of new and possibly more appropriate materials difficult.

How does someone without extensive experience make quantifiable decisions in the area of selection without a senior or expert resource person at their disposal? Application of a selection methodology using material indices for comparative purposes has proven effective in providing the necessary quantitative justification required in making this important decision.

The concept of material indices has existed for many years, possibly being first introduced by the Monsanto Chemical Company as a sales and marketing tool for plastic materials in the 1940s. Monsanto offered tensile strength-to-weight ratios for common engineering materials of the day, including its new plastic formulations. More recently, Dr. David Ashby of Cambridge University has extended this methodology to include not only material selection but also process selection.

The selection of adhesives can be considerably more challenging than the selection of homogeneous metals or plastics owing to their application-specific behavior as well as their sensitivity to users' handling and surface preparation. Their bulk properties are typically well defined as they can be readily quantified under laboratory conditions. Adhesives become most challenging to the design engineer after the selection is made because their successful deployment is highly dependent on process-related activities, over which engineers often feel they have little or no control. This situation can be improved by imposing copious requirements in the engineering drawing or related documentation, but the actual results will only be evident in the performance of the final product. In-process inspection and monitoring can help but, unfortunately, adhesives require destructive inspection techniques to verify their proper application and performance.

Dr. Ashby's material selection methods are best used during the early stages of the design process when one has the most flexibility to adjust the design to accommodate material-specific considerations. With an up-to-date materials database the exercise of exploring the available materials using the indices of interest can be enlightening to the senior engineer and an invaluable tool for the student. The methodology bolsters what may often be a selection based solely on past experience or the use of industry-specific materials most often selected because of their known performance. The hidden danger of this method lies in users' ability to fully define the use environment and apply the appropriate indices for the evaluation. It is often the case that an industry-specific material is used on a regular basis but the actual properties that make it successful are not fully realized by the design engineer. For example, the use of PVC in engineering applications can be successful if it is used judiciously in design configurations that place the material in compression. Environmental effects on PVC in conjunction with its brittle properties need to be closely considered.

Some of the more common material indices include axial tensile, torsion, beam, column, and plate loading cases (Table 1.3). The indices are provided for both specific strength and stiffness for each of the load cases, representing 10 different material indices to choose from, which assists in the material selection during the design phase. The derivation of these and other indices is covered by Ashby in several of his texts, so readers are encouraged to refer to them if they are interested in obtaining other indices. For example, indices involving fracture toughness, thermal expansion, or cost per unit weight might be required, depending on the design objectives.

When making use of Ashby's selection charts, the most generic charts are those representing yield strength versus density (specific weight) and Young's modulus versus density (specific stiffness). These charts are plotted using a log–log scale due to the range of information presented. Material yield strength is a property that is commonly used as a comparative measure between materials in general and specifically within a given material family. The basic nature of this property is the material's resistance to internal movement, the mechanics of which vary between material families. The deformation of metals involve the metallic bond and lattice dislocations;

**Table 1.3**  Combined Properties to Maximize Efficiency or Performance Indices

| Component Shape and Mode of Loading | For Stiffness | For Strength |
|---|---|---|
| *Bar-Tensile Axial Loading* Load, stiffness, length specified, section area variable | $\dfrac{E}{\rho}$ | $\dfrac{\sigma_{ys}}{\rho}$ |
| *Torsion Bar of Tube Torque* stiffness, length specified, section area variable | $\dfrac{G^{1/2}}{\rho}$ | $\dfrac{G_{ys}^{2/3}}{\rho}$ |
| *Beam* Externally loaded or by self-weight in bending; stiffness, length specified, variable section area | $\dfrac{E^{1/2}}{\rho}$ | $\dfrac{\sigma_{ys}^{2/3}}{\rho}$ |
| *Column–Axial Compression* Elastic buckling or plastic compression; compression load and length specified; variable section area | $\dfrac{E^{1/2}}{\rho}$ | $\dfrac{\sigma_c}{\rho}$ |
| *Plate* Externally loaded or by self-weight in bending; stiffness, length, and width are specified, thickness is variable | $\dfrac{E^{1/3}}{\rho}$ | $\dfrac{\sigma_{ys}^{2/3}}{\rho}$ |

plastics involve weaker Van der Waals bonds and polymer chain slippage; ceramics are rigid and brittle because of their stronger covalent and ionic bonds. A review of the strength versus density chart reveals the wide range of engineering materials available while allowing the design engineer to gain a general sense of a material's performance relative to that of other candidates (Figure 1.3). Similarly, the plot of Young's modulus versus density provides an excellent insight into a material's stiffness characteristics relative to its weight, providing a measure of specific stiffness, which can be used to support the selection process when the stiffness efficiency is an important design consideration.

**Figure 1.3** Young's modulus, $E$, plotted against density, $\rho$. This method of data presentation provides graphical regions of like properties for comparison purposes.

*Source*: Reprinted with permission from Materials Selection in Mechanical Design, Michael F. Ashby, p. 37, Fig. 4.3, 1999, with permission from Elsevier.

The required material indices are used in the form of lines plotted on the material property charts. Because the charts are of log–log form, the slope of the line is the inverse of the exponent for the $y$-axis parameter which forms a line of constant value. This guiding line represents the same value for all materials it passes through, with those materials lying above the line exceeding the requirement and those below not meeting the requirement. Additional constraints can be

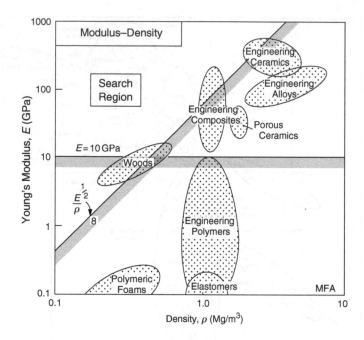

**Figure 1.4** Material selection using the beam stiffness material indice for $E$ versus $\rho$ properties.

*Source*: Reprinted with permission from Materials Selection in Mechanical Design, Michael F. Ashby, p. 82, Fig. 5.11, 1999, with permission from Elsevier.

applied in the form of maximum and minimum values of the plotted parameters, which would be in the form of vertical or horizontal lines, effectively bracketing the solution (Figure 1.4).

For example, a selection based on beam stiffness, $E^{1/2}/\rho$ involves a selection line of slope 2 and this could be plotted for various values of the material index. Based on the requirements of this example, $E^{1/2}/\rho = 8$ is chosen and plotted (Figure 1.4). The resulting line passes through points representing materials that have equal values of this index, meaning the represented materials perform equally well against this index for a light, stiff beam design. To further narrow the selection, a horizontal line restricting the lower limit of the modulus is also plotted. The search area then contains an even shorter list of candidates for review, which can be expanded or contracted by altering the value of the index line to suit the design requirements.

This method can be easily repeated for other material indices using a chart with greater resolution to actually identify specific materials. Further analysis can be performed that involves the consideration of the section shape of the load member and its efficiency in supporting the design loads. This form of analysis is characterized by a review of the section shape and moments (area, bending, torsion, etc.) to determine the most efficient shape for a given material and is covered in great detail by Ashby.

And lastly, a consideration of the structural index can also be included in the optimization of a design and related material selection. Determining these is outside the scope of this text but is no less important to the overall success of a design.

## SELECTION OF METALS

It is often early in the material selection process that the determination of the material family is made. This determination is typically driven by the perceived design requirements but, at times, can lead to the selection of a metallic material, owing to the notion that metals are better than plastics. A careful consideration of all the materials that meet acceptable material index values coupled with experience is needed, and an open mind should prevail.

Metals divide into two basic families: ferrous and nonferrous. Within these families are specific formulations that offer a wide range of material properties and cost. Ferrous metals are typically denser and offer greater overall strength and superior corrosion resistance when used without post-processing. In contrast, nonferrous metals are typically less dense (there are exceptions) and provide greater load-carrying efficiency (specific strength) and excellent corrosion resistance when post-processing is used to seal the material surface. Determining which material is most appropriate is best done after fully defining the design requirements and related processing activities and then applying the criteria as discussed above to develop a candidate list. Further review of these candidates in terms of material availability, processing requirements, and other lifecycle costs often yields a shortlist that provides alternatives and second sources.

The Young modulus for metals ranges from 10 Mpsi for aluminum to near 60 Mpsi for tungsten. The upper useful tensile strength (without becoming too brittle) is 300 ksi. These two basic properties exhibit the range and versatility offered by metals, which, when coupled with their chemical and other mechanical properties, allows them to be used in many applications.

## SELECTION OF PLASTICS

Selection of plastic materials can be considerably more involved than the selection process for metals due to the often overwhelming number of available materials. Although plastics can be grouped into generic categories, there are many possible variants to the base material that possess significantly different properties, bluring the performance characteristics across several material families and resulting in the need for significantly more research to discern the most appropriate candidates. Use of the above-mentioned tools can significantly reduce the selection time while providing a thorough review of the available materials.

Plastics are available in two basic forms: thermoplastics and thermosets. These differ in their response to heat, with thermoplastics exhibiting the ability to be reformed by the application of heat while thermosets, true to their name, remain "set" after their initial processing and do not respond to subsequent applications of heat. Along with this obvious property difference between these two families there are also differences in chemical, physical, and mechanical behavior. As a general rule, thermosets are more expensive than thermoplastics with correspondingly better performance with respect to chemical, physical, and mechanical properties. This tends to be the case also with regard to manufacturing processing characteristics, with thermosets requiring more costly processing techniques and tooling.

Thermoplastics are most often selected for commercial applications because of their lower cost and common processing requirements, whereas thermosets are more often found in applications requiring higher performance. The elastic modulus of plastics ranges from 100 to 1000 ksi (1 Mpsi), with unmodified materials typically not exceeding 500 ksi. The inherent ability of plastics to resist chemical attack often makes them attractive materials for applications in which metals would corrode. In addition, the lower density of plastics offers the advantage of reduced weight for these lower load applications. Dimensional stability in the form of thermal expansion and creep can be a problem for plastics depending on the application. Careful consideration for these properties as well as environmental behavior must be part of the selection process.

## SELECTION OF ADHESIVES

Adhesives have continued to gain importance as engineering materials because of their use in fastening and joining dissimilar materials. They also reduce or eliminate the stress concentration induced by a mechanical fastener, while reducing overall system weight. Adhesives work best when they are subjected to near-constant loading conditions and are often used in conjunction with mechanical fastening of some type; this provides a means to fixture the joint while it is curing and also protects the joint against overloading. Compared with other joining methods, adhesives are much more sensitive to the application process, making them appear to be less robust and more costly. Great care should be taken to document the process used to properly assemble the joint during the development phase so that it can be replicated in production. All too often, highly skilled technicians participate in the early development, performing tasks with excellent workmanship, only to have the design moved on to a production environment where it is found not to be practical to apply the same, or similar, methods meticulously. This leads to a significant reduction in joint performance or failure. Along with creating the needed documentation, the design engineer must follow the design through manufacturing by participating in the manufacturing design review (MDR) and the initial production builds.

There are many families of adhesives, each of which offers a wide range of standard formulations as well as the ability to create custom formulations to address a unique set of design requirements. Epoxies tend to exhibit the greatest flexibility, as alterations to the volume, type of fillers, and cure method are very common. Silicones and pressure-sensitive adhesives (PSAs) are also offered in numerous forms, each of which addresses specific bonding requirements. The remaining choices offer solutions for specific joint configurations and requirements. For example, anaerobic adhesives are commonly used in threadlocker applications to secure fasteners in a specific position when operating in a dynamic environment. Another example is room temperature vulcanized (RTV) silicones, which provide a unique combination of joint-sealing (gap-filling) properties and strength with a relatively flexible adhesive.

Adhesive selection is typically more influenced by joint design and application environment than the materials being joined. Although the material surface can have a significant impact on the overall joint performance, there are numerous surface preparation techniques that improve adhesion. Joints that put the adhesive in shear and allow for the least amount of gap are the most efficient as they minimize the effect of the adhesive material properties on the design. For example, a joint that unnecessarily allows for a thicker bond line will also suffer from greater strain at the joint due to the lower material modulus, which could adversely affect the design.

The load-carrying capability of adhesives varies widely by adhesive, environment, and joint design. Surface preparation and application methods also have significant influence on the behavior of the resulting joint. Epoxy adhesives are typically the most robust and offer the highest lap shear strengths, which can exceed 10 ksi under tightly controlled conditions. Greater overall joint strength and design margin (factor of safety) can be added by simply increasing the area of the joint and by controlling the application techniques to include proper surface preparation, product mixing (if necessary), and joint design.

## UNDERSTANDING DESIGNS AND THEIR BEHAVIOR

The material selection process is only the beginning of the lifecycle of a product, but it is very critical in its future performance. Understanding the impact selection has on a product's future is essential to making a successful selection and requires the design engineer to consider specific details of the implementation. For example, the selection process should include not only consideration of material properties but also the resulting total part cost. This cost is the sum of the mate-rial costs and resulting fabrication and processing necessary to produce the finished part. Fabrication methods for some materials could require costly machining equipment or close tolerances to achieve the desired part geometry. In addition, post-processing may be necessary to achieve the proper corrosion protection or part stability.

Material availability may also be a factor in part cost, at times requiring the selection of a material that is more readily available but has properties that may be somewhat greater or less than desired depending on the engineer's judgment and compensation in other areas of the design. For example, a material of slightly lower strength may be used if it is sized to result in an appropriate stress level. Material availability may also be driven by quality certifications or propriety grades that increase cost and often delivery times.

Production quantities may also influence the selection of materials as there are grades of materials and processes that lend themselves to a range of quantities. Machining methods may be sufficient for low-volume applications whereas injection molding or stamping may be appropriate for higher-volume products.

Failure modes and field service considerations also are important in the selection of materials for engineering applications. These two factors are very much related as it is often the "off-nominal" use that results in product failure and the need for field service. The knowledge that a material will fail in a predictable fashion can be used to the advantage of the designer by allowing for it in the design. For example, if the elongation of a specific component is sufficient to prevent further operation then it can be designed in an appropriate way to prevent further malfunction. Predictable failure modes are related to field service as a way of determining service intervals as well as diagnosis and prevention of catastrophic failure. Field service is also important when considering what it may take to replace the failed component. The need for special tools or tightly controlled processes makes field service an impractical task at best and at other times just impossible. For example, if the materials are selected to be joined by adhesives it may be impractical to repair them in the field or it may require a higher level, more costly assembly to be replaced.

And lastly, the need to allow for controlled disassembly and disposal can place further restrictions on the materials and processes available to the design engineer. These restrictions are most likely to involve surface treatments or materials that do not lend themselves to reprocessing, such as thermoset plastics or adhesives. The environmental and regulatory requirements of recycling will continue to gain influence in the selection of materials as governments and social concerns address the limited nature of our natural resources.

# 2

# Structure of Materials

Materials possess numerous properties and types of behavior that often become evident only during the selection and application process. Metals and polymers have significantly different physical properties that are defined by their atomic structure. To understand these atomic structures it is important to review the concepts of the atom and its components.

## ATOMIC ELEMENTS AND STRUCTURES

The atom and its subatomic particles are a form of energy. Modeling these particles in sufficient detail to describe the observed behavior has challenged scientists and physicists for centuries. In some aspects of the analysis it is desirable to model matter as particles and in other cases it is more helpful to consider it as quanta of energy which obey the laws of wave mechanics rather than those of particle mechanics. For the purpose of this discussion it is sufficient to use the particle representation and this will be the approach throughout the text.

|

**Table 2.1**   Atomic Size Comparison

|  | Electron | Proton | Neutron |
|---|---|---|---|
| Charge (Coulombs) | $-1.69 \times 10^{-19}$ | $1.69 \times 10^{-19}$ | 0 |
| Mass (kg) | $9.1 \times 10^{-31}$ | $1.67 \times 10^{-27}$ | $1.67 \times 10^{-27}$ |
| Radius (cm) | $4.6 \times 10^{-13}$ | $1.4 \times 10^{-13}$ | $1.4 \times 10^{-13}$ |

## The Atom and Subatomic Particles

The atom is made up of electrons, protons, and neutrons, which are referred to as subatomic particles. It is important to note that other subatomic particles exist but these three are sufficient to describe atomic properties as they apply to engineering materials and for the purpose of their selection.

The electron is the unit quantity of electrical energy. It can be considered as a minute particle or as an energy wave of negative charge. The electrical energy we use every day is a flow of electrons each having a charge of $-1.69 \times 10^{-19}$ C. If we think of the electron as a particle its rest mass is $9.1 \times 10^{-31}$ kg with an effective radius of $4.6 \times 10^{-13}$ cm (Table 2.1).

The proton is a heavy particle of matter; it is 1836 times heavier than the electron. It carries a positive charge which is equal to the charge of an electron but opposite in sign. The rest mass of a proton is $1.6726 \times 10^{-27}$ kg with an effective radius of $1.4 \times 10^{-13}$ cm.

The neutron is slightly heavier than the proton and is 1838 times heavier than the electron and has a rest mass of about $1.6749 \times 10^{-27}$ kg. The effective radius of the neutron is almost the same as that of the proton but it is not a charged particle, and its role is to only contribute to the mass of the atom without any effects on bonding. This is seen in isotopes, which are atoms of same element with the same number of protons and electrons but different number of neutrons.

Ernest Rutherford described the atom as a positively charged nucleus consisting of its mass at the center with negatively charged electrons clustered around it. This is more commonly referred to as the planetary model (Figure 2.1).

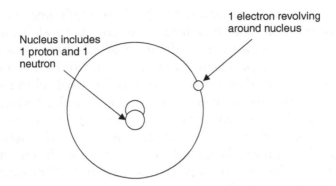

**Figure 2.1**   Planetary model of hydrogen atom.

**Table 2.2**   Periodic Table of the Elements

| IA | | | | | | | | | | | | | | | | | | VIIIA |
|---|---|---|---|---|---|---|---|---|---|---|---|---|---|---|---|---|---|---|
| H | | | | | | | | | | | | | | | | | | H |
| 1 | | | | | | | | | | | | | | | | | | 2 |
| 1.01 | IIA | | | | | | | | | | | IIIA | IVA | VA | VIA | VIIA | | 4.00 |
| Li | Be | | | | | | | | | | | | B | C | N | O | F | Ne |
| 3 | 4 | | | | | | | | | | | | 5 | 6 | 7 | 8 | 9 | 10 |
| 6.94 | 9.01 | | | | | | | | | | | | 10.81 | 12.01 | 14.01 | 16.00 | 19.00 | 20.18 |
| Na | Mg | | | | | | | | | | | | Al | Si | P | S | Cl | Ar |
| 11 | 12 | | | | | | | | | | | | 13 | 14 | 15 | 16 | 17 | 18 |
| 22.99 | 24.31 | IIIB | IVB | VB | VIB | VIIB | VIIIB | VIIIB | VIIIB | IB | IIB | | 26.98 | 28.09 | 30.97 | 32.07 | 35.45 | 39.95 |
| K | Ca | Sc | V | Ti | Cr | Mn | Fe | Co | Ni | Cu | Zn | Ga | Ge | As | Se | Br | Kr |
| 19 | 20 | 21 | 23 | 22 | 24 | 25 | 26 | 27 | 28 | 29 | 30 | 31 | 32 | 33 | 34 | 35 | 36 |
| 39.10 | 40.08 | 44.96 | 50.94 | 47.88 | 52.00 | 54.94 | 55.85 | 58.93 | 58.69 | 63.55 | 65.39 | 69.72 | 72.61 | 74.92 | 78.96 | 79.90 | 83.80 |
| Rb | Sr | Y | Zr | Nb | Mo | Tc | Ru | Rh | Pd | Ag | Cd | In | Sn | Sb | Te | I | Xe |
| 37 | 38 | 39 | 40 | 41 | 42 | 43 | 44 | 45 | 46 | 47 | 48 | 49 | 50 | 51 | 52 | 53 | 54 |
| 85.47 | 87.62 | 88.91 | 91.22 | 92.91 | 95.94 | (97.9) | 101.07 | 102.91 | 106.42 | 107.87 | 112.41 | 114.82 | 118.71 | 121.76 | 127.60 | 126.90 | 131.29 |
| Cs | Ba | La | Hf | Ta | W | Re | Os | Ir | Pt | Au | Hg | Tl | Pb | Bi | Po | At | Rn |
| 55 | 56 | 57 | 72 | 73 | 74 | 75 | 76 | 77 | 78 | 79 | 80 | 81 | 82 | 83 | 84 | 85 | 86 |
| 132.91 | 137.33 | 138.91 | 178.49 | 180.95 | 183.85 | 186.21 | 190.2 | 192.22 | 195.08 | 197.97 | 200.59 | 204.38 | 207.2 | 208.98 | (209) | (210) | (222) |
| Fe | Ra | Ac | Rf | Db | Sg | Bh | Hs | Mt | | | | | | | | | |
| 87 | 88 | 89 | 104 | 105 | 106 | 107 | 108 | 109 | | | | | | | | | |
| 223.02 | 226.03 | 227.03 | (261) | (262) | (263) | (262) | (265) | (265) | | | | | | | | | |

| Ce | Pr | Nd | Pm | Sm | Eu | Gd | Tb | Dy | Ho | Er | Tm | Yb | Lu |
|---|---|---|---|---|---|---|---|---|---|---|---|---|---|
| 58 | 59 | 60 | 61 | 62 | 63 | 64 | 65 | 66 | 67 | 68 | 69 | 70 | 71 |
| 140.12 | 140.91 | 144.24 | (145) | 150.36 | 152.97 | 157.25 | 158.93 | 152.50 | 154.93 | 157.26 | 168.93 | 173.04 | 174.97 |
| Th | Pa | U | NP | Pu | Am | Cm | Bk | Cf | Es | Fm | Md | No | Lr |
| 90 | 91 | 92 | 93 | 94 | 95 | 96 | 97 | 98 | 99 | 100 | 101 | 102 | 103 |
| 232.04 | 231.04 | 238.03 | 237.05 | (240) | 243.06 | (247) | (248) | (251) | 252.08 | 257.10 | (257) | 259.10 | 262.11 |

The number of protons and neutrons in the nucleus is the basis of chemical identification of an atom. The number of protons in the nucleus is known as the *atomic number* $(Z)$. The periodic table is a guide to understanding the order of atomic numbers. It displays the collection of atoms of one type (known as elements) according to increasing atomic number (Table 2.2).

The periodic table begins with the lightest element (1 proton) and ends with the heaviest element (highest number of protons). It provides a classification of all chemical elements along with the appropriate element symbol and sorts them into periods (horizontal rows) and columns (vertical columns). Most chemical elements are metals. These elemental metals are malleable, ductile, and generally denser than other elemental substances. Some of the most common elements used in fabricating engineering materials include iron (Fe), aluminum (Al), and copper (Cu) (for base material) and chromium (Cr), nickel (Ni), zinc (Zn), and lead (Pb).

The mass of an atom is given as *relative atomic mass*. It is the mass in grams of $6.023 \times 10^{23}$ atoms (Avogadro's number) of that element. Usually the mass of the element is located below the atomic symbol in a periodic table. The mass of the carbon atom with six protons and six neutrons is the reference for atomic masses. The *atomic mass unit* is defined as $\frac{1}{12}$ of the mass of this carbon atom.

## The Quantum Numbers

To understand how electrons fill orbits around the nucleus and effect of this arrangement on the chemical properties of an element, it is necessary to understand quantum numbers. Niels Bohr, in 1913, concluded that of all possible orbits, only certain orbits were permissible at any given time. The stable orbits were characterized by the angular momenta of the electrons in the orbits defined by the expression $nh/2\pi$, where $h$ is *Planck's constant* and $n$ can only take integer values ($n = 1, 2, 3$ etc.).

From later developments in atomic theory it became clear that the classical laws of particle dynamics could not be applied to describe particle motion. In classical dynamics, it is a prerequisite that the position and momentum of a particle be known. However, in atomic dynamics only one of them can be determined accurately, leaving the other an uncertain quantity. This is known as Heisenberg's uncertainty principle. The consequence of this principle is that we can no longer think of an electron as moving in a fixed orbit around the nucleus but must consider the motion of the electron in terms of a wave function. This function specifies only the probability of finding one electron having a particular energy in the space around the nucleus. The situation is made more complex by the fact that the electron is not only revolving around the nucleus but also spinning around its own axis. So instead of specifying the motion of an electron by a single integer $n$ as Bohr suggested it is necessary to specify the electron state by four numbers. These numbers are known as the electron quantum numbers, these are $n$, $l$, $m$, and $s$, where $n$ is the principal quantum number, $l$ is the orbital quantum number, $m$ is the magnetic quantum number, and $s$ is the spin quantum number.

With the introduction of the four quantum numbers it is necessary to state the Pauli exclusion principle, formulated by the physicist Wolfgang Pauli in 1924. It states that no two electrons in the same atom can have the same four quantum numbers. Or, to put it simply, no more than two electrons can occupy any one orbital at any instant in time.

The most important quantum number is the principal quantum number because it is responsible for determining the energy of the electron. Electrons having a principal quantum number $n$ can take up all those integral values of orbital quantum number $l$ lying between 0 and $(n-1)$. The orbital quantum number is associated with the angular momentum of the revolving electron and determines what would be regarded in non-quantum terms as shape of the orbit. The remaining two quantum numbers $m$ and $s$ are concerned with the orientation of the electron's orbit around the nucleus and the direction of the spin of the electron, respectively. For a given value of $l$, an electron may have any value of $m$ from $+l$ through 0 to $-l$. The energies of electrons having the same values of $n$ and $l$ but different $m$ are the same provided there is no magnetic field. For the electron, the spin quantum number $s$ can have only one of the two values $+\frac{1}{2}$ and $-\frac{1}{2}$.

## Atomic Bonding

Matter can exist in three states: solid, liquid, and gaseous. Its state will depend on the type of bonding between the atoms

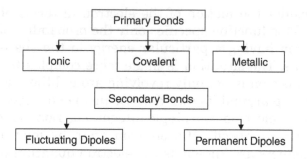

**Figure 2.2**   Major classes of atomic bonding.

and consequentially how those bonds allow the movement of independent atoms. Atoms in gases move more freely, indicating that the bonding between them is very weak when compared to that of solids, where the bonding is considerably stronger.

Atomic bonding occurs when there is a lowering of energy in the bonded state of the atoms relative to when they were independent. The two major classes of atomic bonds are primary and secondary bonds (Figure 2.2). Primary bonds are generally one, or more, order of magnitude stronger than secondary bonds. The reason for this is they involve relatively large interatomic forces created between them. The three major types of primary bonds are ionic, covalent, and metallic bonds. All primary bonds involve either the transfer of electrons from one atom to another or the sharing of electrons between atoms. The formation of primary bonds allows the atoms to achieve a noble gas configuration. Secondary bonds are also called weak bonds and they involve weak interatomic and intermolecular forces. They do not involve any electron transfer.

Prior to discussing the details of each type of bond, it is important to consider the concept of electronegativity (EN), which is an important factor in determining the type of bond that an atom will form. The electronegativity of an element is defined as the relative tendency of that element to gain, or attract, an electron. Electronegativity generally increases across the periodic table from left to right. Hence, in the second row it starts with lithium having an electronegativity of 0.98, carbon with an electronegativity of 2.55, and fluorine with an electronegativity of 3.98. Elements with high values of electronegativity are said to be electronegative and the ones with low electronegativity are called electropositive. While forming a bond, the difference in the electronegativities of the two atoms is also crucial to what type of bond will form.

## Ionic Bonding

An ionic bond is the result of electron transfer from one atom to another. This transfer is from an electropositive (metallic) atom to the electronegative (nonmetallic) atom. A high difference of electronegativity between the two bonding atoms favors the formation of an ionic bond. A common example of the ionic bond involves table salt (sodium chloride) (Figure 2.3). The transfer of an electron from sodium is favored because the result is a more stable electronic configuration, $Na^+$, which has a full outer orbital. Similarly, the chlorine atom readily accepts the electron because the resulting Cl ion has a full

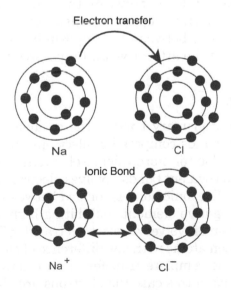

**Figure 2.3** Ionic bonding of sodium chloride.

outer orbital. The resulting species $Na^+$ and $Cl^-$ are called ions and hence result in an ionic bond. The positively charged ion, $Na^+$, is called a cation and the negatively charge ion, $Cl^-$, is called an anion. It is important to note that the ionic bond is nondirectional. A positively charged $Na^+$ will attract any adjacent $Cl^-$ equally in all directions.

The ionic bond is a result of the electrostatic, or coulombic, force of attraction of oppositely charged ions, because there is a net decrease in the potential energy of ions after bonding. The coulombic forces draw the ions together until their filled electron shells begin to overlap. Until the point of overlap the electrons associated with the cation are independent of those of the anion. When the electron shells of the ions begin to overlap their electrons begin to interact and can no longer be considered independent. We know according to Pauli's exclusion principle that no two electrons can have the same four quantum numbers, so some of the electrons will need to be promoted to a higher orbital, requiring an increase in energy. To avoid this, repulsive forces develop which will work against the overlapping of the electron shells of adjacent ions. When the attractive coulombic forces equal these repulsive forces, there will be no net force between the ions and they will remain at an equilibrium separation distance. Hence, we can say that the net force between the oppositely charged ions is equal to the sum of the attractive and repulsive forces.

## Covalent Bonding

The second type of primary bond is the covalent bond, which is formed in compounds composed of electronegative elements, most notably, with four or more valence electrons. The difference between ionic and covalent bonds involves electronegativity and electron transfer. The ionic bond is formed between electropositive and electronegative elements exhibiting high difference of electronegativity and involves an electron transfer. The covalent bond forms between atoms with low difference of electronegativity and involves no complete transfer of an electron from one atom to another, as in this case the electrons are shared rather than exchanged. The covalent bond can be both directional and nondirectional, producing significantly different structures. Metals are generally nondirectional and nonmetals directional. The nature of nondirectional covalent bonding results in the ability to conduct heat and electricity. Directional bonding produces a very strong bond or "directional relationship" between the atoms as found in the carbon bonds in diamonds.

The name covalent derives from the cooperative sharing of valence electron between two adjacent atoms. In a single covalent bond, each of the two atoms contribute one electron to form an electron pair which will be shared, and the energies of the two atoms associated with this bond will be lowered because of this electron-pair sharing. Covalent bonds can also occur with the same or different type of atoms with one or more electron pairs shared (Figure 2.4).

It should be clear that the shared electrons can only be accurately depicted by an electron cloud density. The probability of finding the shared electron pair is greater between the two nuclei because that is where the two outer orbits interact.

Covalent bonds are most common in polymeric materials where the backbone is mostly made of carbon atoms. Carbon has the ability to form single, double, and triple bonds with itself. Carbon can make up to four covalent bonds of equal strength making it a very essential element in polymers. For

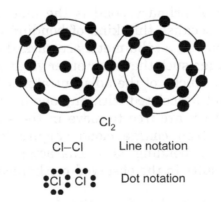

$Cl_2$

Cl–Cl        Line notation

Dot notation

**Figure 2.4** Covalent bond. When present between two atoms of chlorine it causes a molecule of chlorine gas to form. Different notations can be used to show the covalent bond.

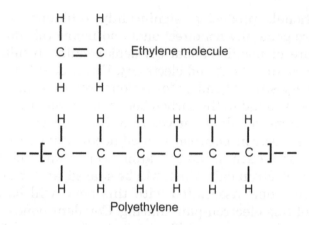

**Figure 2.5**   Ionic bond. Polyethylene is made by breaking the carbon–carbon double bond of the ethylene molecule and polymerizing it.

example, the common plastic polyethylene is made by breaking the carbon–carbon double bond of the ethylene molecule and polymerizing it (Figure 2.5). The bonds between C and H and C and C in the polyethylene molecule are covalent.

## Metallic Bonding

The third type of primary bond is the metallic bond. It is called metallic because this kind of bonding occurs in solid metals. The atoms in a solid metal are packed very closely, in an orderly manner, such that their outer valence electrons are attracted by the nuclei of the adjacent atoms. These valence electrons are not associated with any one nucleus, or pair of nuclei, rather, they are free to move in the structure in the form of an electron charge cloud (Figure 2.6). This type of nondirectional bonding is a characteristic of metallic bonding and results in the observed mechanical and thermal properties.

Solid metals consist of positive ion centers with their valence electrons dispersed in the form of a charged cloud. These electrons, known as "free electrons," are very weakly bonded to their nuclei and are free to move about in the structure.

Positive ion centers

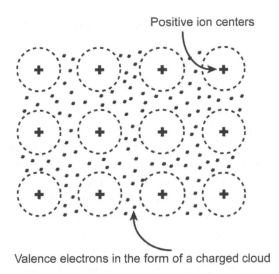

Valence electrons in the form of a charged cloud

**Figure 2.6** Metallic bond. These valence electrons are not associated with any one nucleus, or a pair of nuclei, rather, they are free to move in the structure in the form of an electron charge cloud.

The high electrical and thermal conductivity of metals is associated with this free movement of the electron cloud. The formability and ductility of metals is also explained on the basis that the atoms can slide with respect to each other without "breaking" these metallic bonds.

Metallic bonds happen in solid electropositive elements containing three or fewer valence electrons. These are generally nondirectional bonds as there are no electron-pair restrictions. Like ionic and covalent bonds, the metallic bond attempts to achieve a lower energy state. The atomic holding force is the attraction between the positively charged ion-centers and negatively charged electron cloud. The electrons are shared but they are not spatially localized.

As in previous bond types, Pauli's exclusion principle results in a repulsive force that becomes significant when the filled electron shells associated with the ion cores begin to overlap. The sum of the attractive and repulsive forces leads to bond-force and bond-energy curves, both similar to those of ionic and covalent bonds. All metals exhibit this type of bonding in their solid state, e.g., iron, steels, aluminum, their alloys, etc.

## Secondary Bonds

These are weak bonds with bond energies of 4–42 kJ/mol compared to the primary bond energies of 200–700 kJ/mol. These bonds involve no electron transfer or sharing and are sometimes referred to as physical bonds or Van der Waals forces. The mechanism of secondary bonds is somewhat similar to that of the ionic bond, because it involves attraction of opposite charges but without any electron transfers. This attraction depends on the asymmetrical distribution of positive and negative charges within each bonding atom or molecule. Such charge asymmetry is known as a *dipole*. Hence, the driving force of the secondary bond is the attraction of the dipoles contained in an atom or molecule.

There are two main forms of secondary bonds involving dipoles between atoms or molecules: those involving fluctuating dipoles and those involving permanent dipoles. Fluctuating dipoles can develop between the atoms of noble gas elements which have complete outer valence electron shells. These forces arise because, at any instant, the distribution of the electron charge cloud can be asymmetrical and can cause a dipole. In other words, at any given time the electron cloud has a higher probability of being on one side of the atom than another, resulting in a changing electron cloud that causes a fluctuating dipole. Fluctuating dipoles of adjacent atoms can attract others causing weak secondary interatomic, nondirectional bonds. Noble gases including helium, neon, and argon exhibit fluctuating dipole bonds.

Permanent dipoles, as the name suggests, are contained in a molecule which is covalently bonded as a result of their directionality. These dipoles are not fluctuating and do not disappear with time. As long as the parent covalent molecule is present these dipoles will remain. Consider the bond between adjacent water molecules. It occurs as a result of the directional nature of electron sharing in the covalent O–H

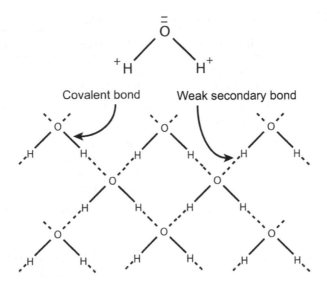

**Figure 2.7** Secondary bond. Adjacent water molecules have a secondary bond due to the permanent dipole between O and H of neighboring molecules.

bonds in which the H atoms become positive centers and O atoms become negative centers for the water molecule. The adjacent water molecules will have a secondary bond due to this permanent dipole between O and H of neighboring molecules (Figure 2.7). A summary of bond types can be found in Table 2.3.

## Mixed Types

The bonding methods described above do not necessarily occur independently. Atoms and ions can coexist in a mixture of primary bonds and secondary bonds. Water, as mentioned in the previous section, is also a mixture of covalent and secondary bonds. Also, primary bonds of any type can occur with other types of primary bonds. The following combinations can occur:

- *Mixed ionic–covalent bonding.* Occurs in both ionic and covalent molecules and ions with no hard transition

**Table 2.3**   Summary of Bonding in Materials

| Bond Type | Bonding Forces | Properties | Example |
|-----------|----------------|------------|---------|
| Ionic | Electrostatic attraction | Hard, high melting point | NaCl |
| Covalent | Shared electrons | Very hard, high melting point | Diamond |
| Metallic | Electron charge cloud | Hard to soft, high electrical and thermal conductivity | Cu, Fe |
| Secondary | Van der Waals forces | Soft, low melting point | Ice, Sugar $(C_{12}H_{22}O_{12})$ |

but rather a gradual one. The difference in electronegativities between the two atoms or molecules is an indication of the extent of ionic bonding. Many semiconductor compounds such as GaAs and $SiO_2$ exhibit this kind of mixed bonding.

- *Metallic–covalent bonding*. Occurs commonly in transition metals. Iron and titanium show this kind of bonding.
- *Metallic–ionic bonding*. Occurs in intermetallic compounds where the difference of electronegativity is high. Intermetallics like AlLi, $Ni_3Al$, $Al_3V$, AlSb, and CuZn exhibit either metallic ionic or metallic covalent bonding depending on the difference in electronegativity.

## Crystal Structure

Once there is an understanding of atomic structure and how atoms bond to other atoms, one must consider the nature of the arrangement of atoms, ions, and molecules that form a structure. Upon casual observation of a material we seldom think how the individual atoms are arranged, rather, we think of it in terms of its properties such as color, hardness, strength, ductility, etc. These properties are dependent on how the atoms, ions, or molecules are arranged inside the material. Carbon is a common example, as it can exist in the form of diamond or graphite. Both are crystalline forms of carbon and consist only of covalently bonded carbon atoms, but this is where the similarity ends. Diamond is hard and brittle and often used as a cutting tool. Graphite is black, very soft, and is often used as a lubricant. How can the same element occur in two drastically different forms? The answer lies within the crystal structure and how the atoms are arranged.

It is necessary to review the concept of short- and long-range molecular order before defining crystal structures. Figure 2.8A depicts a crystal with long-range order, where the pattern repeats itself on a larger scale, with limited discontinuities. Figure 2.8B depicts a pattern of short-range order that repeats itself with the possibility of many discontinuities and is not global in nature. For example, in a liquid, short-range order is present because the distance between molecules is fixed, but this order does not go farther than that, allowing it to be fluid. In a solid, long-range order is observed, with the distance between the molecules remaining fixed, and a pattern which repeats itself throughout the structure is observed.

## Lattice and Unit Cell

Most solids are crystalline and defined by long-range order that requires atoms to be arranged in a repeating three-dimensional (3D) array. This 3D network is known as a *lattice*

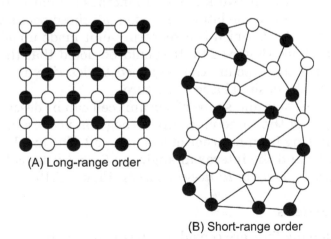

(A) Long-range order

(B) Short-range order

**Figure 2.8** Crystal structures. Long- and short-range molecular order.

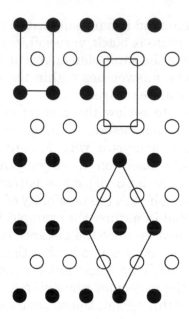

**Figure 2.9**   Unit cell. These repeating units of atoms or molecules are represented by a specific set of axis, edge lengths, and faces to define a crystallographic structure.

and described by a cloud of points in space. In an ideal crystal any point in this network has surroundings similar to those of any other point in the lattice. If we join all these points we can specify a *unit cell* which is a repeating unit made up of points joined by lines (Figure 2.9). It should be noted that there are many ways these points can be joined to describe a unit cell, producing the desired lattice structure.

The 3D size and shape of the unit cell can be described by the length of the three mutually orthogonal vectors **a**, **b**, and **c**, originating from a corner of the unit cell. The axial lengths $a$, $b$, and $c$ and the interaxial angles $\alpha$, $\beta$, and $\gamma$ are called the *lattice parameters* or *lattice constants* (Figure 2.10).

## Crystal Systems

The following four are considered basic unit cells:

- Simple
- Body-Centered

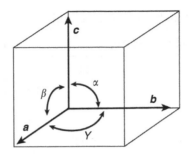

**Figure 2.10** Size and shape of the unit cell. The cell is also described by interaxial angles defining surface orientation.

- Face-Centered
- Base-Centered

There are three variations possible in the cubic lattice system: simple cubic, body-centered cubic (BCC), and face-centered cubic (FCC). The tetragonal system occurs in simple and body-centered variations. The monoclinic system occurs in simple and base-centered types. The orthorhombic system occurs in all four types. Rhombohedral, hexagonal, and triclinic systems occur only in one variation each. Crystallographers have shown, by varying the unit cell's axial lengths and angles, that only seven crystal systems are necessary to define all existing crystal structures (Table 2.4).

Having defined the seven basic crystal systems the question remains as to how many unit cell arrangements can be formed. A.J. Bravais determined that 14 standard unit cells could describe all possible lattice networks (Figure 2.11).

## Miller Indices and Crystallographic Planes

Even in the simplest of structures there is a need for a common language to describe specific points, directions, and planes in crystals. This need is filled by defining any desired point, direction, or plane by means of atomic positions, direction indices, and Miller indices for planes. *Miller indices* are defined as the reciprocals of the intercepts made by the planes on the crystal axes, the $x$-, $y$- and $z$-axis.

**Table 2.4**   Crystal Structure Form Variables

| System | Axial Length and Angles | Unit Cell Geometry |
|---|---|---|
| Cubic | $a = b = c,\quad \alpha = \beta = \gamma = 90°$ | Cubic   BCC   FCC |
| Tetragonal | $a = b \neq c,\quad \alpha = \beta = \gamma = 90°$ | Simple   Body |
| Orthorhombic | $a \neq b \neq c,\quad \alpha = \beta = \gamma = 90°$ | Simple   Base   Body   Face |
| Rhombohedral | $a = b = c,\quad \alpha = \beta = \gamma \neq 90°$ | Simple |
| Hexagonal | $a = b \neq c,\quad \alpha = \beta = 90°, \gamma = 120°$ | Simple |
| Monoclinic | $a \neq b \neq c,\quad \alpha = \gamma = 90° \neq \beta$ | Simple   Base |
| Triclinic | $a \neq b \neq c, \alpha \neq \beta \neq \gamma \neq 90°$ | Simple |

## Lattice Positions

The position of an atom in a unit cell is described by using a rectangular $x$, $y$, $z$ coordinate system which follows the right hand rule (Figure 2.12). To describe the location of an atom at a distance of one unit from zero on the $x$-axis at the base of unit cell, its coordinates will be noted as (1, 0, 0). If the atom is

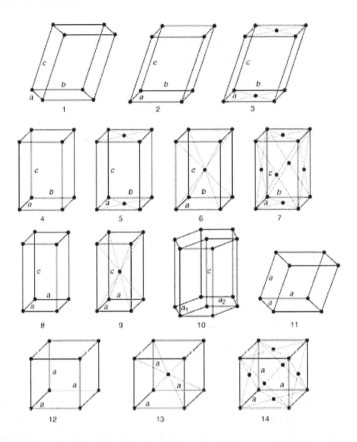

**Figure 2.11** Standard unit cells. The 14 space lattices illustrated by a unit cell of each: (1) triclinic, primitive; (2) monoclinic, pri-mitive; (3) monoclinic, base-centered; (4) orthorhombic, primitive; (5) orthorhombic, base-centered; (6) orthorhombic, body-centered; (7) orthorhombic, face-centered; (8) tetragonal, primitive; (9) tetragonal, body-centered; (10) hexagonal, primitive; (11) rhombohedral, primitive; (12) cubic, primitive; (13) cubic, body-centered; (14) cubic, face-centered.

*Source*: ASM Metals Handbook Desk Edition: Online Desk Edition, Crystal Structure of Metallic Elements, Crystallographic Terms and Basic Concepts, Figure 3.

placed at a distance of one unit on the $z$-axis, its coordinates will be (0, 0, 1). The coordinates for atomic positions in the unit cell is defined by numbers separated by commas. For the position (½, ½, 0), the atom would be placed at the base of the cell in the middle.

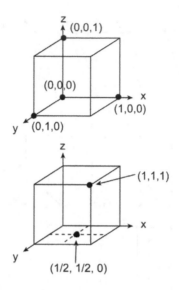

**Figure 2.12**   Lattice positions. Atomic positions are defined using Miller indices in the coordinate system.

## Lattice Directions

It is often necessary to specify directions in a crystal structure to specify properties which are dependent on crystallographic directions. The lattice directions are always expressed as sets of integers. These are obtained by identifying the smallest integer positions intercepted by the line from the origin of the crystallographic axes. The notation used for directions includes integers within square brackets, such as [211]. To draw a given direction, using the example [112], the integer set is divided by the largest integer within the set. For the given example, this would result in ½, ½, 1. Plotting this point within the unit cell of interest and joining it to the origin defines a vector in the direction [112] (Figure 2.13).

## Crystallographic Planes

Crystallographic planes are referred to as Miller indices. Once the reader is familiar with atomic positions and directions,

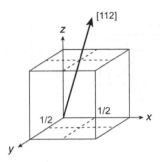

**Figure 2.13** Lattice directions. The position with the lattice structure is defined by reducing the coordinates to the lowest integer vector.

obtaining Miller indices for planes is not difficult. To obtain the indices for a given plane, use the following procedure:

1. Identify the coordinate intercepts of the plane, i.e., the coordinates where the plane cuts the $x$, $y$, and $z$ axes. If the plane is parallel to one of the axes the intercept is taken as infinity. If the plane passes through the origin, then change the location of the origin.
2. Take the reciprocal of the intercepts.
3. Clear fractions but do not reduce to lowest integers.
4. Cite planes in parentheses, for example (hkl), without any commas and place bars over the negative indices.

Figure 2.14 shows several examples of how to convert the intercepts into Miller indices.

## Polymorphism

A change in temperature or pressure, if not accompanied by melting or vaporization, may cause a solid to change its internal arrangement of atoms. This ability to have more than one crystal structure is called *allotropy* or *polymorphism*. The transition from one crystal structure to another at a particular temperature and/or pressure is called a *polymorphic* (multi-shape) change. For example, iron exists at room temperature in BCC

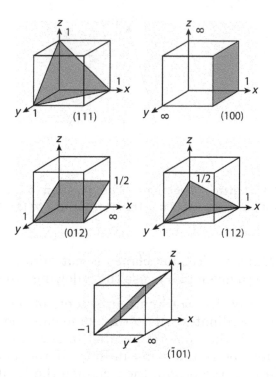

**Figure 2.14**   Miller indices. Converting the lattice coordinate system into Miller indices for several crystallographic planes.

structure ($\alpha$ Fe) and when heated above 910°C changes to a FCC structure (gamma Fe). Upon further heating to 1400°C it reverts back to a BCC structure (delta Fe). At 1540°C iron melts. These changes are reversed upon cooling. The atoms are the same in these different crystal structures but there are differences in physical and mechanical properties. For example, gamma Fe will dissolve up to 2.0% carbon but alpha Fe will only dissolve 0.025% carbon.

Iron, titanium, zirconium, thallium, and lithium exhibit polymorphic transformations from BCC to a close-packed structure (either FCC or hexagonal close-packed (HCP), depending upon the element) on cooling. Iron changes from FCC back to BCC on further cooling to become ferritic. The significance of this behavior depends on the type of solid solution present, interstitial or substitutional. The interstitial forms typically have a significantly greater effect on the material properties than do the substitutional forms. In fact, some of the substitutional forms provide little or no mechanical change but do affect other properties, including corrosion resistance and ductility.

## Main Metallic Crystal Structures

About 90% of elemental metals crystallize into three kinds of crystal structures. These are BCC, FCC, and HCP. Their performance is highly influenced by these structures, so it is important to review them in greater detail.

### Body-Centered Cubic (BCC)

The BCC structure consists of a cubic lattice with one atom at each corner and one atom located at the center of the cube (Figure 2.15). The one atom at each corner is shared by eight neighboring cells, resulting in only one-eighth of corner atom being included in a unit cell. The atom at the center position of (½, ½, ½) is not shared by any other cell. Thus, there are two atoms in each unit cell of BCC.

The *atomic packing factor* (APF) is defined as the volume of atoms in the unit cell divided by the volume of the unit cell and for the BCC structure represents the fraction of the unit

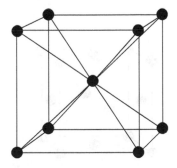

**Figure 2.15** Body-centered cubic. This structure is least resistant to deformation, owing to its more open structure (taken from http://cst-www. nrl.navy.mil/lattice/).

cell occupied by two atoms. For a BCC the APF is 0.68. Typical metals with this crystal structure include alpha Fe, V, Cr, and Mo. An alloy of these metals will also have a dominant BCC structure but the presence of other alloying additions distort and change the structure.

### Face-Centered Cubic

This structure consists of a cubic unit cell which contains eight corner atoms plus six atoms present on each face of the cube (Figure 2.16). As previously reviewed for a BCC, the corner atoms are shared by eight other unit cells, hence only one-eighth of corner atom is included in the cell. Similarly, the face atoms are shared by two cells each, hence half of the six face atoms are included in the unit cell. The total number of atoms included in the unit cell is four: one from corner atoms and three from face atoms.

The APF of FCC is 0.74, meaning it is more densely packed than BCC. In fact, it is known that 0.74 is the highest APF possible for filling a space with equal-sized spherical atoms. Many metals like gamma Fe, Al, Ni, Cu, Ag, Pt, and Au exhibit FCC structure.

### Hexagonal Close-Packed (HCP)

This structure is more complicated than BCC or FCC (Figure 2.17). Three atoms form a triangle in the middle layer,

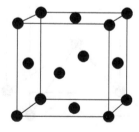

**Figure 2.16**   Face-centered cubic. This lattice form is more resistant to deformation due to its increase structural complexity (taken from http://cst-www.nrl.navy.mil/lattice/).

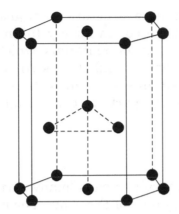

**Figure 2.17** Hexagonal close-packed. This lattice form provides the greatest resistance to deformation, resulting in high yield strength materials.

with six, one-sixth atom sections on both top and bottom layers making an equivalent of two atoms. To complete the structure there is half an atom in the center of both the top and bottom layers, making the equivalent of one more atom. The total number of atoms in one unit cell in the HCP crystal structure is six. The HCP structure has the same APF as the FCC structure which is 0.74. Both these structures have a *coordination number* (the number of nearest neighbor atoms surrounding each atom) of 12 but they differ in their stacking.

The ratio of height $c$ of the cell to its basal side $a$ is called the $c/a$ ratio. An ideal HCP cell with uniform spherical atoms packed tightly should have a $c/a$ ratio of 1.633. However, it typically varies from 1.57 to 1.89. Metals having $c/a$ ratio higher than the ideal ratio have atoms slightly elongated in the structure and metals having $c/a$ ratio lower than the ideal have atoms slightly compressed in the crystal structure. Typical metals with HCP structure include Be, Mg, $\alpha$-Ti, Zn, and Zr.

## Noncrystalline (Amorphous) Structures

Not all materials are crystalline in nature. Two other states of matter, liquid and gas, have a random or noncrystalline structure. Noncrystalline structure is not just limited to gases or liquids, solids also occur in noncrystalline forms. The two most important classes of materials which have noncrystalline states are polymers and glasses. Apart from total crystalline and noncrystalline structures there also exist mixed structures where some parts of a material show order and others do not.

## Structure of Polymers

The word *polymer* literally means "many parts." Each "mer" or part is the repeating unit of the chain. For many plastics the monomer, or repeating unit, can be deduced by deleting the prefix "poly" from its name. As an example, consider polyethylene, a polymer which has ethylene as its monomer or repeating unit (Figure 2.18). The chemical reaction which combines thousands of monomers into a long chain to form a polymer is called polymerization reaction or just *polymerization.*

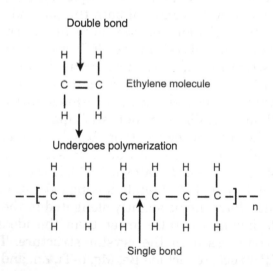

**Figure 2.18**  Monomer structure. The varying monomer chain length affects the properties of the material offering a resistance to its movement, increasing the apparent material strength.

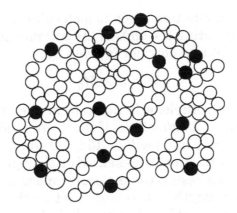

**Figure 2.19** Polymer chain orientation. The random orientation of the chains within these structures are highly dependent on their processing and chain length.

Polymers, also generically called plastics, are more complex in structure than metals. Polymers can be described as large chain-like molecules in which atoms are linked by covalent bonds to form the backbone of the chain. The backbone generally consists of carbon atoms. These long chains are not arranged in a specific fashion, rather, they form random structures, like spaghetti in a bowl (Figure 2.19).

The fact that the chains of a polymer are not arranged in straight lines and occur in a random formation can be described on the basis of atomic bonding. The directionality of the covalent bonds forces the chain to be formed in a zigzag fashion, rather than straight line, while also being flexible enough to form the 3D structure without breaking the bonds. These chains are held together by Van der Waals bonds and mechanical entanglement.

The degree of polymerization of the polymer chain is the number of subunits or "mers" used to form the chain. Again, consider the example of polyethylene, its chain length ranges from 3,500 to 25,000 "*mer*" units, $n$. This value is the product of the degree of polymerization and the molecular weight, $M_0$.

Ignoring the contribution made to molecular mass of polymer by groups terminating the ends of the polymer chain, then the molecular weight of a polymer is given by

$$M = n \times M_0$$

For example, polyethylene is made from the ethylene "mer" which has a $M_0$ of 28 g/mol. So, if there are 6,000 "mers" in a chain the molecular mass will be 168,000 g/mol. Note that the degree of polymerization is presented as a range, meaning $n$ is not constant. As a result, the molecular mass will be an average of the chain length $n$. It is worth noting that as the material's properties are affected by the chain length, all values of these properties will be ranges, dependent on the average chain length in the sample.

## Molecular Chains

The molecular chains in a polymer can be linear, cross-linked, or branched (Figure 2.20). The circles in this figure represent one repeating unit or "mer" of the polymer. Linear chains are

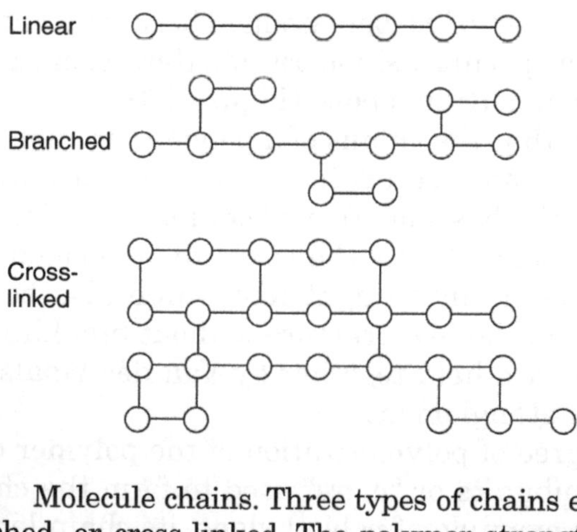

**Figure 2.20**  Molecule chains. Three types of chains exist, namely, linear, branched, or cross-linked. The polymer strength is directly related to the type of chain formation with the cross-linked form being the strongest.

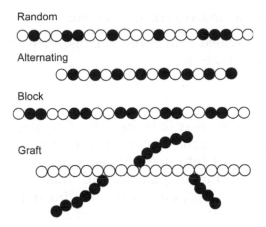

**Figure 2.21** Copolymers. This molecule consists of a long chain formation formed by the reaction of two or more dissimilar molecules.

the weakest form because the chains have no mechanism to prevent relative motion between adjacent chains. Cross-linked chains have this mechanism, limiting motion to result in a more rigid structure.

The above discussion has considered polymers made up of only one kind of repeating unit, a *homopolymer*, such as polyethylene. Polymers which include more than one type are known as copolymers, and these are produced by combining two or more different kinds of monomers into a single chain (Figure 2.21).

A solid polymer may consist of linear or nonlinear chains and be made of a single monomer or multiple monomers with the chains held together by Van der Waals forces. Copolymers, like homopolymers, are made of linear chains that can move easily and slide past each other without much resistance, resulting in more flexible plastic. If the chains are cross-linked or have side branches they will have greater difficulty moving past one another, resulting in a more rigid plastic. This knowledge leads to the conclusion that the properties of a plastic are highly dependent on the type of monomer(s), the chain length, and the chain type.

## Polymer Classification and Molecular Structure

Polymers can be divided into three classes: thermoplastics, thermosets, and elastomers.

### Thermoplastics

Thermoplastics require heat to make them formable and will retain their new shape upon cooling. These polymers can be reheated and reformed into new shapes a number of times without any significant degradation of their properties. Thermoplastics are very popular in the bottling industry due to their recyclable nature and low melting temperature. Thermoplastics consist of polymers with long chain molecules that are either linear chains or ones with short branches. These long chains are mostly made up of covalently bonded carbon atoms. In addition, nitrogen, oxygen, or sulfur atoms are present that are also covalently bonded in the main molecular chain. The length of the thermoplastic chain significantly influences the material and processing properties, because the longer the polymer chain, the more entangled it is, resulting in a more rigid plastic.

### Thermosets

Thermosets or thermosetting plastics are those polymers which are formed into a permanent shape, or cured, by a chemical reaction. Once formed, thermosets cannot be remelted or recycled. Heating a thermoset will cause it to decompose with no softening. Thermosetting plastics can be cured, or set, by providing heat or by a room temperature chemical reaction. The atoms in a thermoset form a 3D structure of chains with frequent cross-links between the chains. The bonds linking the chains are strong and not easily broken, which prevents the chains from sliding over one another. The result is a stronger, stiffer material than thermoplastics. But unlike thermoplastics, they cannot be heated and injected into a mold, rather, they are formed by mixing chemicals in the mold or immediately before placing them there, so that the cross-linking occurs within the confines of the desired shape.

## Elastomers

Elastomers are polymers that can show very large, reversible strains when subject to stress. These materials are perfectly elastic up to twice their original length as defined by ASTM. It is not uncommon for elastomers to be strained up to four or five times their original length. Elastomers have a structure consisting of tangled polymer chains held together by occasional cross-linked bonds. The difference between thermosets and elastomers is that thermosets have frequently cross-linking between chains, while in elastomers it is occasional (Figure 2.22).

### Glass Transition Temperature

The glass transition temperature $(T_g)$ is the temperature where amorphous polymers become soft and flexible when heated. Typically, it occurs above room temperature but this is not a requirement. For example, PVC, which is quite rigid at room temperature, becomes flexible and rubbery when heated to its $T_g$ of 87°C. Above this temperature it stretches considerably, opposed to a very moderate elongation at room temperature (note: considerable care is required when working with

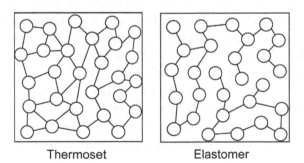

Thermoset      Elastomer

**Figure 2.22** Elastomer structure. The limit to cross-linking in elastomers allows them to recover their original shapes after the application of very high strains.

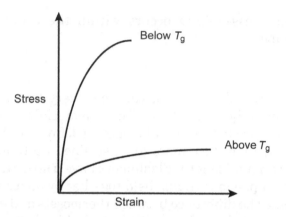

**Figure 2.23**   Glass transition temperature. The stress–strain curves above and below the transition temperature differ significantly due to this property.

PVC at elevated temperatures due to the potential for release of poisonous chlorine gas). This property of amorphous polymers is used to form these polymers into desirable shapes and forms without melting them (Figure 2.23).

Below the glass transition temperature only limited molecular motion is possible, while above it quite large motion can occur. The amount of movement possible depends on the structure of polymer chains and how easily they can slide past each other. The result of this behavior is that linear-chain polymers have a lower $T_g$ and polymers with bulky side groups, branches, and cross-linking have a higher $T_g$. From experimental data it is noted that many polymers which have symmetric monomers, such as $C_2H_2$ and $C_2F_4$, have a $T_g/T_m$ ratio of $\sim\!\tfrac{1}{2}$ ($T_m$ is the melting point). Similarly, many polymers which have unsymmetrical monomers exhibit $T_g/T_m$ ratio $\sim\!\tfrac{2}{3}$. This parameter is a useful tool in evaluating polymers against one another as a relative measure of their molecular makeup.

## MIXED STRUCTURES

Crystalline and noncrystalline structures can occur simultaneously in polymers to produce a composite structure known as semicrystalline polymers. The degree to which this occurs is dependent on numerous factors including rate of cooling during processing and the type of additives present. Rapid cooling generally results in an amorphous-dominated material because the quenching action tends to freeze the state of the material, not allowing it to crystallize fully.

## Crystallinity in Polymers

Generally, polymers are disordered structures in which chain molecules of various lengths form a tangled mass. However, it is quite possible for chain molecules to form regions in which repeating units are aligned in ordered arrays, resulting in local crystalline structures. The resulting structure includes folded chains forming ordered or crystalline regions (Figure 2.24). These regions in polymers are generally small, on the order of 10 to 20 nm in cross-section. In most polymers the degree of crystallization is so small that it can be ignored, however, in some cases such as polyethylene the crystallinity can be as high as 85%.

As a rule, thermosets do not crystallize, while thermoplastics do—but their ability to do so is dependant on several factors. These factors determine the ease with which the molecules can move and be efficiently packed together to form

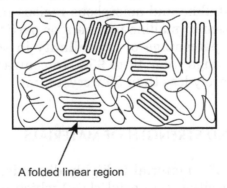

A folded linear region

**Figure 2.24** Polymer crystallinity. Amorphous and crystalline regions coexist within the polymer structure but to varying degrees. Generally, greater cystallinity results in improved material properties.

some order. They include:

1. *The size of side groups*: Polymers with bulky side groups cannot be packed efficiently and therefore have difficulty in forming crystals.
2. *Extent of chain branching*: Polymers with branches, especially longer branches, are difficult to crystallize; the ones with smaller branches and the unbranched ones crystallize easily.
3. *Tacticity*: This is the manner in which side groups are attached to the backbone. Structures where all side groups are on one side of the backbone, called *isotactic* structures, are easier to crystallize, whereas *atactic* polymers with large side groups are difficult to pack in order.
4. *Complexity of monomer*: The more complex the monomer, the more difficult it is to crystallize and vice versa.
5. *Secondary bonds between chains*: If they are present at regular intervals on the backbone due to polar side groups it is easier to align the adjacent chain segments. Hence, presence of polar side groups aid in crystal formation.

Even with all factors present the polymer may never reach 100% crystalline structure due to process variations. Those that have both crystalline and amorphous regions in the structure are known as semicrystalline polymers or partially crystalline polymers. The percentage of crystallinity in semicrystalline polymers can range from 40 to 95%. In contrast, metals and ceramics are rarely less than 99% crystalline.

## STRUCTURE AND STRENGTH OF MATERIALS

The structure of a material is highly dependent on its basic chemical composition (material class) which directly affects its mechanical behavior. Although the actual performance depends on several factors, including manufacturing processing, alloying, and defects, the inherent structure and bonding dictates the nominal mechanical modulus of a material. The bond energy and its chain length define the stiffness of the structure. The higher the bond stiffness, greater is the modulus of the crystal.

This concept gives rise to the idea that if one can manipulate the crystal structure, mechanical properties can be modified. Generally, the strengthening mechanisms and polymer additives are critical to achieving the desired performance, which can involve modifying a specific property by changing the crystal structure or the microstructure of a material. The following key concepts are essential to the understanding of strengthening mechanisms.

### Single Crystals and Alloys

Because most metals and alloys are melted and solidified into desired shapes by the process of solidification, it is crucial to understand the mechanisms involved. During solidification, two basic steps take place: formation of nuclei and growth of nuclei into crystals. Nuclei are the small solid particles formed during cooling and continue to grow until all the liquid is converted into a solid (Figure 2.25). This nucleation process can be of two types: homogeneous and heterogeneous. Virtually all nucleation is heterogeneous, meaning that the walls of container, particles, or impurities can provide nucleation sites for

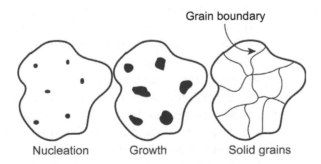

Nucleation      Growth      Solid grains

**Figure 2.25**   Crystal growth. Upon initiation at nucleation sites, the crystals expand until a solid is formed. Controlling this process is critical to the extent of crystallization of the material.

solidification. Because all nuclei grow independent of each other they all have different orientations. While growing bigger, they retain their orientations. The fully grown nuclei, called grains, all have different orientations giving rise to a polycrystalline material.

The growth of nuclei ceases when they meet adjacent growing grains, resulting in the formation of a grain boundary. A grain boundary is the region, where two grains meet forming an imperfect crystal, which separates the two different orientations of two grains in a polycrystalline material. It should be understood that in a polycrystalline material different grains (provided they are in the same phase) have the same crystal structure but each one has a different orientation and size.

The shape and size of the grains depends heavily on thermal gradients while cooling. When grains have similar size they are called *equiaxed* grains. Almost all of the engineering materials and alloys are polycrystalline, with a few exceptions. The most notable exception is silicon, used in the semiconductor industry and commonly grown as a single crystal to obtain the required purity and performance.

To form a single crystal, all solidification must take place around a single nucleus so that there are no other grains and no grain boundaries. The process of growing a single-crystal material is not easy and it requires strict control of temperature and other environmental factors. It also requires highly purified molten material, to ensure that there are no other nuclei formed during solidification. These demanding requirements are the reasons why single crystal alloys are only manufactured when their properties are specifically required. All other materials are polycrystalline with varying degrees of crystallinity. Single crystal alloys provide superior electrical properties because there is no disruption due to grain boundaries. In some cases, they also result in superior mechanical properties because the effect of grain boundaries and defects are eliminated. An important use is in a turbine blade where high-temperature creep is an important consideration and having a single crystal with no grain boundaries is very effective in reducing their disastrous effects.

Overall, polycrystalline materials are easier to fabricate due to easier temperature control and purity control resulting in economical manufacturing. Apart from the economic drivers, the properties offered by the elimination of grain boundaries are not often required to achieve the desired performance, allowing general applications to use lower-cost, polycrystalline materials.

## Strengthening Mechanisms in Metallic Alloys

The need to strengthen metallic alloys is commonly necessary to meet design requirements. Dislocations are the crystal defects which propagate when excessive stress is applied. Most strengthening mechanisms revolve around the concept of restricting dislocation movement. This ability is important for metals because the movement of dislocations is much easier in metals than in ionic and covalently bonded crystals. Dislocations are difficult to eliminate but their movement can be restricted.

### Alloying or Solid Solution Strengthening

The addition of a foreign elements, or atoms, in a lattice generally increases its strength due to the type of solid solutions formed. The solid solution can be either interstitial or substitutional. Interstitial formations are those in which there is an extra atom or foreign element in the lattice while substitutional formations are those in which one of the native atoms is replaced by a foreign one (Figure 2.26).

This introduction of foreign atoms into a crystal results in a minor shifting of the neighboring solvent atoms away from their equilibrium positions and a corresponding increase in the strain energy of the crystal. The region of the material over which the impurity atom exerts its influence is known as its *strain field*. These strain (or stress) fields interact with dislocations and make their movement more difficult, allowing the solid solution configuration to become stronger than the pure metal.

The degree of strengthening is proportional to the difference in the size of the solute and solvent and its disruptive

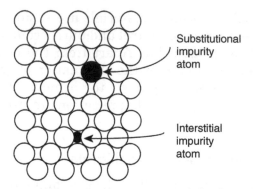

**Figure 2.26**    Foreign elements can be interstitial and substitutional in nature.

effect on the short-range order of the lattice. The greater the size difference between the atoms, the greater is the lattice distortion, which results in greater difficulty in generating dislocation movement. The greatest strengthening effect is achieved when interstitial elements occupy the tetrahedral positions in BCC metals. Substitutional atoms do not produce a great distortion field because their sizes are about the same.

Strain Hardening

When alloying dissimilar atoms, lattice distortions, and the energy required for dislocations to occur, increase. Similarly, an increase in the concentration of dislocations also increases crystal strength. The mechanism is very similar to alloying, with the stress due to many dislocations making it more difficult for a single dislocation to move within the structure. It should be noted that during plastic deformation the number of dislocations in a crystal increases rapidly, causing the increase in flow stress. Therefore, if we continue to plastically deform a material again and again, higher stresses are needed everytime. This increase in flow stress resulting from plastic deformation is known as *strain hardening*. Strain hardening is typically performed through cold working and is one of the most important methods for strengthening some metals. Pure copper and aluminum can be strengthened significantly by cold working.

## Grain Refinement

Most engineering alloys are polycrystalline materials. The presence of defects such as point and line defects as well as grain boundaries impede the movement of dislocations, increasing the stress needed to move the dislocations. Because of this, increasing the number of grain boundaries presents a barrier for dislocation movement, resulting in higher strength. Decreasing the grain size will introduce more boundaries, making dislocation movement correspondingly more difficult. Grain refinement can be achieved by introducing several methods like hot rolling and heat treatment.

## Precipitation Hardening

Strengthening alloys can also be performed by introducing precipitates in the crystal. Dislocation movement is hindered by the lattice distortion in the surrounding precipitate, causing the flow stress to rise. This method is widely used in aluminum alloys, nickel alloys, steels, and many others. Precipitates are usually introduced by heat treatment. Many forms of heat treatment exist and are applied with the objective of altering the microstructure of an alloy for the purpose of strengthening, softening, or hardening it.

## Strengthening of Polymers

The strength of a polymer depends upon its molecular weight, chain length, and material family. As with metals, heat treatment can be used to change the material properties, but often, a chemical reaction is required to obtain the desired properties. The following methods can be used to change the mechanical properties of a polymer:

1. *Increasing the chain length of a linear polymer.* This increases tensile strength and stiffness as chains become more tangled.

2. *Introducing large side groups into a linear chain.* This increases the tensile strength and stiffness because side groups inhibit motion.

3. *Producing branches on a linear chain.* This also increases tensile strength and stiffness as more branches limit movement.

4. *Introducing larger groups into the chain.* Larger groups reduce the ability of the chain to flex and, therefore, increase rigidity.

5. *Cross-linking chains.* Increasing the degree of cross-linking increases rigidity.

6. *Introducing liquids between chains.* The addition of external plasticizers fills up the space between the chains, making it easier for chains to move, providing greater flexibility.

7. *Increasing crystallinity.* If the degree of crystallinity can be increased in linear chains, it increases the density of polymer while also increasing tensile strength and modulus.

8. *Including fillers.* Fillers, such as glass fibers, can increase tensile modulus and strength by disrupting the matrix and preventing chain movement.

9. *Orientation.* Applying stress or strain during formation of polymers can result in orientation of polymeric molecules. This can increase the properties in that particular direction dramatically.

10. *Copolymerization.* Combining two or more monomers in a single chain can also change the properties depending upon the ratio of the mix.

11. *Blending.* This is done in order to obtain a more desirable polymer and is closer to metallic alloying. Two or more polymers are combined and the result depends on the proportion of the polymers.

## Additives in Polymers

Polymers are commonly combined with additives to obtain a particular combination of desired properties. Some of their names have already been mentioned, but the following table is provided to summarize their function (Table 2.5).

**Table 2.5** Polymer Additives

| Additive | Effect on Polymers |
|---|---|
| Stabilizers | Protect from damage by ultraviolet light. Direct sunlight can degrade the properties of a polymer. These additives are ultraviolet light absorbers designed to protect against UV exposure |
| Plasticizers | Increase flexibility. It is commonly a liquid plasticizer dispersed throughout the polymer matrix, filling the voids between the chains. This makes the chain movement easier and hence increases flexibility |
| Flame retardants | Chemicals and fillers are introduced to the material formulation prior to polymerization. Their presence alters the materials flammability by not only increasing the flammability temperature but also by suppressing the tendency to smoke or burn when exposed to high heat |
| Pigments and Dyes | These are added when a change in color is desired. |
| Fillers | Fillers are typically added to either reduce cost of the material by effectively thinning it using a less expensive material or to improve the base material properties. For example, glass fibers or beads are used as fillers to increases tensile strength while mica (clay) is used to improve electrical resistance. Gas can be used as a filler to produce foam |
| Micro-fillers | Nanoclays in platelet form are added to improve stiffness, fire resistance, and as a gas barrier. These forms of fillers offer reduced weight penalty compared to traditional fillers |
| Lubricants and Heat Stabilizers | These are added to facilitate processing and vary according to the process used |

*Note*:  Materials performance can be greatly enhanced by the use of additives.

## IN SUMMARY

The selection of a material for a specific engineering application is fundamentally a question of the material structure. Other considerations are certainly critical to the ultimate success of the design, but the common starting point for the selection is rooted in the material's elemental structure. A material's structure is important to its selection because its basic engineering properties are defined by it. The strength and performance of both metals and plastics are dependent on not only the chemical composition but also the structure of their atomic particles and crystalline lattice. Primary bonds have the greatest influence while secondary bonds have a lesser role in the material properties due to the reduced strength of those bonds. Metals exhibit the highest strengths because their bonds are atomic in nature, meaning that they share or transfer electrons, employing nature's most powerful attractive forces. Plastics do not employ atomic level bonding at the bulk material level, which results in reduced material properties. They also exhibit two structural forms: one which is crystalline and the other which has no crystalline form, but is amorphous. These structures provide insight into the material's behavior and anticipated performance, allowing the design engineer to make a quantitative judgment early in the selection process, without extensive knowledge of materials.

The two basic forms of plastics, thermoplastics and thermosets, each offer unique engineering properties due to their structure. Thermoplastics do not have the interlocking structure of a cross-linked thermoset, preventing it from exhibiting the higher level of strength. Thermoplastics obtain their strength from the entanglement that is a function of the polymer chains, both in length and branching. The third form of plastics is one of a mixed form, involving either the blending or mixing of plastics without the benefit of atomic bonding. These plastics will be reviewed in greater detail in Chapter 6.

Elastomers are a unique class of polymers that have many applications because of their flexible structure. Although they are cross-linked polymers, their glass transition temperatures are below room temperature, resulting in very flexible materials capable of extremely high strain rates.

In general, materials are strengthened by altering their natural atomic formation by the introduction of foreign materials or by forced displacement of the structure. How this is achieved for each material family is discussed in detail in the following chapters.

The challenge of successful material selection is a part of every design, requiring engineers to make informed decisions based on the system requirements and experience. Success can be achieved by applying the underlying principles presented in this chapter as well as the specific material properties presented in the following chapters.

# 3

# Material Properties
# and Behaviors

The first step in selecting an appropriate material is to develop a functional specification of the device that quantifies the operating environment and life expectancy. This specification can then be evaluated against material costs and the necessary processing requirements. In many designs, there are some material properties that are absolutely essential and others that may be somewhat subjective and likely to be selected based on cost and market considerations. For example, a shovel is typically a two-piece device that has a metal end attached to a handle. The metal end must be durable and somewhat corrosion-resistant; selecting a material in this case is fairly straightforward. The handle can be made from numerous materials because there are many that meet the necessary functional requirements, causing the material selection to be driven by marketing requirements rather than engineering necessities. Materials most often selected for this application include wood and composites which differ significantly in raw material and fabrication costs. Since either of these offer a solution to the design problem, the selection becomes based on the desired market and selling price.

The material selection process may also require consideration for the product lifecycle properties. These include the following:

- availability (ease of refinement)
- processibility
- post-process requirements
- short-term properties
- long-term properties
- suitability for recycling
- disposal

The abundance of the required raw materials and the ease of extracting them and producing a useful form determines their general availability. The physical and chemical makeup of the material will subsequently drive the processing requirements and short-term properties. Post-processing of the material will determine not only its long-term properties (corrosion-resistance, etc.) but also its suitability for recycling. For example, most aluminums require post-processing to protect them from corrosion. Anodizing, or the formation of a similar ceramic oxide coating, is commonly used to provide this protection. Recycling these coated materials is difficult because separating aluminum oxide from the base material requires additional processing to recover the aluminum.

Design engineers are increasingly being faced with product designs requiring consideration for material disposal. This end-of-life requirement has been applied to metals for many years, because their material properties lend themselves to reprocessing and it is cost effective to do so. Plastics present a greater challenge. Thermoplastics can be recycled when properly separated and grouped by material type, while thermosets cannot. To some degree, recycling has been part of the injection molding process for many years because of the ability to "regrind" processed material and remix it with virgin material. Thermosets are not recycled because of their cross-linked structure, which does not allow reprocessing by heating. They are most often disposed of with everyday trash.

Material properties are the core of the selection criteria. These involve chemical, physical, mechanical, and dimensional properties (Table 3.1). These can be further subdivided by material family: ferrous and nonferrous; thermoplastics

**Table 3.1** Property Spectrum for Metals, Plastics and Adhesives

| Material | Mechanical | Chemical | Physical | Dimensional |
| --- | --- | --- | --- | --- |
| Metals | Tensile/ compressive properties Toughness Ductility Fatigue Hardness Creep resistance Shear Strength | Composition Microstructure Phases Grain size Corrosion resistance Inclusions | Melting point Thermal Magnetic Electrical Optical Acoustic Gravimetric Color | Available shapes Available sizes Available surface texture Manufacturing tolerances |
| Plastics | Tensile/ compressive properties Heat distortion PV limit Toughness | Composition Fillers Crystallinity Molecular weight Flammability Spatial configuration Chemical resistance | Melting point Thermal Magnetic Electrical Optical Acoustic Gravimetric Color | Manufacturing tolerances Stability Available sizes |
| Adhesives | Tensile/ compressive properties Shear strength Adhesion Toughness Creep resistance Hardness Fatigue resistance | Composition Fillers Molecular weight Chemical resistance Adhesion | Color Optical Thermal Melting point Surface energy | Bond line (thickness) Contact area Surface finish (macro) Surface finish (micro) |

and thermosets; and ceramics (metal matrix composites) and composites (cut fiber, wound, layup).

This text will not discuss ceramics or composites because of the unique nature of their properties and applications. Although they possess many desirable properties, their costs remain high and their lead times are longer than those of "conventional" materials, limiting their use only to those specific design solutions which justify their selection. Designs involving composites continue to expand into many areas including sporting equipment and transportation systems. Ceramics have a limited application base; but with improved processing methods, ceramic materials are being produced that are more tolerant to damage and have shorter lead times, thus improving their appeal.

The following sections describe the mechanical, chemical, and physical properties of metal- and polymer-based materials as they apply to their selection for engineering applications.

## MECHANICAL PROPERTIES

Mechanical properties that influence the success of a design can be classified into three basic groups: structural properties, process-related, and material life (Figure 3.1). The chemical and physical properties are also critical to the design success and are discussed in greater detail in the following sections.

Engineering materials for most designs are selected based on strength criteria which include static or dynamic loading. The candidate list of materials is further reduced by applying additional material property requirements, such as corrosion resistance, failure mode (ductile or brittle), electrical and thermal properties, and processing requirements. The remaining materials can then be sorted by cost, availability,

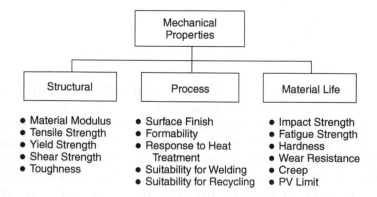

**Figure 3.1**   Mechanical properties important to material selection.

and suitability for recycling prior to making the final selection. Laws regarding recycling continue to evolve and affect product design by requiring disassembly or disposal methods to be part of the product specification.

The mechanical properties of materials are evaluated using numerous tests to quantify their behavior, including AS1391 and ISO6892, for testing tensile strength, to name a few. Using the results of these tests and others, the design engineer can make an informed decision regarding the suitability of the material to a specific design. The most commonly referenced material property is tensile strength (force per unit area), which is a short-term measure of how a material handles a tensile load. Related measures include strain and the material modulus.

Flexural strength is another measure of a material's strength. It is a measure of bending strength and is also measured in terms of force per unit area. This property is significant for materials that exhibit asymmetric properties, such as reinforced plastics and composites.

Toughness, impact strength, notch sensitivity, and hardness are measures of damage-tolerant behavior. It is also important to consider these properties during the selection process, as they are responsible for the intrinsic "robustness" of a design. Products are often subjected to loads and environments that were not taken into account during the original design and development. To compensate for these conditions a safety factor must be applied to the calculated loads to provide sufficient design margin in order to ensure user safety.

The importance of process-related properties varies depending on the design requirements. The process most common to all designs, regardless of the material, is surface finish. Defining the appropriate surface finish for a specific application is important for metals and plastics as well as in the application of adhesives. It can also have an influence on other properties, such as fatigue and wear. Formability can also be applied to both metals and plastics in the context of what it takes to reshape a material. Metal extrusion and forming require unique material properties to facilitate the bulk changes that are a part of the process. This is also true for plastics in the context of injection molding or thermoforming. Each of these processes and their derivatives require unique material properties in order to facilitate production at a reasonable cost.

Product life must also be included in the design specification. It can depend on several important material properties, such as fatigue and creep. The specification of the pressure–velocity (*PV*) limit is unique to plastics and composites as it describes their ability to withstand dynamic wear resulting from simultaneous application of a load and relative motion.

Careful consideration of the mechanical properties discussed below will result in a more effective design solution. It is also often necessary to perform trade-off analyses to obtain the desired properties that best meet the design specification. Recall the example of the shovel and its handle. The most common and inexpensive solution is to fabricate the handle from wood, but wood does not offer the superior strength and damping characteristic of a handle made from a composite. Of course, the composite handle is considerably more expensive and may not be as durable owing to added processing costs and the less damage-tolerant behavior of composites.

## Structural Properties

The concepts of material strength and how a material behaves under load are embodied in numerous properties. Those presented here are only the most common and, typically, span the entire material spectrum. A careful review of these values as they relate to a specific design and marketing environment must be performed if the most appropriate candidate is to be selected. Using selected material indices (which can be found in Chapter 1.) can effectively support these decisions.

### Stress, Strain, and Young's Modulus

Mechanical stress and strain are a function of the material modulus according to the following formula:

$$\text{Stress} = \text{strain} \times E$$

The modulus $E$ is the slope of the elastic portion of the stress–strain curve measured at the 0.2% offset point

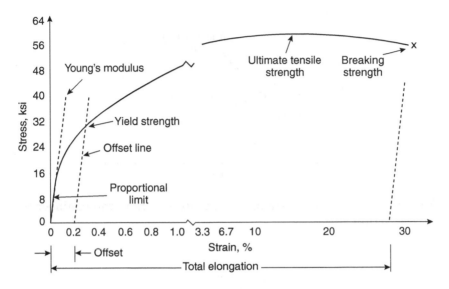

**Figure 3.2** A typical engineering stress–strain curve for a nonferrous metal in tension.

(Figure 3.2). The ultimate tensile strength and the breaking strength are two additional properties that can be recorded during a tensile test and provide some insight into the material toughness under extreme loading (more on this in the section on "Toughness").

The stress–strain curve can be presented in two forms: as a true stress–stain curve or as an engineering stress–strain curve. The engineering stress–strain curve is used for design purposes because it presents an accurate picture of the material strength prior to deformation (Figure 3.2). It is based on the original dimensions of area and gauge length, which, actually, continually change throughout the test. Most designs are not intended to be used past their proportional limit, so the engineering stress–strain curve is acceptable and it is generally used for engineering design. Conversely, manufacturing operations such as forming and machining rely on material deformation to achieve the final form and benefit from the true stress–strain curve data (Figure 3.3).

Two other types of moduli can also be used to quantify the behavior under load for those materials which do not

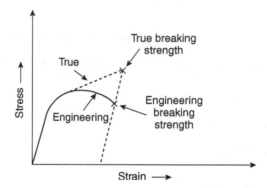

**Figure 3.3**  The true stress–strain curve versus the engineering stress–strain curve.

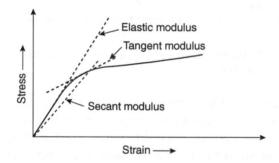

**Figure 3.4**  Three types of modulus ratios that can be derived from a stress–strain curve.

follow Hooke's law (Figure 3.4). These materials exhibit non-linear behavior and do not have a distinct proportional limit. Examples of materials without a proportional limit include gray cast iron and viscoelastic materials such as plastics and elastomers. The tangent modulus is obtained by placing a tangent line to the stress–strain curve and determining its slope. Similarly, the secant modulus is taken from the stress–strain curve by placing a linear line from the origin to the desired stress value, using the resulting slope as the modulus. Selecting the region of interest for calculating these values is based on design information required to meet the application specifications.

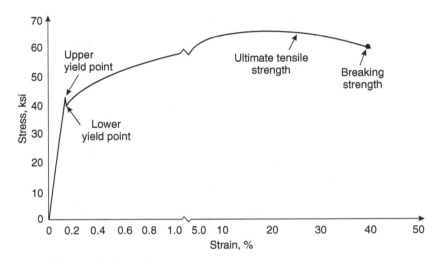

**Figure 3.5** Typical tension stress–strain curve for iron containing 0.15% C.

*Source:* ASM International, Online Metals Handbook, Desk Edition, Properties of Metals: Mechanical Properties, Figure 4.

Stress–strain curves provide unique insight to the material composition as exhibited by iron containing small amounts of carbon (Figure 3.5). The initial displacement or yield point for this material is followed by a reduction in apparent strength, as the stress for plastic deformation is less owing to the formation of dislocations. After this region, the curve is similar to one produced by a pure metal. This distinct behavior may be undesirable in an application in which a gradual transition is required. For example, a heavy press fit part relies on the smooth transition from proportional limit to the yield point to maintain a uniform clamping force on the mated parts.

Stress–strain curves are obtained by tensile testing, which involves the axial loading at a constant rate of a defined test specimen (Figure 3.6). The shape of the sample and the specific geometries of the mechanism are defined by specific standards based on the material. These tests provide the designer valuable information regarding the material's response to axial loading, such as the yield and tensile strength. Yield strength is determined at the 2% offset point while tensile strength is the

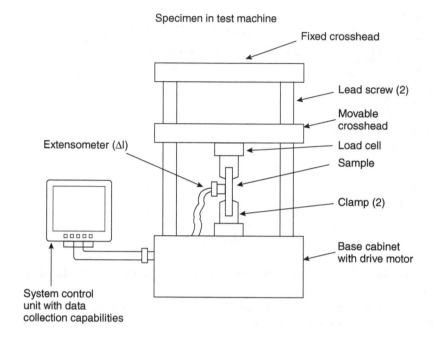

Specimen in test machine

Fixed crosshead

Lead screw (2)

Movable crosshead

Load cell

Sample

Extensometer (Δl)

Clamp (2)

Base cabinet with drive motor

System control unit with data collection capabilities

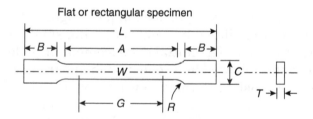

Flat or rectangular specimen

**Figure 3.6** Tensile test specimen and associated instrumentation setup for tensile test.

maximum value reached during the test, often referred to as the ultimate strength. Tensile strength is the maximum load divided by the original cross-sectional area with units of load per area such as psi, ksi, Mpa, or $N/m^2$.

Obtaining stress–strain curves for brittle materials offers a unique challenge because these materials are sensitive to gripping mechanism and shape, often failing early and not

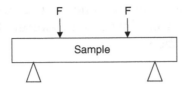

**Figure 3.7** Tensile testing method for brittle materials is actually performed in bending because of the difficulty in fabricating and holding a form based on a typical metallic sample.

providing an appropriate measure of strength. To overcome these shortcomings, brittle materials are tested by bending them rather than by applying tension. A simple rectangular bar shaped sample is used, which includes two support points located near the sample ends and two loading points on the top (Figure 3.7). The disadvantage of this test is that the stress distribution is nonuniform in the sample which can result in an overestimation of strength. Brittle materials have no significant region of plastic deformation in their stress–strain curves.

When considering material strength criteria, the yield strength is used to define the maximum load carrying capability of the device or structure. Although this value represents the maximum allowable stress, the actual stress multiplied by an appropriate factor of safety must be calculated for the design, and this must be less than the yield strength.

## Flexural Strength

Tests configured for bending are also used for plastics, reinforced plastics, and composites to measure another important property known as flexural strength. This property is similar to a tensile test as it measures the ability of the material to resist deformation under load. It is measured by using a three point test, similar to the four point test used for brittle materials but with only one pressure point rather than two. The load at yield, typically measured at 5% deformation/strain of the outer surface, is reported as the flexural strength. This test is used for high-modulus plastics, composites, and electrical insulating materials. For materials that fail in the outer surface within the 5%

strain limit, the four point bend test is used. ASTM D790 describes the testing of plastics for flexural strength using the three point test method.

## Toughness

The engineering stress–strain curve also provides a measure of the material toughness or the ability of a material to elastically absorb impact energy without failure. It is quantified by the area under the stress–strain curve (Figure 3.8). Those materials which are considered tough are ductile and exhibit greater elongation prior to failure while brittle materials have little or no measurable elongation. Applications involving impulse or shock loading benefit from these tough materials as they absorb the impacts, reducing the likelihood of a catastrophic failure while providing some isolating capability to the associated structure.

## Fracture Toughness

The fracture toughness of a material is its ability to resist internal crack growth, by limiting crack propagation. Material composition, heat treatment, and ambient temperature have a significant influence on this property. The crack serves as a method of relieving the internal material stresses, but fracture

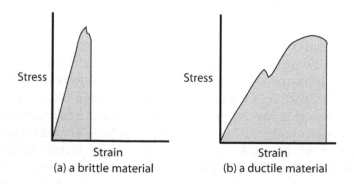

**Figure 3.8**   Material toughness is a function of the area under the stress–strain curve.

toughness quantifies the tendency of the material to halt or resist further crack growth. The parameter used to quantify this behavior is known as the crack tip intensity factor, K. ASTM E-399 describes test procedures and methods for estimating the fracture toughness of metallic materials.

## Notch Sensitivity

Impact testing serves as a measure of a material's notch sensitivity or behavior toward stress concentration. Notches, or stress concentrations, are important design considerations as they are often present as a result of manufacturing operations or use. Material failures frequently originate at a discontinuity in the structure, such as a hole, thread, notch, groove, or scratch. When designing load-bearing parts, it is essential to know how the material will respond to load-concentrating discontinuities. Proper design can eliminate or reduce stress concentrations and therefore minimize potential failure sites. Notch sensitivity is defined as the extent to which the sensitivity of a material to fracture is increased by the presence of a surface discontinuity such as a notch, a sudden change in section, a crack, or a scratch. Brittle materials usually have a high notch sensitivity as opposed to the low values of sensitivity observed in ductile materials.

## Process Properties

The ability to manufacture a successful design involves the process capabilities and processing properties of the chosen materials. Manufacturing capabilities are driven by a company's resources, which are, typically, related to the users market and cost structure. For example, the aircraft industry is required to produce a safe and efficient product, and, therefore, it can bear the high cost of materials and their related processing in the quest for a safe and reliable product. Alternatively, the toy industry must also build safe products, but it is also highly cost-sensitive, owing to its elective nature (one does not have to have a toy). Toy manufacturers cannot justify high material or processing costs, so product design must include a rigorous review of the process properties.

The most common process-related properties for all designs are surface finish and formability. Surface finish involves not only the machine or molded finish but also includes post-processing, which is intended to improve product life by protecting the surface, either by altering the surface structure or by application of a surface coating.

Formability involves not only the bulk material forming properties required for stamping or cold drawing but also includes machining characteristics. These latter characteristics influence processing times and surface finish.

## Surface Finish

The material finish of every design requires consideration, as it is related to the desired material performance. Mechanically machined materials have been the traditional area of concern and are the basis for the specification and inspection methods employed in the area. Surface finishing techniques including grinding, buffing, and polishing are also included. Nontraditional machining techniques have become more cost-effective for many manufacturing processes as well as a necessity for others and includes laser machining, electrical discharge machining (EDM), and chemical machining. How the resulting surface finish for these processes are defined and quantified becomes a significant consideration for the success of many designs, requiring the design engineer to place specific instructions on the engineering drawing in conjunction with the applicable specifications.

The fundamental concept involving surface finish is driven by the need to control the microscopic surface shape (Figure 3.9). The ability to quantify the surface condition is necessary in order to consistently obtain the desired form and can be defined by several different methods, each selected based on the design requirements (Figure 3.10). A low $R_a$ surface finish is also important for designs involving cyclic loads, as small surface imperfections can be stress concentration points, which can initiate a fracture.

The surface finish symbols used on engineering drawings and in related specifications and processing documents are

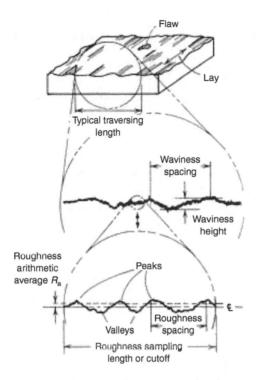

**Figure 3.9** Schematic of roughness and waviness on a surface with unidirectional lay and one flaw. See Figure 3.10 for definition of $R_a$ and waviness height.

*Source*: Reprinted from ASME B46.1-1985 and Y14.36-1978, by permission of The American Society of Mechanical Engineers. All rights reserved.

defined in great detail in order to provide clarity to the requirements and allow for manufacturing repeatability (Figure 3.11). Generally, the drawing title block defines a standard requirement with these symbols only added when the requirements are different from the standard. For example, a finish of 8 RMS may be placed on a piston bore as an overriding requirement because of the need to produce the appropriate finish for sealing the piston seal with the bore.

Design requirements may necessitate the orientation of the machine finish to provide the desired surface finish (Figure 3.12).

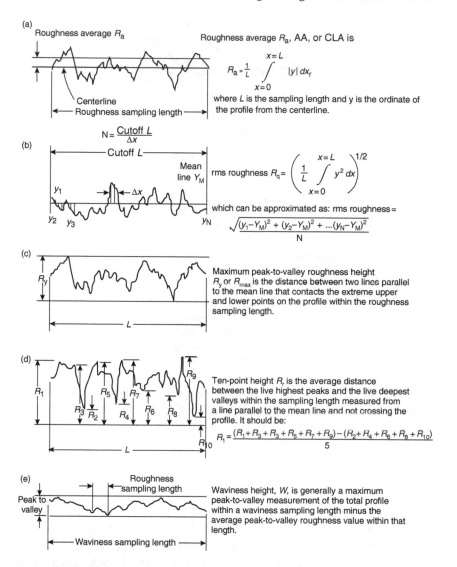

**Figure 3.10**  Some commonly used designations of surface texture. (a) $R_a$. (b) $R_q$. (c) $R_y$ or $R_{max}$. (d) $R_z$. (e) $W$.

*Source*: Reprinted from ASME B46.1-1985 and Y14.36-1978, by permission of The American Society of Mechanical Engineers. All rights reserved.

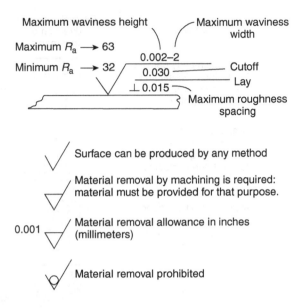

Maximum waviness height / Maximum waviness width
Maximum $R_a \longrightarrow 63$
Minimum $R_a \longrightarrow 32$
0.002–2
0.030 —— Cutoff
⊥ 0.015 — Lay
Maximum roughness spacing

Surface can be produced by any method

Material removal by machining is required: material must be provided for that purpose.

0.001 — Material removal allowance in inches (millimeters)

Material removal prohibited

**Figure 3.11** Surface texture symbols used for drawings or specifications. In this example, all values are in inches except $R_a$ values, which are in microinches. Metric values (millimeters and micrometers) are used on metric drawings.

*Source*: Reprinted from ASME B46.1-1985 and Y14.36-1978, by permission of The American Society of Mechanical Engineers. All rights reserved.

These are obtained by noting the surface "lay" or direction of the tool marks in the material surface. For example, machine tools very often use an angular or radial lay to produce a surface that can retain lubricants and reduce wear and tear.

Manufacturing operations involving material removal can be quantified in terms of surface roughness (Figure 3.13). Most operations can produce a range of finishes depending on the machine settings, tools, and the specific material properties. The design engineer needs to take these methods into consideration during the material selection process as well as work closely with a knowledgeable materials engineer and supplier representative.

| Lay symbol | Meaning | Example showing direction of tool marks |
|---|---|---|
| = | Lay approximately parallel to the line representing the surface to which the symbol is applied | 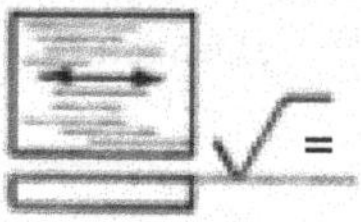 |
| ⊥ | Lay approximately perpendicular to the line representing the surface to which the symbol is applied |  |
| X | Lay angular in both directions to line representing the surface to which the symbol is applied | 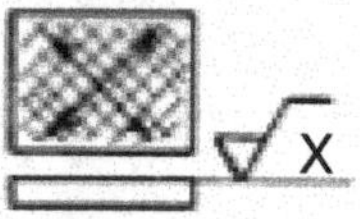 |
| M | Lay multidirectional | 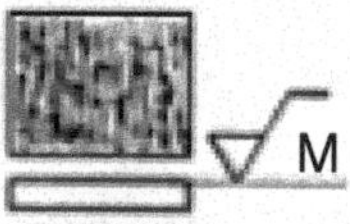 |
| C | Lay approximately circular relative to the center of the surface to which the symbol is applied | 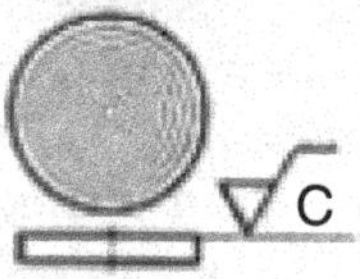 |
| R | Lay approximately radial relative to the center of the surface to which the symbol is applied | 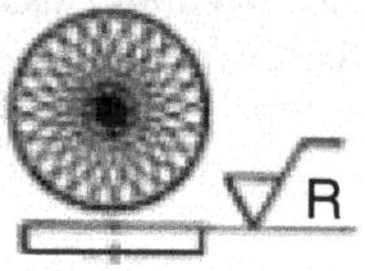 |
| P3 | Lay particulate non directional, or protuberant |  |

**Figure 3.12**   Symbols used to define lay and its direction.

*Source*: Reprinted from ASME B46.1-1985 and Y14.36-1978, by permission of The American Society of Mechanical Engineers. All rights reserved.

Each of the finish classifications also affect the producible tolerance range for the machined part (Table 3.2). The ability to obtain these finishes and tolerances are highly dependent on the material properties and may require additional surface

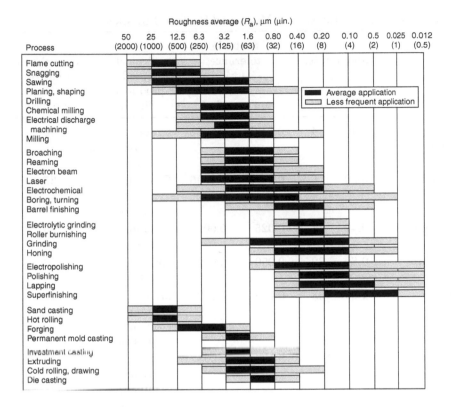

**Figure 3.13** Surface roughness produced by common production methods. The ranges shown are typical of the processes listed. Higher or lower values can be obtained under special conditions.

*Source*: Reprinted from ASME B46.1-1985 and Y14.36-1978, by permission of The American Society of Mechanical Engineers. All rights reserved.

preparation. For example, alloyed aluminum is too soft to produce a highly polished, durable finish, as might be required for a piston sealing surface, as described above. To obtain the required tolerances and surface finish, the aluminum is anodized which can then be finished to a low surface roughness.

## Heat Treatability

The response of a material to thermal treatment is used to alter properties such as hardness, strength, or microstructure

**Table 3.2**   Classification of Machined Surface Finishes

| Class | Roughness, $R_a$ | | Suitable tolerance ($\pm$) | | Typical method of producing finish | Approximate relative cost to produce |
|---|---|---|---|---|---|---|
| | μm | μin. | mm | in. | | |
| Super finish | 0.10 | 4 | 0.0125 | 0.0005 | Ground, microhoned, lapped | 40 |
| Polish | 0.20 | 8 | 0.0125 | 0.0005 | Ground, honed, lapped | 35 |
| Ground | 0.40 | 16 | 0.025 | 0.001 | Ground, lapped | 25 |
| Smooth | 0.80 | 32 | 0.050 | 0.002 | Ground, milled | 18 |
| Fine | 1.60 | 63 | 0.075 | 0.003 | Milled, ground, reamed, broached | 13 |
| Semifine | 3.2 | 125 | 0.100 | 0.004 | Ground, broached, milled, turned | 9 |
| Medium | 6.3 | 250 | 0.175 | 0.007 | Shaped, milled, turned | 6 |
| Semirough | 12.5 | 500 | 0.330 | 0.013 | Milled, turned | 4 |
| Rough | 25 | 1000 | 0.635 | 0.025 | Turned | 2 |
| Cleanup | 50 | 2000 | 1.25 | 0.050 | Turned | 1 |

*Source*: ASM International, Online Metals Handbook Desk Edition, Surface Finish and Surface Integrity in Machining: Surface Finish or Surface Texture, Table 3.

in an attempt to prepare it either for additional processing or for placement in its final use location. Heating and cooling the material at various rates and over specific temperature ranges is commonly done in order to produce specific material conditions (Figure 3.14). To be useful, this processing must be completed without cracking the material or an excessive change in its size.

Formability

The ability to change the shape of a material after it has been produced is the basis for most manufacturing. Two basic methods are typically employed: mechanical or chemical machining and forming. The composition, temperature, strain rate, and previous fabrication history have an affect on the ability of a material to be formed. The concepts of ductility and plasticity quantify the formability of a material, providing an insight into the material's response to loading extremes.

### Ductility

This is the ability of a metal to plastically deform without breaking or fracturing. It is measured as the percentage reduction in area or elongation that occurs during a tensile test. It is important both for processing and as a design requirement. Ductility is an important characteristic to consider when selecting a material, because it represents the ability of a material to fail in yield rather than in brittle fracture. This does not mean that a material with higher ductility will never fail in a brittle manner. Ductile metals can fail by brittle fracture if they are at a temperature below their ductile-to-brittle transition temperature (DBTT) or in a stress–corrosion environment. Ductility is also a property that may not be beneficial when it is greater, because it may render the device unusable. During the selection process the design engineer must evaluate this material property and determine its effect on the design's performance. Metals such as platinum, steel, copper, and tungsten have high ductility.

### Plasticity

Ductility is influenced by a material's plastic deformation behavior. Metals and plastics both exhibit this behavior, but each involve different mechanisms during the process. Metals plastically deform when there is motion along crystallographic planes. Superplasticity of metals occurs in certain grades and at specific strain rates, resulting in significantly greater deformation than

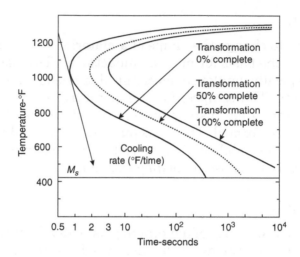

**Figure 3.14**   The time-temperature-transformation (TTT) chart provides insight into the material compositions that are possible during cooling.

is commonly observed and improved material properties. In contrast, plastics deform due to their chain-like structure which uncoils as the polymer deforms and the molecules become aligned. Continued loading past the alignment stage results in an observed increase in modulus as the bonds of the polymer are stressed and ultimately fail. The difference between metals and plastics is significant (Figure 3.15).

### Machinability

The ease with which a material can be machined to the desired shape is an important consideration in material selection because it influences manufacturing costs. Machinability is not a property of the material, but an attribute that quantifies the machining process. Machinability directly affects surface finish and dimensional accuracy, which are important factors for any engineered part and are included on the engineering drawing. Tool life and cutting speeds affect production rates and cost, making them valuable considerations to be included during the design process.

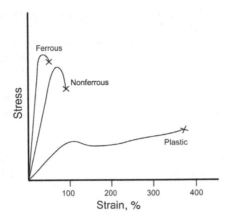

**Figure 3.15** Typical stress–strain curves for engineering materials.

A material is considered to be good for machining if the tool wear is low, the pressures applied are low, and the chips break into small pieces. Machinability is influenced by material strength, the presence of lubricants such as lead, sulfur, phosphorous, and graphite. The presence of abrasive constituents such as carbides reduces the machinability. Tool geometry as well as processing conditions such as cutting speed and lubrication impacts machinability. In practice, AISI 1112 carbon steel is assigned a machinability rating of 100 at a cutting velocity that provides for a 60 min tool life. Considering this in terms of a production rate it provides for a tool life of 60 min when machined at a cutting speed of 100 ft/min. The machinability ratings or comparisons with other materials provides a relative measure of their processing behavior (Table 3.3). Extensive listings can be found in material manufacturers' and industry related literature.

### Weldability and Joinability

The ability of any material to be welded to itself and other similar materials with ease and accuracy is important to consider when selecting a material. The strength of a weld should be no less than the parent metals joined together by it. Unfortunately, the area adjacent to the weld, known as the HAZ (heat affected zone), produces an sensitized zone that lacks the

**Table 3.3**  Comparative Measure of Material Machinability

| Material | Machinability Rating |
|---|---|
| Free cutting brass | 300 |
| Wrought Al | 200 |
| AISI 112 | 100 |
| Pearlitic gray cast iron | 70 |
| 3140 steel | 55 |
| 17–7 PH stainless steel | 20 |

parent metal properties. The weldability depends on the type of welding process used, the type of filler material used, environment, oxide formation, and surface conditions. Most ferrous metals have good weldability while nonferrous metals are generally more difficult to weld.

A measure of whether a steel can be easily welded is to determine the cabon equivalent ($C_{equiv}$ or $C_{equiv}^D$) of the steel. There are a number of different equations that can be used to determine this, the most appropriate choice depends on the type of steel being considered. For low alloy C-Mn steels with a ferrite + pearlite structure the appropriate equation is:

$$C_{equiv} = C + \frac{Mn}{6} + \frac{Cr + Mo + V}{5} + \frac{Ni + Cu}{15}$$

(alloying additions in wt%). For modern low carbon steels the equation becomes:

$$C_{equiv}^D = C + \frac{Si}{25} + \frac{Mn + Cr}{16} + \frac{Cr + Ni + Mo}{20} + \frac{V}{15}$$

(alloying additions in wt%). For the steel to be easily weldable $C_{equiv}$ needs to be less than 0.4 or the $C_{equiv}^D$ needs to be less than 0.25[1]. There are a number of different welding techniques that can be used for steels. Selecting the most appropriate technique depends upon the final product, plate thickness, and geometry of the weld required.

On a more generic level it is important to understand how to join materials, either to each other or to other similiar materials, using various processes (Table 3.4). The specifics for each design case are best addressed by following standard industry practices as well as the manufactures recommendations when appropriate. For example, when welding aluminum there are standard filler metals that are to be used based on the materials being joined. Similarly, when bonding with adhesives there are standard surface preparation techniques as well as specific recommendations from the adhesive manufacture.

### Castability

The ability to cast a material into a useful engineering shape (one that meets specific structural requirements) is influenced by many factors, such as the materials liquid–metal transition properties and how it solidifies. These results are influenced by the material's fluidity, shrinkage, and resistance to cracking. The fluidity, which measures how well the material flows into the mold, is influenced more by surface tension than viscosity as well as by the presence of insoluble inclusions. Pure metals generally have better fluidity than alloys. Also, the shape of the mold may have significant implications on how the mold fills. Shrinkage, porosity, and cracking of the material while it is being cast affect the usefulness and accuracy of the casting, requiring that material selection be closely coupled to the mold design.

### Suitability for Recycling

Consideration for product disposal is a design requirement that is becoming more prevalent due to the environmental impact of improper disposal. The ability to separate the various components of a product to allow the material families to be reprocessed makes for a more challenging design effort, adding time and cost to the product design. The European automotive industry has taken the lead in this area of "green" engineering by requiring that a vehicle be scraped by disassembly into its component parts and recycled.

**Table 3.4** Weldability and Joinability of Common Materials

| Material | Arc welding | Oxyacetylene welding | Electron beam welding | Resistance welding | Brazing | Soldering | Adhesive bondig |
|---|---|---|---|---|---|---|---|
| Cast iron | 7 | 10 | 1 | 1 | 3 | 1 | 7 |
| Carbon Steel, low-alloy steel | 10 | 10 | 7 | 10 | 10 | 3 | 7 |
| Stainless steel | 10 | 7 | 7 | 10 | 10 | 5 | 7 |
| Aluminum | 7 | 7 | 7 | 7 | 7 | 1 | 10 |
| Magnesium | 7 | 7 | 7 | 7 | 7 | 1 | 10 |
| Copper, Copper alloys | 7 | 7 | 7 | 7 | 10 | 10 | 7 |
| Nickel, Nickel alloys | 10 | 7 | 7 | 10 | 10 | 5 | 7 |
| Titanium | 7 | 1 | 7 | 7 | 3 | 1 | 7 |
| Lead | 7 | 7 | 1 | 3 | 1 | 10 | 10 |
| Zinc | 7 | 7 | 1 | 3 | 1 | 7 | 10 |
| Thermoplastics | 10[a] | 10[b] | 1 | 7[c] | 1 | 1 | 7 |
| Thermosets | 1 | 1 | 1 | 1 | 1 | 1 | 7 |
| Elastomers | 1 | 1 | 1 | 1 | 1 | 1 | 10 |
| Ceramics | 1 | 1 | 7 | 1 | 1 | 1 | 10 |
| Dissimilar metals | 3 | 3 | 7 | 3 | 3–7 | N/A | 10 |

*Note:* 10 = Excellent, 5 = Fair, 1 = Seldom/never used. N/A = not available.
[a] Heated tool; [b] Hot gas; [c] Induction.

*Source:* http://www.efunda.com/processes/metal_processing/welding_table.cfm

## Life Properties

The material selection process must also take into consideration the type of loads the part or structure must bear throughout its lifecycle. These loads may be static, impact or cyclic in nature. Life properties such as fatigue and creep take into consideration long-term cyclic loading and constant loading, respectively. Knowing a material's response to these loading conditions is necessary to ensure that the design can survive its intended life. The surface condition also has an affect on life limiting wear behavior, requiring consideration for hardness and finish.

### Impact Strength

The ability of a material to withstand impact loading conditions is quantified by an impact strength test which determines a material's ability to absorb energy over a short period of time. The impact property is proportional to the area under the stress–strain curve.

In general, the more brittle the material the lower the impact strength. Materials can be evaluated by using either the Izod or Charpy impact test methods which are fully defined in ASTM D 256. These tests both involve a swinging pendulum impacting a notched sample, with the Izod test method impacting the notched side and the Charpy test impacting the non-notched side. The energy consumed in breaking the test specimen is measured in Joules. A material's sensitivity to impact loading in part dependent on environmental conditions as it evident can change rapidly due to its ductile-to-brittle transition temperature. This characteristic can be more in some materials because it coincides with ambient or operating conditions. If a material exhibits this behavior the conceivable operating temperature range, the impact characteristics should be rigorously defined by temperature-controlled impact testing to fully characterize the material performance.

### Hardness

The hardness of a material is a measure of its resistance material to indentation or abrasion. In a hardness test a load is

Hardness Testing Methods

| Test type | Penetrator type | Application |
|---|---|---|
| Brinell |  10 mm ball | Nonferrous and softer ferrous metals |
| Rockwell C, A, N |  120° Diamond | Ferrous metals |
| Rockwell B,T |  $\frac{1}{16}$" Diameter ball | Nonferrous and softer ferrous metals |
| Shore A and D |  20° Needle | Plastics and elastomers |

**Figure 3.16** Typical hardness penetrators used to determine a material's hardness.

placed on an indenter, which is driven into the specimen. The degree to which the indenter can penetrate is the measure of the material's ability to resist plastic deformation or penetration (Figure 3.16). The commonly used test methods for hardness include the Brinnell, Rockwell, and Vickers method (Figure 3.17). This is a relatively simple and inexpensive test, which, for many materials, can be correlated to property data such as tensile strength, wear resistance, and fatigue strength, without having to perform individual tests for each (Table 3.5 and Table 3.6). Hardness data is used most often for two purposes: to select a material and its appropriate material condition for a specific application; as a field test to determine the general

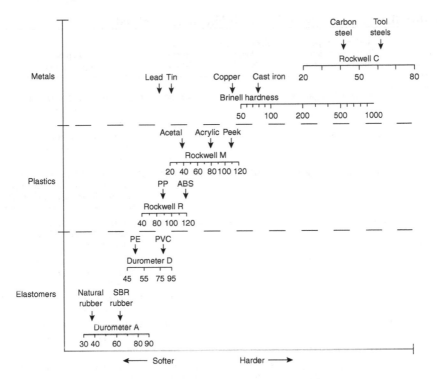

**Figure 3.17** Comparative hardness scales for common measuring methods.

characteristics of a material without having to perform a detailed chemical or mechanical analysis.

## Fatigue

The long-term effect of cyclic loading on a material's performance is highly dependent on the material family, material conditions and the amplitude of loading (Figure 3.18). Fatigue strength is defined as the maximum stress that can be endured by a material, for a specified number of cycles, without failing. Low cycle fatigue strength approaches the static strength. When the number of cycles exceeds one million, the fatigue strength typically reduces to a fraction of the static strength. Ferrous materials exhibit a finite fatigue life, meaning that there is a limit to which the cyclic loading affects the material

**Table 3.5**   Summary of Rockwell Hardness Methods

| Scale symbol | Indenter type (Ball dimensions indicate diameter) | Preliminary force N(kgf) | Total force N(kgf) | Typical applications |
|---|---|---|---|---|
| *Regular Rockwell Scales* | | | | |
| A | Spheroconical Diamond | 98.07 (10) | 588.4 (60) | Cemented Carbides, thin steel, and shallow case hardened steel. |
| B | Ball - 1.588 mm (1/16 in.) | 98.07 (10) | 980.7 (100) | Copper alloys, soft steels, aluminum alloys, malleable iron etc. |
| C | Spheroconical Diamond | 98.07 (10) | 1471 (150) | Steel, hard cast irons, pearlitic malleable iron, titanium deep case hardened steel, and other materials harder than HRB 100. |
| D | Spheroconical Diamond | 98.07 (10) | 980.7 (100) | Thin steel and medium case hardened steel, and pearlitic malleable iron. |
| E | Ball-3.175 mm (1/8 in.) | 98.07 (10) | 980.7 (100) | Cast iron, aluminum and magnesium alloys, and bearing metals. |
| F | Ball-1.588 mm (1/16 in.) | 98.07 (10) | 588.4 (60) | Annealed copper alloys, and thin soft sheet metals. |
| G | Ball-1.588 mm (1/16 in.) | 98.07 (10) | 1471 (150) | Malleable irons, copper-nickel-zinc and cupro-nickel alloys. |
| H | Ball-3.175 mm (1/8 in.) | 98.07 (10) | 588.4 (60) | Aluminum, zinc, and lead. |
| K | Ball-3.175 mm (1/8 in.) | 98.07 (10) | 1471 (150) | Bearing metals and other very |
| L | Ball-6.350 mm (1/4 in.) | 98.07 (10) | 588.4 (60) | soft or thin materials. |
| M | Ball-6.350 mm (1/4 in.) | 98.07 (10) | 980.7 (100) | Use smallest ball and heaviest |
| P | Ball-6.350 mm (1/4 in.) | 98.07 (10) | 1471 (150) | load that does not give anvil |
| R | Ball-12.70 mm (1/2 in.) | 98.07 (10) | 588.4 (60) | effect. |
| S | Ball-12.70 mm (1/2 in.) | 98.07 (10) | 980.7 (100) | |
| V | Ball-12.70 mm (1/2 in.) | 98.07 (10) | 1471 (150) | |
| *Superficial Rockwell Scales* | | | | |
| 15N | Spheroconical Diamond | 29.42 (3) | 147.1 (15) | Similar to A, C and D scales, but |
| 30N | Spheroconical Diamond | 29.42 (3) | 294.2 (30) | for thinner gage material or |
| 45N | Spheroconical Diamond | 29.42 (3) | 441.3 (45) | case depth. |
| 15T | Ball-1.588 mm (1/16 in.) | 29.42 (3) | 147.1 (15) | Similar to B, F and G scales, |
| 30T | Ball-1.588 mm (1/16 in.) | 29.42 (3) | 294.2 (30) | but for thinner gage material. |
| 45T | Ball-1.588 mm (1/16 in.) | 29.42 (3) | 441.3 (45) | |
| 15W | Ball-3.175 mm (1/8 in.) | 29.42 (3) | 147.1 (15) | Very soft material. |
| 30W | Ball-3.175 mm (1/8 in.) | 29.42 (3) | 294.2 (30) | |
| 45W | Ball-3.175 mm (1/8 in.) | 29.42 (3) | 441.3 (45) | |
| 15X | Ball-6.350 mm (1/4 in.) | 29.42 (3) | 147.1 (15) | |
| 30X | Ball-6.350 mm (1/4 in.) | 29.42 (3) | 294.2 (30) | |
| 45X | Ball-6.350 mm (1/4 in.) | 29.42 (3) | 441.3 (45) | |
| 15Y | Ball-12.70 mm (1/2 in.) | 29.42 (3) | 147.1 (15) | |
| 30Y | Ball-12.70 mm (1/2 in.) | 29.42 (3) | 294.2 (30) | |
| 45Y | Ball-12.70 mm (1/2 in.) | 29.42 (3) | 441.3 (45) | |

*Source*: National Institute of Standards and Technology (NIST), Special Publication 960–5, Rockwell Measurement of Metalic Materials, Table 1, page 5.

**Table 3.6** Summary of Common Hardness Measurement Methods

| Hardness test | Application | Indentor | Load range |
|---|---|---|---|
| Knoop (microhardness) | Soft steels to ceramics | Diamond | 1 g–2000 g |
| Rockwell A | Cemented carbides | Diamond | 50 kg |
| Rockwell C | Hardened metals, thick | Diamond | 150 kg |
| Rockwell N | Hardened metals, thin (sheet metals) | Diamond | 15, 30–45 kg |
| Rockwell B | Soft steels and nonferrous metals | Ball | 100 kg |
| Brinell | Soft steels and nonferrous metals | Ball | 500–300 kg |
| Rockwell R | Polymers and hard Elastomers | Ball | 10 kg |
| Shore Durometer D | Soft Polymers and hard elastomers | Needle | Spring |
| Shore Durometer A | Soft elastomers | Needle | Spring |

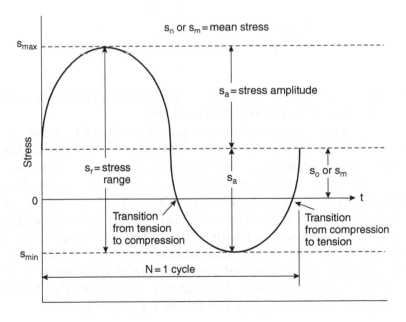

**Figure 3.18** Fatigue analysis criteria.

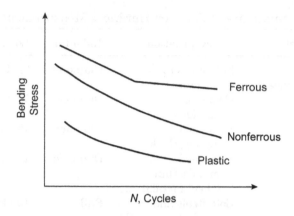

**Figure 3.19** Fatigue properties of metals and plastics differ greatly in their ability to handle cyclic loading.

**Table 3.7**   Dynamic Loading Design Criteria

| Load case | Equation |
|---|---|
| Stress range | $S_r = S_{max} + S_{max}$ |
| Stress amplitude | $S_a = \frac{1}{2}(S_{max} - S_{max})$ |
| Mean stress | $S_m = \frac{1}{2}(S_{max} + S_{max})$ |
| Stress ratio | $R = S_r = S_{max}/S_{max}$ |

strength, resulting in a point at which the material life is unaffected by the cyclic loading (Figure 3.19).

Fatigue testing involves the application of cyclic stress over an extended period of time at defined values. The following summarizes the nomenclature used to define the loading (Table 3.7).

Data from the test is plotted as stress, S, versus the logarithm of the number of cycles to failure, $N$. These curves are popularly known as *SN* curves (Figure 3.19). If the curve flattens out, or becomes horizontal, the specimen is said to have reached its fatigue limit. The fatigue limit is the maximum stress which can be applied over an infinite number of cycles and for ferrous metals is typically 35 to 60% of the tensile strength and estimated by the following equation

$$\text{Fatigue Limit} \approx \frac{1}{2}S_{ut}$$

Conversely, nonferrous metals and plastics do not exhibit this behavior of infinite life, resulting in a steady decline in material strength at increasing cycles. Their *SN* curves do not flatten out, indicating they do not have a fatigue limit.

## Creep

Creep is a process in which a material exhibits permanent deformation when subjected to a constant load over a long period of time. It is also defined as a time-dependent strain occurring under a constant stress. Creep is a property most often associated with ferrous metals subjected to constant loading at elevated temperatures, but it is also a phenomena which occurs in polymeric materials. The stresses involved in this form of loading are far below the material tensile strength. Generally, creep is considered a thermally activated process, meaning that the rate of creep or permanent deformation is increased at elevated temperatures.

Creep testing is performed under a constant load and a constant temperature over an extended period of time with the data plotted as strain versus time (Figure 3.20). The results exhibit three stages of strain that occur prior to failure. Initially, there is an instantaneous strain defined as

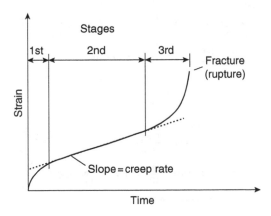

**Figure 3.20** The three stages of a creep curve as a function of strain and time.

Stage 1, followed by an increase in strain with time but with a reduction in slope of the curve. This stage is known as primary creep. Stage 2 creep is defined as the region where the slope of the strain rate is nearly constant and known as secondary creep. Following Stage 2, the strain rate increases sharply and ends in fracture. Stage 3 is called tertiary creep.

The effects of creep become a design concern with metallics when the operating environment is greater than one-third of the melting temperature because of the thermal influence on the process. For ferrous metals the region is typically above 400°C, but it is much lower for nonferrous alloys. Creep in plastics can occur at or above room temperature, and for elastomers creep begins below room temperature. These behaviors make creep a critical consideration for material selection, requiring adequate definition of the operating environment and loading.

Creep behavior is also extremely sensitive to material composition, microstructure, and prior processing history.

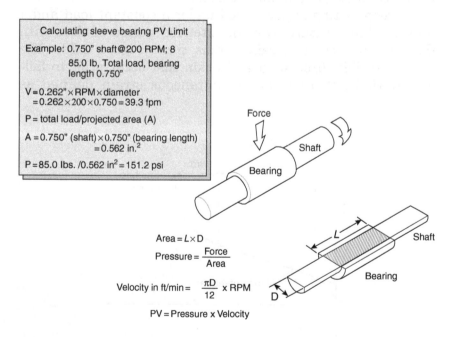

**Figure 3.21**   Example of the PV limit as calculated for a rod bearing.
*Source*: Polygon Corporation. Used with permission.

Knowing this history is important to the selection process and it is best to obtain it from the material supplier or their representative.

## PV Limit

The PV limit applies to plastics placed in sliding wear applications. The product of the pressure, P, and the surface velocity, V, provides a quantitative limit for the materials involved (Figure 3.21). The concern for this type of design is the affect of frictional heating that is taking place in a localized environment that can cause premature failure. The failure involves the gradual degradation of the surface with a corresponding loss of mechanical and physical properties.

## CHEMICAL

The response of a material when exposed to unintended environments can result in product failure, while extended exposure to even the intended environments can produce unanticipated degradation. Generally, all materials have some form of response to chemical exposure. For example, when exposed to ambient conditions, untreated aluminum produces an oxidized coating that helps retard further oxidation, but which is not useful or desirable for most applications. Secondary processing is typically required to produce the desired finish. A careful review of a materials response to chemical exposure, including ambient environmental conditions, must be formulated and confirmed through testing and close field monitoring of prototypes.

### Corrosion Resistance

Corrosion is defined as the gradual chemical or electrochemical attack on a metal by atmosphere, moisture, or other agents. The extent to which a material can be shielded against corrosion depends on the type of corrosion occurring, the service environment and material composition. There are several types of corrosion which may occur with respect to metallic or plastic materials: atmospheric corrosion, pitting, crevice, galvanic, stress corrosion cracking, intergranular attack, dealloying, and high temperature corrosion.

Corrosion protection can often be achieved by several methods that range in cost and application time. Consider the applications where you have to select corrosion protection for two different environments. One is a steel structure frame on an oil rig in the ocean and the other is steel beams used in the structure of a building. The environments for both of these applications are vastly different. On the oil rig the structure is attacked from seawater and moist sea air, as well as also from oil and other chemicals. In the second application, the steel is confined within a concrete structure. The kind of protection required for these applications are vastly different, requiring careful analysis on the part of the design engineer in order to quantify the environment prior to making a selection on the protection method (Figure 3.22).

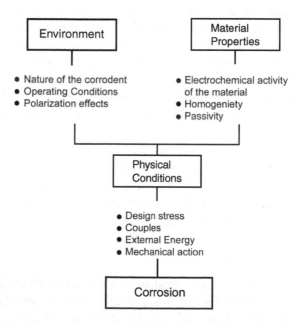

**Figure 3.22**   Factors affecting corrosion originate with the design environment and material properties and are further influenced by the physical conditions applied to the design.

**Table 3.8**   Galvanic Series in Seawater

| | |
|---|---|
| 1. Magnesium (most active) | 18. Inconel (active) |
| 2. Magnesium alloys | 19. Hastelloy B |
| 3. Zinc. | 20. Brasses |
| 4. Aluminum 100 | 21. Copper |
| 5. Cadmium | 22. Bronzes |
| 6. Aluminum 2017 | 23. Copper-nickel alloys |
| 7. Steel (plain) | 24. Titanium |
| 8. Cast iron | 25. Monel |
| 9. Chromium iron (active) | 26. Silver solder |
| 10. Nickel cast iron | 27. Nickel (passive) |
| 11. 304 stainless (active) | 28. Inconel (passive) |
| 12. 316 stainless (active) | 29. Chromium iron (passive) |
| 13. Hastelloy C | 30. 304 stainless (passive) |
| 14. Lead-tin solders | 31. 316 stainless (passive) |
| 15. Lead | 32. Silver |
| 16. Tin | 33. Graphite (least active) |
| 17. Nickel (active) | |

*Source*: ASM International, Online Metals Handbook Desk Edition, Corrosion Characteristics of Carbon and Alloy Steels, Corrosion in Seawater: Galvanic Corrosion, Table 5.

Corrosion protection can be provided by passive coatings, paints, cathodic, or anodic protection or by providing a sacrificial anode (Table 3.8). Metals offer a greater challenge as plastics are inherently non-corrosive. Corrosion control can be addressed at the part level, assembly level or system level (Figure 3.23). That is to say that parts and assemblies can be protected by the previously mentioned methods while at the system level corrosion can be addressed by the operational characteristics or methods. For example, it may not be possible to alter the fluid velocity in a mass transfer system but it would be possible to select a material or coating to limit the rate of errosion of critical components.

The composition of a material is critical to the material selection process. It is not necessary to know the chemical composition in great detail, but the design engineer should know the bulk composition or key constituents that differentiate the material from similar grades. For example, the addition of lead to copper produces a free machining material such as yellow brass; reducing the carbon content of 303 produces

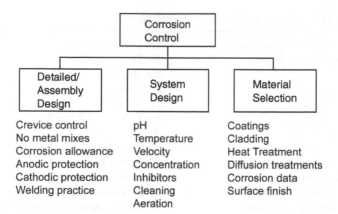

**Figure 3.23**  Corrosion control through design. How the detailed parts and assembly are designed controls the corrosion characteristics of the system.

304 stainless steel, which is better suited for welding because of its reduced sensitivity to carbide precipitation. Detailed information regarding the material composition can be obtained from industrial reference sources such as ASM International and manufacturers.

## Microstructure

Material microstructure is critical to many properties that can influence material selection. A materials microstructure can have a significant effect on its machinability, tensile strength, fracture behavior, and creep. Information regarding microstructure and process history can typically be obtained from industry sources and manufactures.

## Crystal Structure

A material's crystal structure is one of its basic properties, similar composition. The crystal structure dictates most of its mechanical properties, including modulus, toughness, and ductility. Most engineering materials are polycrystalline materials, meaning they are made from numerous crystals packed in a manner specific to the material type. It is not crucial to know the exact details of the structure of the alloy but it is reasonable to know what major crystal system is present in it. The crystal system will allow you to predict certain behaviors in term of ductility and formability as well as hardness.

## Molecular Weight

Exclusively for polymers, the molecular weight is an important consideration during material selection. A polymer's molecular weight is important because it determines many of its physical properties including transition temperature from elastomer to solid as well as mechanical properties such as stiffness, strength, viscoelasticity, toughness, and viscosity. If a polymer's molecular weight is too low, the transition temperature and the mechanical properties will generally be too low for it to have any useful commercial applications. For a polymer to be useful it must have a transition temperature above room temperature and must have mechanical properties that are suitable for bearing designed loads. Unlike small molecules, however, the molecular weight of a polymer is not one, unique value. Rather, a given polymer will have a distribution of molecular weights from which an average can be obtained for use in a specification. The distribution will depend on the way the polymer is produced.

## PHYSICAL

Engineering materials are most commonly referred to by their physical properties because they can be readily observed. A materials mass or surface finish can often convey basic performance expectations, although the design engineer needs to be careful in not allowing these initial expectations to cloud their rational decisionmaking process or to pass over a material all together. Physical properties not only affect how a design performs but also how it is perceived. The look and feel of something made from metal conveys a sense of strength and durability, while the same item, if it were made from plastic, may not, even though both perform equally in the same application.

Overcoming these perceptions while designing an economical and effective product can be difficult, offering a significant challenge to even the most experienced design engineers.

## Mass Properties

Density and specific gravity are the mass properties most commonly referred to. Density is defined as mass per unit volume and is often considered in conjunction with a materials strength. The ratio of strength to density is known as specific strength, $Sy/\rho$. This is a useful measure when searching for a material that is both strong and light, such as those required by transportation systems. Conversely, high density and strength may be the desired material for designs involving momentum, such as projectiles and flywheels. Materials including tungsten and heavy metals fit these requirements.

Specific gravity is a numerical value representing the weight of a given substance compared with the weight of an equal volume of water, making the specific gravity of water 1.0.

## Physical Properties

Poisson's ratio, water absorption, and color are physical properties that must be considered in the material selection process. Naturally, their importance will be dependent on the application.

Poisson's ratio is the ratio of the transverse strain to the corresponding axial strain in a body subjected to uniaxial stress measured in the elastic region. For example, when a square bar is stressed axially in a tension testing machine such that the length increases, a contraction in each orthogonal direction occurs which causes a decrease in bar crossection. The ratio of the contraction at right angles to the extension is known as Poisson's ratio. Its value in steel is on the order of 0.283. It is defined by

$$\nu = -\, \varepsilon_{trans}/\varepsilon_{longitudinal}$$

Water absorption is defined as the amount of water absorbed by a material and is expressed in percent weight change.

A sample is weighed, immersed in room temperature water (typically for 24 hours or until saturation), then reweighed. The percentage of water absorbed by the material is

$$\left(\frac{\text{Wet weight}}{\text{Dry weight}} - 1\right) \times 100$$

Water absorption can be an important consideration for many applications because of the potential for part distortion that could lead to failure due to binding or excessive wear.

Material color is an elusive property which is open to much interpretation because of its dependency on the illuminating light spectrum, surface finish, and viewing angle, as well as other factors specific to the application. ASTM standard D1729 requires visual examination in a standard, controlled environment that defines the light source and ambient lighting. Color may be important in post-processing stages during material finishing applications, such as painting and anodizing, to enhance aesthetics or provide corrosion protection.

## Thermal Properties

The thermal performance of a material can play a significant role in the success or failure of a design. Thermal expansion and conductivity are the two most common properties to be considered. Material behavior at elevated temperatures also affects creep properties and glass transition temperature.

Thermal expansion is an important property for both metals and polymers as it provides an understanding of the physical changes occuring in the material within the range of system operating temperatures. The thermal characteristics of metals are fairly uniform within a given material family, but do vary between families. For example, carbon steels expand similarly to stainless steels, while being notably different from copper or aluminum. The response of polymers to thermal heating is greater than metals by an order of magnitude; greater attention has to be paid to this physical attribute of polymers for designs involving them that experience a wide range of temperatures. Plastics can have a wide range of expansion coefficients depending on the type of bonding, crystallinity, extent of polymer-chain branching or cross-linking and the amount of additives/fillers.

Thermal conductivity characterizes a materials ability to transfer heat and is defined by the relationship $q = -k(dT/dx)$ which is a measure of heat flux or flow per unit time per unit area ($W/m^2$). The transfer area and material thickness perpendicular to this area determine the available conduit for the heat transfer. The thermal conductivity, k, is in W/m $\times$ K while $dT/dx$ is the temperature gradient through the material. Metals generally have the greatest conductivities, but a significant range exists within this classification. Typically, nonferrous metals have greater conductivity than ferrous metals. Plastics do not exhibit significant thermal conductivity due to the nature of their chemical composition. Additives and fillers incorporated into the polymer can significantly alter their thermal behavior to predictable ranges. This ability to tailor their performance is used extensively within the electronics industry to obtain the desired electrical and thermal dissipation qualities.

Heat capacity is defined as the amount of energy required to produce a unit rise in temperature. It provides a measure of the materials ability to absorb heat from the surrounding environment and is expressed by $C = dQ/dT$. Transient temperature applications which are sensitive to thermal expansion should employ materials with low heat capacity values to limit distortion.

The glass transition temperature ($T_g$) is the temperature below which molecules have very little mobility. Generally, polymers are rigid and brittle below their glass transition temperature and elastic above it. $T_g$ is usually exhibited by amorphous materials, including glasses and plastics. At a molecular level, $T_g$ requires an understanding of the mechanical loss mechanisms within the material's specific functional groups and molecular arrangements; a discussion of this is beyond the scope of this text. A well known design example involves the loss of the space shuttle, Challenger, which involved a rubber O-ring that was operating below its glass transition temperature, preventing it from adequately flexing to form a proper seal within one of the two solid rocket boosters.

The subsequent leak of hot gases resulted in a structural failure and the loss of the spacecraft.

The melting point is the temperature where solid and liquid phases are in equilibrium or, in other words, a solid transforms into a liquid. Melting points are of great interest in metals used for casting. The boiling point is the temperature at which a substance changes its state from liquid to gas. A stricter definition of boiling point is the temperature at which the liquid and vapor (gas) phases of a substance can exist in equilibrium. Boiling points are more important in the case of gases and liquids. Metals do not reach their boiling points during processing or general fabrication methods.

## Electrical Properties

Electrical properties of a material depends on its atomic and microscopic structure. While mechanical properties of a material such as elastic modulus, yield strength, and tensile strength, describe a material's response to the application of external loads, the electrical properties describe a materials response to external electrical fields.

If an electric voltage $V$ is applied to a material of length $L$ and cross-sectional area $A$, the current flow as given by Ohm's law is

$$V = IR$$

where $I$ is the current in amperes and $R$ is the resistivity in ohms. The opposite of resistively is conductivity and is given in units of $ohm^{-1}$. Electrical conductivity is the product of mobility and charge density with the total conductivity being the sum of the conductivities all carrier types. In many materials, such as metals, the electrical charge is carried by electrons only. In ionic solids there is another charged species which can contribute to electrical conduction, the ions. There can be more than one type of ion in an ionic solid and because of this the term "sum of all carriers" is used when defining the conductivity of such a solid.

The other term used is "charge mobility," which is the ease with which a charge carrier can move through the lattice. Electrons are smaller in size than ions and can move through the material more easily. Their movement depends on the average drift velocity and how few collisions they encounter along their path. The more collisions, the less the current flow. The detailed models of conduction and carrier movement require the knowledge of quantum mechanics, and is beyond the scope of this text. This text will simply consider electrical current or thermal effects due to the flow of electrons and ions.

If the potential difference between two plates separated by a dielectric is too high or the thickness of the dielectric is too small, the dielectric breaks down and the electrical charge can move between the two plates. *Dielectric strength* is the voltage required for an insulator to breakdown and conduct electricity; it is inversely proportional to material thickness.

Dielectric strength = breakdown voltage/insulator thickness

The units of dielectric strength are volts per meter or kilovolts per millimeter.

The *dielectric constant*, also called the permittivity, is the ratio of the capacitance of an insulator to the capacitance of a vacuum or dry air (the dielectric constant of a vacuum is 1.00000, dry air is 1.00054) and denoted by DK or Er. Capacitors release their charge when a circuit is broken. Capacitance is the ratio of charge absorbed to the potential (voltage applied). If an insulator has a dielectric constant of 3, it means that it will absorb 3 times more electrical charge than vacuum will. The value is affected by temperature, voltage frequency, and humidity. The higher the dielectric constant, the better capacitor a material will be.

*Volume resistivity* is the ratio of the electrical resistance through a cross-sectional area divided by the length through which the current flows. It is usually measured in ohms-cm or ohms-inch and is a property particularly important in polymer-based materials. Measurements of volume resistivity are affected by temperature and humidity and must be quantified when comparing like performance for different materials. Volume resistivity is important when selecting plastics as insulators in electrical appliances and other insulating applications.

The material *dissipation factor* is another important measure for quantifying material capacitance and alternating current (AC) loss. Also called the *loss tangent*, it is the percentage of electrical energy absorbed and lost when current is applied to an insulating material. Most of the absorbed energy is dissipated as heat. If the material has a dissipation factor of 0.05 this means that 0.05% of the energy being stored (capacitance) is lost. The dissipation factor is the ratio of the resistive component of a capacitor to the capacitive reactance of the capacitor. The lower the dissipation factor, the higher the material capacitance. At elevated temperatures, the material dissipation factor is reduced.

*Arc resistance* is another electric property that applies primarily to plastics. It is the ability to withstand exposure to an electric voltage without arcing. This is the total time, in seconds, for which an intermittent arc may play across a plastic surface without rendering the surface conductive. The arc value is the known voltage and arc resistance is measured in units of time.

The property which defines whether a material is a conductor or an insulator is the *electrical resistivity*. It is the electrical resistance to the flow of current offered by a material times the cross-sectional area of current flow and per unit length of current path. It is also called resistivity or specific resistance. Apart from thermal vibrations, irregularities on a microscale also present difficulties for the charge carriers to move. These defects can include vacancies, dislocations, and grain boundaries. In the case of pure metals only thermal vibrations contribute to the resistivity. This is the reason why alloys with interstial and substitutional solid solutions have higher resistivities than pure metals.

A common example involving electrical power lines may help to explain material selection considerations and necessary trade-offs. Consider power transmission lines that connect across long distances, transferring high-voltage power from one region to the next. Two conditions must be satisfied in the application: conductivity and strength. A pure metal would ideally satisfy the first requirement but not the second as it lacks sufficient strength. An alloy provides a reasonable compromise between high conductivity and sufficient strength to support the weight of the transmission line. The strength requirement is indirectly a density requirement, because the material must be able to adequately support its own weight. In the case of this design, aluminum provides the required conductivity and strength–weight ratio for the design to be economical.

## Magnetic Properties

Ferrous materials exhibit magnetic characteristics that are specific to their material family and are known as ferromagnetic materials. Nonferrous materials and plastics are generally not magnetic, but this can be altered by the use of ferromagnetic fillers and binders. The specific properties of magnetic materials are influenced by chemical composition, microstructure, and crystallographic phase or lattice structure. The magnetic field strength, $H$, is measured as a function of the flux density, $B$ by the relationship

$$B_0 = \mu_0 H$$

where $\mu_0$ is a constant called permeability of free space or magnetic susceptibility. If air is used as the core, the term $\mu_r$ is used, instead of $\mu_0$, which is called relative permeability. It is given by

$$\mu_r = B/B_0 = B/H\mu_0$$

The material parameters such as grain size, dislocation density, and presence of precipitates can have a significant influence on the strength of the field. The maximum value of $B$ is known as the saturation point and is noted by the unchanging flux density when plotted against the magnetic field strength, $H$ (Figure 3.24).

An important property in the application of magnetic materials is the *Curie point* or *Curie temperature*. This material characteristic occurs in certain metals, alloys, and rare earth elements and involves its ability to remain magnetic at elevated temperatures. A material may be capable of retaining a magnetic field at room temperature, but at some elevated temperature loses its ability to do so, effectively making it only

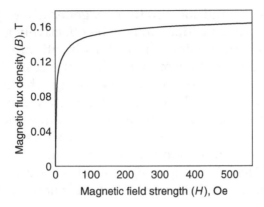

**Figure 3.24** An example of the magnetization curve for an iron based material.

a paramagnetic material, or one that can exhibit a magnetic characteristic only in the presence of a magnetic field. The loss of magnetism can occur at relatively low temperatures, as low as 200°F, when compared to the other components within the operating system, making it critical to understand the material selection requirements. This property is important to electrical motor design and similar applications requiring permanent magnets that operate at elevated temperatures.

## Radiation Properties

A material's ability to absorb radiation may be an important consideration in the selection process for applications involving ionizing radiation sources such as X-rays and gamma rays. It may also be important for nonionizing radiation sources, including lasers. Materials can be adversely affected by these conditions and would be unable to perform as intended. For example, glass fiber products are generally not good candidates for these types of applications because of the organic binder that is used to hold the matrix in place. Glass fibers resist radiation because of their amorphous structure, which does not distort the atomic ordering, but the binder may breakdown in the presence of high levels of radiation. Lead is commonly known for its ability to absorb radiation but due to its high density and low modulus it generally requires additional structural considerations to be effectively applied.

Ferrous and nonferrous materials all are effective at absorbing radiation to some level, it is the task of the design engineer to properly determine the design thickness for the intended application. Material transparency and absorption properties are of concern with plastics used in nonionizing radiation applications because the design requirement is either based on operation or safety with the former requiring transmission and the latter requiring absorption or reflection to control the radiation path. Careful selection is necessary to achieve the required result.

Radiation transmission is determined by the inverse square law which involves the attenuation of an energy source equally in all directions without limit (Figure 3.25). The intensity of the radiation at any given radius $r$ is the source strength divided by the area of the sphere. Being strictly geometric in its origin, the inverse square law applies to diverse phenomena including radiation, gravity, light, sound, and magnetic fields. Point sources of energy obey the inverse square law.

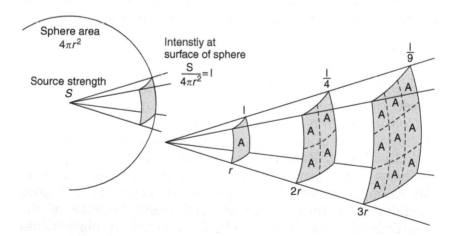

**Figure 3.25**  Inverse square law. The energy twice as far from the source is spread over four times the area, hence one-fourth the intensity.

*Source*: HyperPhysics by Dr. C.R. Nave, 2005. Used with permission.

## Optical Properties

Selection of a material based on its optical properties is a challenging task, requiring knowledge of some important properties specific to how light interacts with plastics and ceramics. The type of chemical bonding involved in a polymer dictates the refractive index, dispersion properties, and the color. Processing methods also are important considerations because of their effect on surface finish and resulting reflectivity as well as visible properties such as transmittance and haze. Understanding the concept of each of these properties while working closely with the material supplier's technical staff will be an important component in the success of a design.

### Refractive Index

When light passes from one medium to another, reflection and refraction occur at the interface. Perhaps the simplest way to define refractive index is as the ratio of the speed of light $(v)$ in vacuum to that in the material

$$n = (v_{vac})/(v_{mat})$$

Or more simply, the ratio of the incidence angle and refraction angle also describes this phenomena. This ratio is the basis for Snell's law. Consider light passing through medium A to medium B having an incidence angle $i$ (as measured from the surface normal) and a refraction angle $r$. Then the refractive index $n_{AB}$ is given by Snell's law as

$$n_1 \sin \theta_1 = n_2 \sin \theta_2$$

Using Snell's law and the material properties on either side of the interface, the incidence and refraction angles can be determined (Figure 3.26). Plastics and ceramics are the materials of interest for this property. As an example, consider light passing through an acrylic plastic. The light frequency (or source) has an impact on the index of refraction. This property is known as *dispersion* and is highly dependent on the material structure and is an important optical property when considering focusing applications because of the frequency range of the light involved (Figure 3.27). Temperature and material strain also can affect the index of refraction. The

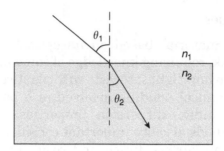

**Figure 3.26**  The different refractive index between two materials bends or alters the light path through them as defined by Snell's Law.

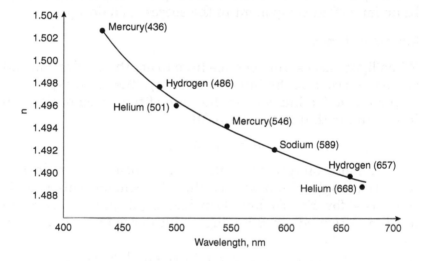

**Figure 3.27**  Dispersion and its affect on the refractive index as a function of light source wavelength.

*Source*: ASM International, Online Engineered Materials Handbook Desk Edition, Optical Testing of Polymers: Refractive Index, Figure 4.

numerical range for this property is from 1.00 for vacuum to 2.42 for diamond.

Transmittance

Another important optical property is transmittance. It is the amount of electromagnetic radiation (visible, infrared, and ultraviolet) which is allowed to pass through a material. Transmittance is measured in percent and plotted against wavelength. The most common use of transmittance data is in making filters for certain wavelength.

Each material has its own unique transmittance pattern and suitable ones can be used to block out unwanted wavelengths. Commercially available long- and short-wave filters made from fused silica and specially coated to block unwanted wavelengths are two examples of application specific designs (Figure 3.28). Both filters are successful in blocking certain wavelengths of the spectrum.

A transmittance graph of an acrylic plotted over a larger range exhibits the variability of the property (Figure 3.29). It can be seen that it allows over 90% transmittance in the visible range but certain wavelengths in infrared region have only 15% transmittance. Glasses usually have a transmittance

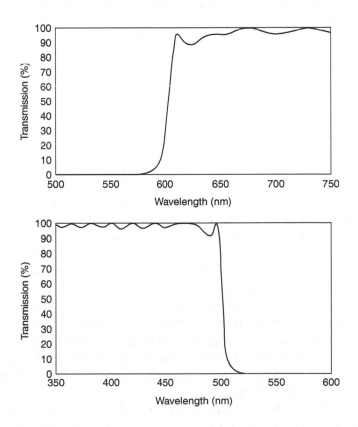

**Figure 3.28** Examples of high-pass and low-pass optical filters.

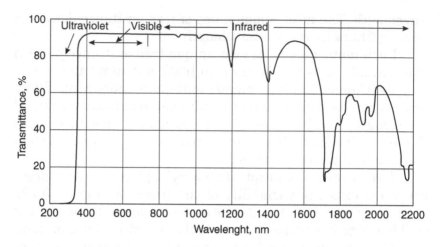

**Figure 3.29**   Spectrophotometric transmission of acrylic.

*Source*: ASM International, Online Engineering Materials Handbook Desk Edition, Optical Testing of Polymers: Transmission and Haze, Figure 1.

over 90% in the visible light range with the remaining 10% being reflection loss. As measured in accordance with ASTM D 1003.

Optical quality is a generic term for glass or plastic clarity that has a transmittance of >90%.

Haze

Haze is an important property when designing for a transparent "sight" application in which observers must be able to see inside or through a part easily and clearly. Haze is defined as light scattered at greater than 2.5° and is a result of material variations in properties such as density, fillers, pigments, and voids. If a material has a high haze value, it will have decreased transparency, making it more difficult to see inside or behind the "sight" part. A part with a high haze value will still transmit light, but images may appear foggy or blurred. Typically, polycarbonate resins have a haze value of about 0.5 to 2.0%. Haze value ranges for other transparent materials are also available (Table 3.9).

Clarity is also a measure of haze. As with haze it is associated with polymers. Clarity is dependent on the nature of chemical bonds inside the polymer.

**Table 3.9**  Haze Values for Selected Materials

| Material | Haze % |
|---|---|
| Polystyrene | 0.1–3.0 |
| Styrene acrylonitrile (SAN) | 0.6–3.0 |
| Polymethylmethacrylate (PMMA) | 1.0–3.0 |
| Cellulose acetate (CA) | 0.5–5.0 |
| Glass | 0–0.17 |

## Reflectivity

This property involves the material surface and is dependent on the temperature and wavelength of the incident radiation. It is a dimensionless value and measured as the fraction of the incident radiation that is reflected from the body. Surface finish plays an important role in the reflectivity of a material. Smooth surfaces reflect radiation specularly, therefore they are not good absorbers or emitters. Rough surfaces which reflect diffusely are better absorbers and also better emitters than smooth surfaces.

## Form and Finish Properties

Many materials are only available in specific sizes, shapes, heat treat or finishes because of production limitations or demand. Considering these limitations and their impact on the suitability for manufacture of the design as it relates to the material form is a necessary part of the design process. It should not be postponed until or transferred to the fabricator but rather indentified prior to submitting a part for fabrication. Shape factors include round, square, rectangular, and hexagonal, rolled or formed shapes. In addition, extruded shapes can be numerous and highly specialized depending on the application and usage. The scope of each of these offerings will vary and are dependent on the foundry capabilities and the market economics.

Material condition as received from the mill or supplier is often the most common condition to satisfy a wide range of needs. Hardenable materials such as tool steels are provided in a nonhardened state to allow for machining, after which they are processed to obtain the desired material characteristics. Conversely, nonferrous metals are available in a specific hardness such as "half hard" or "full hard" which describes the degree of work hardening or cold working that was performed during its foundry processing. This is possible because their machining characteristics are not as demanding as ferrous metals.

Surface finish, including coating or treatments, vary widely depending on the type of material. Ferrous and nonferrous materials can be provided with foundry finishes which vary depending on the processing methods. For example, ferrous steels can be hot or cold rolled, which result in very different finishes. Hot rolled materials typically have a surface scale that must be removed for most applications while cold rolled finishes do not as they impart a usable finish. Other finishes are certainly possible but will be highly specific to the application and commonly applied after final machining.

## IN SUMMARY

Material selection requires the design engineer to understand the mechanical, chemical, and physical nature of the material under consideration. Careful review and understanding of these properties is the basis for all selections and is something obtained by researching the specific properties as they apply to the intended design and becomes less of a task as one obtains greater experience. As this chapter illustrates, the scope of the available materials information is considerable, making selecting one over another a challenging task for even the most competent design engineer. Careful review of the mechanical, chemical, and physical material properties in the context of the application specifications is an important element of the design process and this should not be sacrificed to satisfy project schedules or costs if one is to avoid future product liability.

## REFERENCE

1. http://www.matter.org.uk/steelmatter/manufacturing/6_3.html

# 4

---

# Ferrous Metals

Ferrous metals are those iron-based materials which are modified for various properties using carbon and selected alloys. The classification consists of five general categories: carbon steels, alloy steels, stainless steels, tool steels, and cast iron steels (Figure 4.1). The primary constituent of ferrous metals is carbon and the amount present dictates whether the material is steel or cast iron. Those with a carbon content <2% are considered steels while those with >2% are considered cast irons. Steels are produced in wrought form and then processed into stock shapes for use and can be further altered by additional processing such as machining, welding, or forming. Irons are most commonly cast into net shape, requiring minimal amount of additional machining to clean up critical surfaces for final use and also into wrought forms which can be machined.

These wrought forms can also be classified according to their applications and material structures, (Figure 4.2).

Carbon, alloy, stainless, and tool steels all have a carbon content less than 2% with their qualities are enhanced in various ways to provide the desired material properties. Carbon steel that has been alloyed with only manganese (1.65% max.),

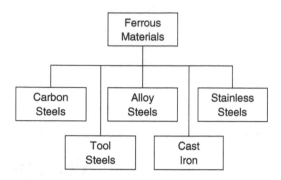

**Figure 4.1**   Ferrous metals commonly used in mechanical design.

copper (0.60% max.), and silicon (0.60% max.) is often referred to as plain carbon steel. It is available in three generic forms: low (<0.25% C), medium (0.25 to 0.55% C), and high carbon (>0.55% C) content. The presence of copper and silicon is unintentional and, generally, a byproduct of the foundry process, which introduces impurities originating from the wrought iron ore or the recycled scrap metals that are included. The primary difference among these materials is in their mechanical strength, which can range from 30 ksi to over 100 ksi depending on the material grade and processing methods, such as heat treatment and hot or cold roll forming.

Plain carbon steels can be improved greatly by adding alloying elements, such as chrome, nickel, molybdenum, and manganese (1.65%), to create alloy steels. They are subdivided into low-alloy and high-alloy steels which signifies an alloy content of ≤8% and >8%, respectively. This can be broken down further by varying the amount of carbon as in the plain carbon steels to create grades of low-alloy/low-carbon or high-alloy/high-carbon or similar combinations. Examples of these grades include high-strength low-alloy (HSLA) steels such as ASTM A242 which has a low carbon content and tools steels which are highly alloyed.

Alloy steels are also produced in various grades that further enhance their properties. These grade designators take the

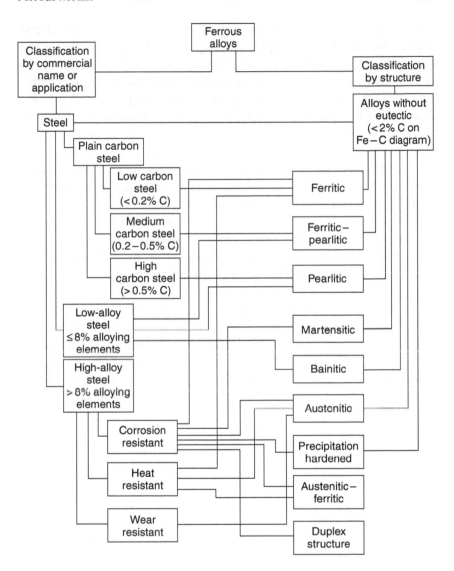

**Figure 4.2** Ferrous alloys classification flow chart showing the relationship between commercial name and material structure.

*Source*: ASM International, ASM 2002 Advanced Materials & Process Guide, December 2001, Vol. 159, No. 12, p. 51.

form of prefixes or suffixes which indicate the nature of the enhancement. For example, placing the letter B in front of the material designator signifies that the material has been modified by adding Boron to improve its strength without significantly altering its other properties. Although this alteration may seem minor, it is significant enough to warrant alternate processing and added costs, so if it is not essential to the performance of the material in a given application, a cost savings can be realized by the selection of a grade which does not include Boron. Many other grade designators are used and they are discussed in detail in the following sections.

Corrosion-resistant steels are generally those materials that have a chromium content of greater than 10.5% and various amounts of nickel, ranging from 1% to as high as 35% in special grades. A general sense of the relationship of chromium to nickel can be seen in Figure 4.3.

The most commonly used class of corrosion-resistant steels are stainless steels. They are available in five generic classes, three of which are designated based on their lattice structure, a fourth, which is a combination of two, and a fifth, designated by its hardening mechanism. The austenitic, ferritic, and martinsitic alloys are the primary candidates, when

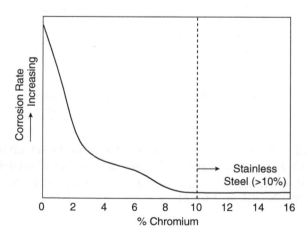

**Figure 4.3**   Corrosion rate as a function of chromium content.

considering stainless steel alloys, because of their reasonable costs, while duplex alloys (a combination of austenitic and ferritic) are less common but provide a solution for those applications requiring greater corrosion-resistance and strength. The final designation, precipitation hardening (PH) stainless steels, are popular materials because of their high strengths and moderate corrosion-resistant properties. They can be processed by heat treatment and also have reasonable wear-resistance, although not equal to other stainless steels.

Steels that provide extreme wear and strength properties are known as tool steels. These steels are selected when extreme conditions involving wear, toughness, and resistance to softening at elevated temperatures are involved. No tool steel provides a solution for all these environments, but by careful selection and processing, an optimum combination can be produced for the intended application. There are four generic classes of tool steels to select from when designing for demanding tooling requirements: cold worked, hot worked, high speed, and special purpose grades. These steels almost always require some form of heat treatment and final finish operation in order that the desired shape and material properties be obtained. Hardness values can vary widely, ranging from 35 to 65 $R_c$, depending on the heat treatment. Thermal distortion during heat treatment, through both growth and warpage, has to be taken into consideration for these materials when designing to close tolerances. Material manufacturers offer this information in detail, including changes in size throughout the range of processing temperatures. Lastly, corrosion-resistance is considered to be fair to poor for tool steels when compared to other steels, requiring that they be used in environments that are inherently noncorrosive such as machine tools that are generously lubricated with oil.

Cast iron is a ferrous metal alloyed with carbon in amounts greater than 2% (Table 4.1). This amount exceeds the solubility limit of the mixture, resulting in the precipitation of graphite formations within the metal matrix. The alloy also contains silicon, which affects the solidification temperature range as well as the casting and mechanical properties. Cast irons are a family of materials which are defined by their

**Table 4.1**   Carbon Content of Unalloyed
Cast Iron

| Type of iron | Carbon (%) |
| --- | --- |
| White | 1.8–3.6 |
| Malleable | 2.2–2.9 |
| Gray | 2.5–4.0 |
| Ductile | 3.0–4.0 |
| Compacted graphite | 2.5–4.0 |

microstructure and resulting graphite formation characteristics, the major groups being gray, malleable, white, and ductile. Gray and ductile cast irons are the most commonly used of this ferrous material because of their low cost and mechanical properties such as damping and impact resistance.

Traditionally, ferrous metals are identified by the assignment of a number designation as defined by one of several organizations (Table 4.2). Two dominant organizations that provide composition guidelines for many ferrous materials are the American Iron and Steel Institute (AISI) and the Society of Automotive Engineers (SAE). A third numbering system in use is the Unified Numbering System (UNS), which was jointly developed within commercial industry and was documented in American Society for Testing Materials (ASTM) and SAE specifications for the purpose of providing a collective listing of all materials from one source. The UNS numbering system is an identification system rather than a specification. It provides for an efficient method for indexing, cross referencing, data storage and retrieval and avoidance of using the same number. It correlates designations from organizations such as AISI, ASTM and SAE as well as independent producers.

There are a significant number of steels that do not follow SAE or ASTM specifications because they are proprietary formulations developed by steel producers to solve a specific design problem. These nonstandard products are often the only solutions to specific design problems, but understanding their compositions and location in the hierarchy of ferrous metals can be a challenge. For example, Hastaloy® C-22® is a

**Table 4.2** Material Specification and Numbering Systems

| Designation | Name | Representation |
|---|---|---|
| AISI | American Iron and Steel Institute | AISI and SAE work closely together to maintain the most widely used numbering system for designation of ferrous materials. AISI designations are not provided in specification form and defer to SAE specifications for procurement purposes |
| SAE | Society of Automotive Engineers | SAE and AISI work closely together to maintain the most widely used numbering system for the designation of ferrous materials. Specifications are generally adequate for procurement purposes |
| ASTM | American Society for Testing and Materials | The most widely used specification system for engineering materials. These specifications are adequate for procurement purposes |
| ASME | American Society of Mechanical Engineers | Provides unique and generic specifications. Specifications commonly incorporate those of other organizations with little or no modification |
| AMS | Aerospace Materials Specifications | Specifications are geared toward the aerospace community and are generally adequate for procurement purposes under government contacting guidelines |
| UNS | Unified Numbering System | Developed primarily by ASTM and SAE with support from other relevant societies and trade associations |

high-chromium steel with excellent corrosion-resistance which is used extensively in highly corrosive environments such as the exhaust stacks of coal fired power plants. This trade name does not lend itself to identifying the composition or family the material originates from, making it difficult for a design engineer to ascertain its generic properties. It also makes finding these variants somewhat problematic as they will not always be identified when searching using generic descriptors based on AISI or SAE specifications. Hastaloy® C-22® is an alloy of nickel, chromium, molybdenum, and tungsten; it has outstanding resistance to pitting, crevice corrosion, and stress corrosion cracking.

The property that differentiates ferrous metals from all others is their ability to be hardened. Controlling the hardness of a material not only affects wear, but also impacts other important properties such as tensile and fatigue strength. There are three principal methods of hardening ferrous metals: transformation hardening (heat treating), work hardening (cold working), and precipitation hardening. The degree to which a ferrous metal responds to these methods strongly depends on its microstructure and processing. Heat treating requires a composition which includes sufficient carbon to produce a hardened microstructure. Generally, the more the proportion of carbon (up to 2%), the greater the degree to which it can be hardened. Surface preparation is the most important aspect that affects all thermal treating applications. Cleaning the surface before heat treating is necessary to prevent smutting on the surface and unintentionally causing the resulting carbon to diffuse into the surface. Work hardening forces lattice displacements that improve the hardness but also induces residual stresses that can be detrimental to processing, such as machining and forming operations. The capacity of a ferrous metal to be hardened by precipitation methods depends on its composition and is related to the material family. The most notable PH ferrous metal is 17-4PH stainless steel, which can be hardened to greater than 43 $R_c$.

Laser technology is relatively new to the materials industry and provides for material processing methods that may not

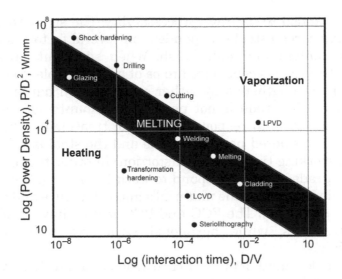

**Figure 4.4** Laser technology offers many solutions to material processing ranging from bulk heating to material vaporization.

*Source*: Laser Material Processing by W.M. Steen, Springer-Verlag, New York, 2003, p. 228, Fig. 6.1.

be possible using other technologies. These processes include heating, melting, or vaporizing the material to transform the surface, weld, or cut/drill, respectively (Figure 4.4). The energy required to perform these operations can vary widely depending on the material and its absorptive characteristics.

Commonly used lasers include Nd-YAG (3 kW range), $CO_2$ (25 kW range), and Excimer (150 W range). They can be used to process most alloys, stainless steels, and cast irons.

## CLASSIFICATIONS

### Carbon Steels

Carbon steels, or plain carbon steels as they are often referred to, are only iron alloyed with carbon and up to 1.65% manganese. Small amounts of up to 0.6% silicon and copper are also added to improve processability. The strength of the resulting carbon steel is dependent on several factors, such as carbon content and mill finish. When sufficient carbon is added, the steel is considered to be of a hardenable grade and it can be strengthened by either solid solution hardening or quenching. If carbon is not present in amounts sufficient to enable the steel to respond to thermal treatments, then the grade is considered nonhardenable and the steel is subjected to cold working (mechanical distortion) in order to harden it. Not all grades of steel respond to work hardening, because is highly dependent on the material's microstructure and ability to flow plastically. The FCC and BCC metals have good flow characteristics, while hexagonal structures have limited formability.

The designation of most carbon steel bar stock is covered by the AISI–SAE system of numbering that employs a four digit format. The first two digits define the grade of steel and the corresponding amounts of manganese, silicon, and copper, while the last two digits represent the percentage of carbon, when divided by 100. The carbon content ranges are typically defined by low, medium and high (Table 4.3).

The designation of sheet, strip, wire, and structural steels are covered by ASTM specifications as are their possible applications — as building beams, columns, and related equipment such as pressure vessels and piping. For carbon and alloy steels, ASTM designations use the AISI/SAE designations as their grade designator in an attempt to bring some uniformity to the numbering process. For example, consider 1020 carbon steel, the 10XX series designates a plain carbon steel, with the "XX" signifying the amount of carbon, as a percentage, when divided

**Table 4.3** Typical Carbon Content Ranges

| Designation | Range (%) |
|---|---|
| Low-carbon | <0.25 |
| Medium-carbon | 0.25–0.55 |
| High-carbon | >0.55 |

by 100, indicating a 0.2% carbon content. The UNS method of numbering employs a similar strategy by also using the AISI/SAE designation in their number. For the same 1020 designation, the UNS number is G10200.

Processing considerations for plain carbon steels are numerous and include not only thermal treatments but strengthening and surface treatments as well. Carbon steels are processed hot or cold, depending on the composition and desirable material condition.

Hot finished carbon steels have the advantage of better weldability and machinability, while being of lower cost because of the simpler processing required. This process does not produce residual stresses because it is performed at temperatures that are above the recrystallization temperature of the steel, which allows the distorted crystalline structure to heal itself, not producing the grain boundary distortion that characterizes cold finished products. The most significant disadvantage of this process is the mill finish, which can be undesirable for many applications and may require additional processing to obtain the desired condition. The hot finished process is most commonly applied to structural steel shapes and thick section plates. These items not only are difficult to process using cold finishing, because of the high forces involved, but also benefit from the improved machining and weldability.

Cold finished or cold rolled steels obtain their strength properties from the cold working effects of this process. Their mechanical properties are defined by the percentage of cold working that is applied to the material and are proportional to it. The greater the tensile strength, the greater the hardness, but both of these are obtained at the expense of toughness. These materials can be heat treated, welded, or formed, as necessary, depending on the grade. When machining cold finished carbon steel, it can be difficult to hold close tolerances because of the residual stresses within the material; because of this, cold rolled products may require stress relieving after initial machining to stabilize the material prior to completing the final machining operation. Different applications are subjected to different finishing process (Table 4.4).

**Table 4.4**   Plate, Bar, Strip, and Sheet Applications

| Product Form | Finish Method | Application |
|---|---|---|
| Plate | Hot rolled | General purpose<br>Pressure vessels,<br>Structural<br>Aircraft |
| Bar | Hot rolled | General purpose<br>Structural<br>Automotive<br>Aircraft |
| | Cold rolled | General purpose<br>Aircraft<br>Rifles<br>Barrels |
| Strip | Hot rolled | General purpose<br>Structural |
| Sheet | Hot rolled | General purpose<br>Structural |
| | Cold rolled | General purpose<br>Structural |

## Properties and Designations

### Numbering

Plain carbon steels are identified using a four digit numbering system where the first two digits identify the unique alloy addition made to wrought iron while the last two digits signify the nominal or middle range of weight percent of carbon added (Table 4.5). For example, 1010 carbon steel has a carbon range of 0.08 to 0.13% with 0.10% being the nominal value that the foundry attempts to achieve.

In addition to the numbering system the SAE J412 further categorizes SAE–AISI plain carbon steels into four groups (Table 4.6). These steels are so grouped because their processing characteristics (hardenability, formability, machinability, and weldability) are similar.

Standard grades are identified as described above, while additional designators are included to the number indicating variations the standard. An "M" prefix designation is used for a grade known as "merchant quality" steels. This grade is

**Table 4.5**  Carbon Steel Designations

| Designation | Significant constituents |
| --- | --- |
| 10XX | Nonresulfurized carbon steel grades, manganese 1.00% max |
| 11XX | Resulfurized carbon steel grades |
| 12XX | Rephosphorized and resulfurized carbon steel grades |
| 15XX | Nonresulfurized carbon steel grades, manganese 1–1.65% |

produced with wider ranges of carbon and manganese than standard grades, thereby reducing the cost and quality of the material. Merchant quality steels are intended for noncritical parts of bridges, buildings, ships, agricultural implements, road-building equipment, and general machinery because they may contain a significant amount of porosity, chemical variations, or other irregularities making them unsuitable for heat treating or forging. Heat analysis is performed by a product analysis on a sample from the ladle or furnace and is the only method used to confirm the material composition, making the analysis less than exact but sufficient for this low-grade product. These grades are generally not produced in units having diameters greater than 3 in.

Additional special-purpose elements such as boron and lead are also added to carbon steels in order improve specific material properties: boron added in small amounts ranging from 0.0005 to 0.003% improves hardenability, while lead, in amounts ranging from 0.15 to 0.30% improves machinability. These grades are indicated by inserting the letter "B" or "L" between the alloy designation and carbon content (e.g., 10B10 or 10L10).

### Strength Properties

Carbon steels are the most commonly used engineering materials because of their high strength-to-cost ratio (strength/cost). They are also among the stiffest engineering materials, with an elastic modulus of 30 Mpsi, making them attractive in terms of both price and performance. The yield strength of carbon steels varies widely, ranging from 25 to

**Table 4.6**    Four Groups of Carbon Steel Designations

|  | Grades in the group | Carbon range (%) | Properties |
|---|---|---|---|
| Group I | 1005, 1006, 1008, 1010, 1012, 1013 | less than 0.15 | Relatively low tensile values Susceptible to strain age embrittlement Have excellent surface appearance and used in automobile panels and appliances |
| Group II | 1015, 1016, 1017, 1018, 1019, 1020, 1021, 1022, 1023, 1025, 1026, 1029, 1513, 1522, 1524, 1526, 1527 | 0.15–0.30 | Have increased strength and hardness and reduced cold formability compared to group I. Used for numerous forged parts. In general they are suitable for welding before carburizing |
| Group III | 1030, 1035, 1037, 1038, 1039, 1040, 1042, 1043, 1044, 1045, 1046, 1049, 1050, 1053, 1536, 1541, 1548, 1551, 1552 | 0.30 to less than 0.55 | Used for higher mechanical properties. They are frequently hardened and strengthened by heat treatment or cold forming. They are used in a wide variety of automotive applications. They are good for normal machining and welding |

**Table 4.6** (*continued*)

|  | Grades in the group | Carbon range (%) | Properties |
|---|---|---|---|
| Group IV | 1055, 1059, 1060, 1065, 1069, 1070, 1074, 1075, 1078, 1080, 1085, 1086, 1090, 1095, 1561, 1566 | 0.55 to less than 1.0 | Higher carbon grades are used for applications where improved wear characteristics are required. Cold forming is not done on these grades. Used in washers, flat springs, stamped parts, blades, disks, and plow beams |

180 ksi, depending on the applied heat treatment and extent of cold working, as well as the behavior of the physical alloying elements and the material's form factor. Having to consider all these factors makes the selection of a specific type of carbon steel a complicated process while it also allows for compromises between different types owing to the considerable overlap in their performance characteristics.

The yield strengths of mill finished bar stock in quenched and tempered forms usually range from 25 to 90 ksi, depending on whether they are hot or cold finished. Application of specific secondary processes, such as aggressive quenching followed by tempering, can improve the yield strengths significantly, usually obtaining values above 100 ksi. Tensile strengths are proportionally higher and are, typically, listed with the yield strengths, giving the design engineer a sense of the available margin when they are considered along with the percentage elongation and area reduction.

### Heat Treatability

One of the most attractive properties of carbon steels, which can significantly enhance their performance, is that their properties can be altered by heat treating. The carbon content of the material is the primary factor that determines the heat treatability of any carbon steel. Generally, the lower the carbon content, the harder it is to obtain a significant hardness. Low-carbon steels are generally cold finished and can be case-hardened using carburizing techniques to increase the surface carbon content and resulting hardness, producing a product with good wear-resistance, while preserving the inherent toughness of the base material. Medium-carbon steels can be subjected to direct hardening, using an oven, flames, or induction techniques, to produce the necessary depth and grade of hardness appropriate for the application. High-carbon steels are not extremely common, as the higher carbon content is usually found in combination with other alloys, but they are available and heat treatable much like the medium-carbon grades (Figure 4.5).

### Fatigue Properties

It is well documented that ferrous metals exhibit a fatigue limit (also known as the endurance limit) below which they are considered to have infinite life. The stress level, $S$, and number of cycles, $N$, at which this occurs on the $S–N$ plot depends on the material composition and loading history. It is common practice to use $\frac{1}{2}S_{ut}$ as the fatigue limit for quenched and tempered steels when actual data are not available. The carbon content influences the fatigue limit. With no carbon present, the wrought iron does not exhibit a fatigue limit, behaving much like a nonferrous metal. This behavior changes once carbon is added.

Adverse conditions, such as overloading and corrosion, will affect the fatigue performance of carbon steels. It has been shown for periodic overloading conditions that the fatigue limit behavior disappears, removing the lower cycle limit which provides infinite life. It is important to take this behavior into account in the case of dynamic systems that require careful consideration of the maximum design loads. Corrosion is also a source of significant concern when designing for dynamic applications involving corrosive environments because the fatigue limit is also eliminated by the destructive effects of this type of environment. To preserve the desired material

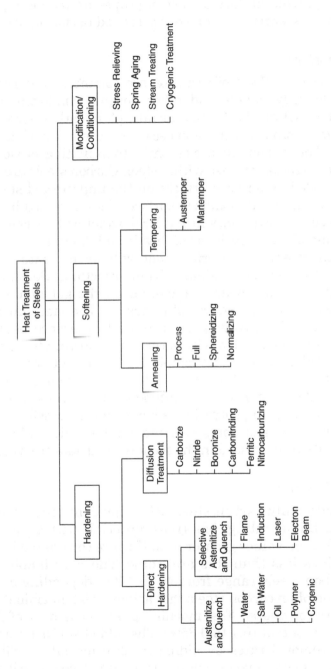

**Figure 4.5** Heat treatment of steels can be categorized into three main types: hardening, softening and conditioning.

properties, including fatigue, use of proper protection against corrosion is essential to achieve the desired performance.

### Corrosion

The visible effects of corrosion are commonly observed in automobile body panels and other designs employing plain carbon steel. Of all the ferrous metals, plain carbon steels are the ones most prone to corrosion because of their lack of alloying elements. These elements are essential to inhibiting, or slowing down, the corrosion process within steel. Carbon steels are, typically, corroded by uniform attacks on the unprotected surface, resulting in an evenly oxidized surface. They can also be subjected to crevice corrosion, which is local oxidation between pairs of joined metal parts, such as bolt heads and flanges.

When selecting materials that will be in contact it is necessary to consider possible galvanic reactions. Selection of materials which are near one another on the galvanic series eliminates a potential source of corrosion. The mechanism of galvanic corrosion involves the potential difference between two metals when coupled with a corrosive electrolyte, such as seawater. When a galvanic couple forms, one of the metals becomes the anode and corrodes faster than it would by itself, while the other material is the cathode and corrodes slower or not at all. Controlling corrosion is key to the successful implementation of plain carbon steels and can be achieved by various methods, as discussed in the section on surface treatments.

### Machinability

Traditionally, carbon steel 1212 represents the baseline for all machining ratings of ferrous materials with a rating of 100. Machining values greater than this are considered better, while values less than this are worse. The machinability of plain carbon steels range from 40 to 170% depending on the alloy. Those plain carbon steels that achieve high machinability ratings are considered "free machining" steels because of their high sulfur content (0.1 to 0.35%). The added sulfur improves material removal rates by aiding in chip formation. Silicon adversely effects machinability (but improves hardness);

therefore carbon steels that have good machinability ratings are generally not deoxidized using silicon but with aluminum. Silicon is used as an alloying element in the necessary proportions to provide the desired properties which typically includes hardness.

### Weldability

As a general rule, carbon steels are readily weldable and tolerant of many variabilities in the welding process, for instance, the joining of steels of different grades. The ability to weld different materials depends on the solubility of the materials into one another. A greater understanding of the process can be obtained by consulting the respective phase diagrams, or better still, a qualified metallurgist. The properties that allow carbon steels to be heat treated also affect the welding process because it subjects the material to similar thermal conditions. This can result in inconsistent material properties in and around the weld that can reduce the performance of the material. This affected area is known as the heat affected zone or HAZ. For example, high hardness adjacent to a weld, caused by the thermal conditions during welding, can act as a stress riser, reducing the fatigue life of the combined structure.

Free machining steels have high sulfur content, which promotes cracking in and around the HAZ; therefore, these steels are not considered good for welding. This condition is observed when welding 303 stainless steel, but not when welding 304L stainless steel, the latter being formulated specifically for welding. Contamination also affects a material's ability to be joined properly by welding. Proper surface preparation followed by immediate welding is important in order to obtain a quality weld.

### Unique Properties

The unique appeal of plain carbon steels is their high stiffness and low cost, making them cost-effective solutions for numerous applications. The high stiffness-to-cost ratio and the ability to control the material's hardness by altering the carbon content are also important in understanding the range of applications in which this material can be used.

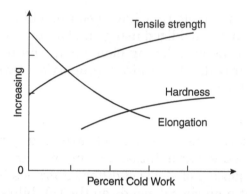

**Figure 4.6** The generic relationship between tensile strength, hardness and percent elongation as a function of cold work for ferrous materials.

The ability of carbon steels to be work hardened is another unique property, which allows carbon steels to be formed into many shapes; not all steels respond to this form of processing. This process yields, along with improved strength, a corresponding increase in hardness and a reduction in ductility, evidenced by the reduction in percent elongation during tensile loading tests, indicating a loss of toughness (Figure 4.6).

The combination of high values of thermal and electrical conductivities of plain carbon steels also sets them apart from other ferrous metals and nonmetals. Nonferrous metals generally have the highest values for these parameters owing to their atomic properties.

**10XX:** There are more than 50 grades of steel available in this series with the range and composition of each grade depending on the product form. The carbon content varies from 0.06 to 1.03% with not greater than 1.0% manganese. Phosphorus and sulfur are limited to 0.04 and 0.05%, respectively. A carbon steel designation with letter B inserted between the second and third digits indicates that the steel contains 0.0005 to 0.003% boron. Likewise, the letter L inserted between the second and third digits indicates that the steel contains 0.15 to 0.35% lead, for enhanced machinability. Mechanical properties such as strength and hardness increase with the carbon content; thermal and electrical conductivities decline, magnetic permeability decreases drastically, and corrosion-resistance also decreases. Adding manganese to the steel increases hardenability but adversely affects cold forming. High-carbon and high-manganese grades are used for greater wear resistance and improved hardenability requirements. Machinability of materials of this series, like their mechanical properties, depends on their carbon content and also on the work hardening to which they have been subjected. Because of the large range of grades available in this series, these steels have widespread applications; they are used, for example, in making forged parts, small shafts, plungers, gears, hand-held tools, agricultural machinery, pressure vessels, and a number of automotive applications. They are available in most wrought forms, including bar, sheet, plate, pipe, and tubing. A sheet is, primarily, a low-carbon steel product, but virtually all grades are available in bar and plate forms.

**11XX:** This series includes resulfurized carbon steels, which are produced for applications requiring good machinability. These materials provide improved chip formation with increased sulfur content, which can be as high as 0.33% in some grades. Machinability improves as sulfur content increases. Sulfur combines mostly with the available manganese and precipitates as sulfide inclusions. These inclusions favor machining by providing a built-in lubricant that prevents the chips from sticking to the tool and undermining the cutting edge. Higher carbon grades in 11XX are widely used for parts where a large amount of machining is needed, for example, 1137 is widely used for nuts, bolts, and studs with machined threads. Lower carbon grades of 11XX are used where combination of good machinability and response to heat treatment is needed, for example, 1117 and 1118, which can be oil quenched after case hardening treatments, as they have more manganese.

**12XX:** These are resulfurized and rephosphorized carbon steels; intended for use where improved machinability is desired, compared to carbon steels of similar carbon and manganese content. Phosphorus additions increase the strength and reduce the ductility of ferrite; hence chips break more easily. The use of other additions such as lead, bismuth, or selenium has declined due to environmental restrictions. Sulfur and phosphorus negatively affect weldability, cold forming, and forging. The low-carbon grades can be used for case hardening operations, while the grades containing more than 0.30% carbon can be quenched and tempered or induction hardened. However, steels high in phosphorus are notoriously notch-sensitive; an excess of phosphorus can raise hardness and strength to levels where it can impair machinability. Hence the 12XX series is limited in phosphorus content to 0.04, 0.09, or 0.12%, as well as in carbon content, up to 0.13%.

**15XX:** These are steels with nominal manganese content of between 1.0 and 1.65% with only residual alloying elements. With over 20 grades available, these were originally a part of 10XX series which was later separated on the basis of higher than normal manganese content. Like the 10XX series they share the varying mechanical properties with the carbon content (0.1 to 0.71%). Lower carbon grades offer better cold formability, while higher carbon grades deliver higher mechanical properties and wear resistance. Hardenability is increased in all grades owing to higher manganese content, increasing the range of hardening operations and quenching media. Steels with less than 0.3% carbon in this series have improved machinability over the lower manganese grades with same carbon content. Their applications range from forged parts, shafts, and gears to tools and farm machinery. They are common in applications that require high strength and better uniformity and are available in bar, rod, tubing, plate, and strip shapes.

**H steels:** Certain steels have hardenability requirements in addition to the limits and ranges of chemical composition. They are distinguished from similar grades that have no hardenability requirements by the use of the suffix H.

## Fabrication Processes

All ferrous metals begin their production in the blast furnace, a large complex system where iron ore, coke, and limestone are combined to produce pig iron, a carbon-rich (typically 4 to 5%) mixture of raw iron. This iron is then either passed in its molten state to the steel-making furnace or cast into "pigs" or billets for processing into steel at a later time or location.

There are two primary processes for producing carbon steels, the basic oxygen furnace (BOF) and the electric arc furnace (EAF) methods. The BOF method involves the introduction of large amounts of oxygen into the molten steel to react with carbon, and other impurities present, to produce the steel. This method is the preferred method of producing the initial batch of most carbon and alloy steels and is the primary method for high-tonnage production. The EAF method produces steel in a similar fashion but passes an electrical current through the material to heat and process the molten steel. This process is most commonly used for lower tonnage production and for secondary refining of specialty steels such as alloy-, tool-, and stainless steels.

A common term used to describe a process applied to both methods is "killing." "Killed steels" are those that are processed by adding aluminum or silicon to the melt to remove the included oxygen by effectively "killing" the bubbling action. This can also be accomplished by vacuum treatment of the molten steel. Using silicon is the least expensive way; it should be noted, however, that the method of oxygen removal can have a significant impact on the quality and cost of the end-product. Vacuum arc remelting (VAR), is a costly alternative to using silicon or aluminum, but it produces a high-grade steel and is typically reserved for high quality alloys and tool steels.

Production of steel has migrated, in recent years, from high-tonnage production facilities to lower tonnage operations commonly called "mini-mills," short for miniature steel mills. These scaled-back operations are intended to be more responsive to industry needs by providing cost-effective quantities in minimal lead times, while still providing high-quality products. Most steels are continuously cast into the desired net shape, rather than through the intermediate step of producing an ingot that must be later reheated and formed. Many of these mini-mills use large amounts of recycled scrap to produce their steels using the EAF process, rather than ore as used in the blast furnace production method, making the operation more flexible and less reliant on large amounts of raw materials. The stigma attached to this process, of using scrap steel in the production of new steel has been eliminated by the success of these mini-mills in producing consistent chemical compositions across a wide range of material grades.

The final form of the steel as produced by the foundry, regardless of the tonnage of the operation, is of two generic forms: net shape or machining/sheet stock. These are then found to be in one of two conditions: cold worked or hot worked. The commonly available forms are presented below.

## Raw and Stock Finishes

Carbon steels are produced in one of two conditions: hot or cold finished. Hot finished products are processed at temperatures above the recrystallization temperature, which allows the material to recrystallize immediately after deformation, relieving any work hardened stresses. For carbon steels this temperature is between 1500 and 2300°F. Processing at these temperatures produces a mill surface finish that is less than desirable and often requires a surface cut to remove the residual scale and impurities embedded into the surface. Materials processed in this manner do benefit from greater stability when machined and do not result in further distortion after machining because of reduced internal stresses. The hot finishing process is cheaper than the cold finishing process because it is easier to deform the material when it is red hot than when it is cold. The following plain carbon steel grades are commonly processed using this method: 1020, 1025, and 1030.

Cold finishing involves work hardening of the material by rolling or shaping when it is below the recrystallization temperature. The objective of cold finishing is to increase the strength of the net shape, whether it is bar stock or a structural angle. Increasing tensile strength by more than 100% greater than hot finished products can be obtained because of the dislocations produced during the rolling/forming operations. The crystalline shape has an impact on the strengthening effect of cold working. FCC and BCC metals such as 1006, 1050, 1112, 1117, and other free machining steels have good flow characteristics, while HCP

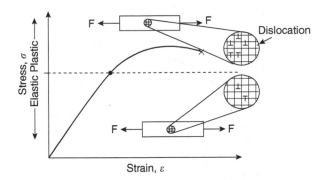

**Figure 4.7** The elastic and plastic regions of the stress–strain curve have uniquely different effects on a materials grain structure. In the elastic region the material returns to its original structure while when deformed plastically it produces an increase in dislocations.

forms, such as titanium, have limited dislocation motion and therefore do not respond to the cold finishing process. This process is not usually applied to high-carbon or alloy steels because the strengthening effect is not as pronounced and formability is poor (Figure 4.7).

Materials that have been cold finished exhibit poor dimensional stability when machined, because of the stress relieving that occurs during the material removal process.

High-formability grades are available with elongations of >50%. These involve a type of steel known as interstitial-free (IF) steels, which are specifically formulated to be free (or nearly so) of interstitial carbon. This allows them to be cold drawn into thinner sections, reducing their weights while increasing their strengths.

Cold finished plain carbon steel bar and strip stock are produced in five grades of hardness as defined by ASTM A109:

- No. 1 (full-hard)
- No. 2 (1/2 hard)
- No. 3 (1/4 hard)
- No. 4 (skin-rolled)
- No. 5 (dead-soft)

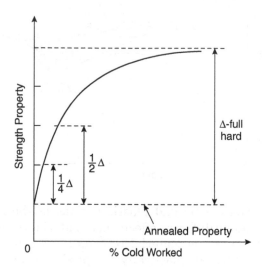

**Figure 4.8**  Temper hardnesses for cold finished plain carbon steel.

The dead-soft condition corresponds to the material's annealed condition while the full-hard condition corresponds to the maximum possible cold worked condition. The maximum condition is determined by the increase in strength as a function of the amount of cold work (Figure 4.8).

Therefore, knowing the material's strength both at the annealed condition and at the maximum cold worked condition then defines it at other conditions. The skin-rolled temper designation signifies that the material has been subjected to a surface yielding process that is intended to eliminate unwanted stretch marks that can be visible when forming dead-soft tempers. These stretch marks are the result of nonuniform atomic slip (uneven localized yielding) in the regions of greatest deformation. Using a skin-rolled tempered product often can eliminate these undesirable effects, while providing an improved surface finish compared with a nonrolled annealed grade.

Sheet stock is not provided in these grades because the extreme forces required when forming make the processing difficult. There are different classes of sheet stock: Class 1 consists of tempered rolled grades and Class 2 consists of steel that is rolled and then annealed to remove residual stresses.

The foundry finished product, whether hot or cold finished, is some net shape that is purchased by the end user for processing. The condition of this material has a significant impact on the effort required to produce the final product. As a general rule, cold finished products are more difficult to process than those that are hot finished because of the work hardening effects present in the cold finished material. The finished shapes for both methods are similar and include plates, sheet, strips, tubing, bars, and structural shapes such as I-beams or C-channels (Table 4.7). These are produced from common plain carbon steels including low-carbon (case hardening), medium-carbon (direct hardening), and high-carbon (case or direct hardening) grades as well as specialty grades intended to improve specific material properties, such as abrasion/impact resistance or tool life.

## Processing Considerations

The physical properties of a carbon steel can be changed by thermal treatment or physical modification. Thermal treatments affect the material's properties by changing the phase or formation of the material's microstructure. These treatments can be applied to a controlled depth or throughout the material, allowing surface-hardening or through-hardening, which produces a material that has increased surface hardness while retaining its toughness or increased strength. Physical treatments involve improving the strength of a material by impeding the movement of its crystalline structure; they modify the bulk of the material to provide greater strength. Surface treatments are the processes most often applied to plain carbon steels, more so than to other ferrous metals; they are applied to reduce or limit corrosion, which is a problem with carbon steel in most environments because of their reactive natures. Without a surface treatment carbon steels will oxidize (rust) in typical atmospheric conditions.

### Thermal Treatments

The hardenability of plain carbon steels is generally considered poor compared with that of alloy and tool steels. There must be a high-enough carbon content to complete the

**Table 4.7**   Common Structural Shapes of Carbon Steel

| Shape | Form | Material | Comments |
|---|---|---|---|
| **Plate** | | | |
| General purpose | Plate | low-carbon: 1015, 1020, 1025<br>medium-carbon: 1045<br>high-carbon: 1060 | General purpose products are available in all three grades: low-, medium-, and high-carbon grades. Regular quality plate — produced to a standard formula, not to specific strength ranges |
| Structural | Plate | A36 | |
| Specialty | Plate | A285 | 0.28°C max. Pressure vessel — ASTM A285 steel w/0.28 carbon max. Abrasion/ Impact resistance — medium-carbon content (0.35–0.50) and moderate hardness (25Rc max.) |
| Sheet and strip | N/a | A569<br>A366 | Alloys |
| **Tubing** | | | |
| Mechanical | Round | 1020, 1026 | Cold drawn grades |
| Mechanical | Square | | |
| Hydraulic | Round | | |
| **Bar** | | | |
| Hot finished | Rounds | Low-carbon: A36, 1018, 1020, 11L17, 1117<br>Medium-carbon: 1030, 1035, 1040, 1045, 1141, 11L41, 1144 | Threaded products are available from hot finished rounds |
| | Squares | Low-carbon: A36, 12L14, 1215<br>Medium-carbon: 1045 | Merchant quality screw machine stock |
| | Flats | Low-carbon: A36, 1018, M1020, 11L17<br>Medium-carbon: M1044, 1045, 1144 | Merchant quality |
| | Hexagons | Low-carbon: 1018, 1117, 12L14, 1215<br>Medium-carbon: 1030, 1035, 1040, 1045, 1141, 11L41, 1144 | Screw machine stock |
| Cold finished | Rounds | 1018, 1050, 1117, 11L17, 1045, 1141, 11L41, 1144CD | Shafting rounds that are turned, ground, and polished are available with diameter tolerances −0.0005 to −0.0015 for 1.5 in. diameter or less. Shafting rounds of greater precision are also available with a diameter tolerance of ±0.0005 |
| | Squares | 12L14, 1215, 1045, 1018, 1117, 11L17 | Screw machine stock |
| | Flats | 1018, 11L17 | |
| | Hexagons | 1018, 1117, 12L14, 1214Bi, 12L14, 1215, 1045, 1117, 1144 | Screw machine stock |
| **Structural** | | | |
| Angles | | A36 | Structural shapes are typically made from low carbon steel (ASTM A36) |
| Channels | | A36 | |
| Tees | | A36 | |
| Beams | | A36 | |

martensite transformation over a volume sufficient to harden moderate sections; generally, a steel with 0.6% wt carbon will produce 100% martensite. Additionally, the time-transformation-temperature (TTT) diagrams confirm just how difficult it is to quench plain carbon steels rapidly enough to produce significant hardness. In heavier sections, only the surface becomes hard, which may or may not be desirable depending on the application. Because plain carbon steels do not through-harden well, they are candidates for flame-hardening and carburizing which initiate surface-hardening. The following heat treatments are commonly applied to plain carbon steels:

- *Hardening*: direct hardening; flame hardening; induction hardening; and laser or electron beam hardening.
- *Softening*: normalizing; annealing; and tempering.
- *Conditioning*: stress relieving; spring aging; and cyrogenic.

Hardening of carbon steels involves two basic considerations: proper thermal cycling and sufficient carbon content. The thermal cycling necessary to produce a hardened carbon steel material includes three stages: heating to the proper austenite temperature range, soaking at that temperature, and then rapid cooling (quenching). This process is only effective in steels with sufficient carbon content, typically >0.6% for plain carbon steels. The methods used in quenching are critical to the success of the process and involve a variety of substances, including water, brine, oils, molten salts, and gases. Each of these has a unique property that controls the rate of cooling to produce the desired material microstructure, with water providing the highest rate of cooling and gases the lowest. Application of all these processes involve the need to agitate the quenching media sufficiently to provide uniform cooling of the material.

The degree of hardness is a product of the material condition and processes applied to the steel (Table 4.8). Direct hardening involves thermal processing of the material in an oven and followed by a controlled cooling to obtain a

**Table 4.8**   Hardenable and Non-Hardenable
Grades of Carbon Steels

| Non-hardenable | Hardenable |
| --- | --- |
| 1006 | 1040 |
| 1010 | 1045 |
| 1020 | 1050 |
| 1044 | 1060 |
| 1112 | 1080 |
| A36 | 1095 |
|  | 1144 (free-machining) |
|  | 1140 (free-machining) |
|  | 1150 (free-machining) |

through-hardened condition. Flame- and induction-hardening are both performed by the direct application of a localized heat source, namely a flame or heated wire coil, to a material with at least 0.4% carbon, followed by a quenching operation. The result is surface- or through-hardening, in the region where heat is applied, from a minimum depth of 0.032 in. to depths that are dependent on the material properties. Induction-hardening, typically, involves reduced cycle times and the ability to handle small, irregularly shaped parts, but with the added penalty of increased capital equipment costs. Although flame-hardening is inexpensive and may require preheating to minimize cracking, it lacks the repeatability of induction-hardening, because of the potential of variability in flame quality. Laser- and electron-beam-hardening result in similar conditions of localized heat-affected zones, but, typically, cover a smaller area to a shallower depth of 0.040 to 0.080 inch (1 to 2 mm). The process can achieve high hardness because of the rapid cooling that takes place due to the local heating and mass of adjacent material that accelerates cooling. Surface reflectivity and spot size can be significant limitations depending on the materials and surface areas involved. Although this method involves an expensive processes, the benefit is in its ability to control the location and hardness when compared with other surface-hardening methods.

Once the desired hardness is attained from the foundry condition, it is often necessary to soften the material to improve its mechanical properties, while also producing changes in other properties or microstructure. The most common methods for doing this are normalizing, annealing, or tempering, which involve reheating the material to an upper critical or subcritical temperature. Normalizing is generally performed at the mill and is intended to obliterate the gain structure which results from hot rolling and forging. Heating the material to 55°C above the upper critical line, or as necessary for the material involved, followed by agitated air cooling produces a finer, more uniform grain structure that will respond more favorably to subsequent heat treating processes. Castings often benefit from the normalizing process.

Annealing is performed by heating above the fully austenite region followed by thermal soaking and oven cooling. The temperature involved is also typically above 800°F but varies widely for each material depending on composition and produces a larger grained material that is softer and weaker. The oven cooling process differentiates it from the normalizing process.

Tempering is a process that is typically performed after quenching but before the material returns to room temperature, and is intended to toughen the material while minimizing the change in hardness. Quenching generally results in a mostly brittle condition that can be treated by tempering the material to regain some of the lost toughness without significant loss in hardness. As a general rule, the hardness loss ranges from 5–10% with a corresponding reduction in tensile strength of 10–30%, which results in an increase in the Izod impact strength of 5–10 times the pretempered condition.

An additional consideration that must not be forgotten is the mass effect of the quench and temper process on the resulting material properties. Because the quenching process is time dependent, the mass (or volume) of the material section generally alters the outcome of the process. Larger sections are prone to reduced strength and softer centers because of the difficulty in removing the heat rapidly enough to meet the TTT diagram requirements (Figure 4.9).

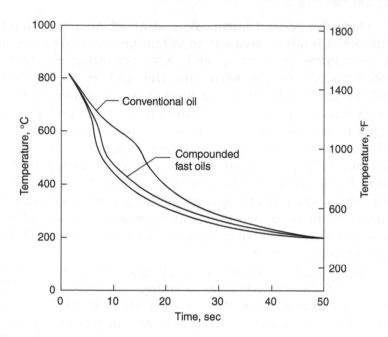

**Figure 4.9**    The time required to quench a piece of steel is dependent on its mass and shape.

*Source*: ASM Metal Handbook: Desk Edition, Figure 4, p. 28–15, ASM International, 1985.

It is required to relieve stresses formed during manufacturing by means of heat treating in order to stabilize materials and to limit potential creep or fatigue failures. This process involves heating the material uniformly to a temperature below the transformation range, typically 1000–1250°F or 100–200°F below the transformation temperature followed by a soaking and controlled cooling cycle. The relieving of stresses induced during manufacturing is a time–temperature phenomenon that is characterized by the Larson–Miller equation:

$$\text{Thermal effect} = T \left(\log t + 20\right)(10^{-3})$$

where the temperature is given by $T$(R) and time by $t$(h). The following figure depicts the relationship as a percentage of stress relief and the temperature required to achieve that level of stress relief (Figure 4.10).

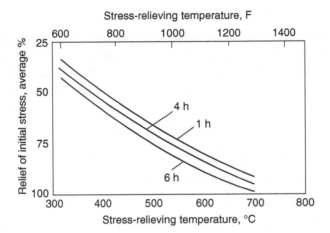

**Figure 4.10** Percent residual stress relief as a function of temperature.

*Source*: ASM Metal Handbook: Desk Edition, Figure 1, p. 28–11, ASM International, 1985.

Although the process does not remove all stresses it does allow for some grain growth which effectively eliminates the majority of them.

A supplemental heat treatment that is becoming more common is the use of a post elevated heat treatment that involves cryogenic temperatures. In this process the material is subjected to temperatures below −300°F at a very slow rate, followed by a extended hold period at that reduced temperature and a slow return to room temperature. This is followed by a mild tempering operation. The entire process can take 24 to 60 h. The effects of these thermal changes are to compress and expand the material in a uniform manner that reduces the internal stresses as well as produces a conversion of softer, retained austenite to harder, martensite. This process improves the wear resistance while also improving the toughness of the material. The merits of this processing are debatable and still being researched as it does not affect all materials in the same manner. Careful material selection and review with a qualified supplier is recommended.

### Strengthening Treatments

Strengthening treatments for plain carbon steels are limited because these materials lack alloying elements. Increasing the carbon content of these materials, in combination with the process of heat treating is the primary strengthening method used. The addition of carbon results in an increase in strength because of the interstitial solid solution interaction with the host iron atoms, preventing dislocations which result in material deformation. Increased carbon content and cold working results in greater tensile strength.

### Surface Treatments

Limiting the exposure of plain carbon steels is critical to their performance as well as overall appearance. If left untreated, they will oxidize at a rate somewhat proportional to the corrosiveness of the environment. For example, when exposed to a dry environment the rate of corrosion is minimal compared with that in a wet or fully submerged condition. Surface treatments commonly consist of three primary methods: coatings, surface diffusion, and electropolishing (etching).

Surface coatings can be of three general forms involving chemical, electrochemical, or physical deposition methods. Chemical methods involve the application of a protective coating by chemical deposition. Conversion coatings and electroless nickel coatings are applied by means of a chemical reaction involving the material being plated, whereas, electrochemical deposition methods involve the application of electrical current in a chemical bath to promote the plating of the submerged material.

Three forms of conversion coatings are commonly applied: chromate, phosphate, and black oxide. Chromate and phosphate coatings are applied through processes involving a heated acidic bath that initiates the consumption of the base material as part of the reaction, allowing the coating to become an integral part of the component being coated. A small amount of part growth is experienced as the coating extends from the surface until it is sealed, halting the reaction. Black oxide coatings are performed in a steam atmosphere (700°F) that produces a very thin oxide coating which is capable of absorbing oils to provide improved corrosion protection. None of the conversion coatings provide a hard surface, as would be found with a nickel or chrome coating, and are only used for their corrosion-resistance properties.

Electroless nickel platings or autocatalytic platings, are also deposited in the absence of an electrical current. The process involves the deposition of a nickel coating by means of a chemically induced reduction of metal ions in a pH-controlled, heated bath. This process differs from conversion coatings in not being limited to a thickness determined by the interaction with the material surface; nickel continues to be deposited until the part is removed from the bath. The coating is hard (48 to 55 $R_c$) and can be finished ground to size for close-fit applications. Additional heat treating can be performed to obtain even greater hardness. Electroless nickel plating also provides lubricity for these close fit conditions as well as toughness or resistance to flaking as a result of the chemical bonds formed during its application. Its application thickness is generally <0.002 in., but it can be applied thicker for those applications requiring grinding to final form. It can also be applied to improve wear-resistance or for surface protection when code-posited with polycrystalline diamond particles. These solutions contain 25 to 30% diamond particulates suspended in the plating solution that are captured by the process to become an integral part of the plating.

An alternative to electroless nickel plating is traditional chrome plating. Hexavalent chromium plating baths have been used extensively for many years, but are becoming less popular due to environmental restrictions imposed on the process. The post-processing required to remove the chromium from the rinse water is costly because of the multiple steps necessary to remove and safely dispose of it. Chrome plating can be applied in soft and hard forms depending on the applied processing. Hard chrome requires higher concentrations, temperatures, and electrical currents, while soft plating does not. Surface hardnesses of up to 72 $R_c$ are possible, with thickness ranging from 0.0002–0.0006 inch (0.00508–0.01524 mm). There are proprietary coatings that can be applied to the surface of the chrome to lower the coefficient of friction, reducing wear. These coatings typically involve PTFE-based cross-linked polymers that are applied at elevated temperatures to the surface of the chrome. The resulting surface is hard chrome with surface impregnation of a lubricious plastic. Recently, the development of a safer and more efficient system based on trivalent chromium provides a cost-effective alternative which has overcome some of the shortcomings of the earlier process, including reduced processing involving the reclamation of the rinse water.

Other methods involve mechanical means to deposit a barrier coating (Table 4.9).

**Table 4.9**　　Barrier Coating Processes for Carbon Steels

| Process name | Method | Thickness (in.) | Cost |
|---|---|---|---|
| Conversion coatings | Phosphate chemical dip | 0.0005 | Medium |
| Electroless nickel | Chemical deposition | 0.00005–0.005 | Medium |
| Chrome plating | Electro-chemical deposition | 0.00005–0.050 | Medium |
| Zinc coating | 1. Hot dip galvanizing<br>2. Electro-galvanizing<br>3. Zinc spraying | >0.0005<br><0.0005<br><0.0005 | Low |
| Aluminum coating | Type 1 – heat resistant<br>Type 2 – corrosion resistant | 0.001<br>0.002 | Medium |
| Tin coating | 1. Hot dip process<br>2. Electrolytic process | 0.015–0.033<br>0.015–0.033 | Low |
| Thermal spray | Mechanical deposition | 0.0005–0.020 | High |
| Welding | Electro-mechanical | As needed | Low |
| Diffusion treatments | 1. Carburizing<br>2. Gas carburizing<br>3. Gas nitriding<br>4. Carbonitriding | <0.025<br><0.010<br><0.0004<br><0.030 | High |
| Electro-polishing | Acid or alkaline solution (material removal) | <0.005 | Medium |

Cold worked steels are typically produced with aluminum, tin, or zinc coatings. These coatings are applied by various means including hot dip galvanizing and electrodeposition methods. Hot dip galvanizing generally allows for coatings exceeding 0.005 in. thick while electrodeposition methods provide greater control over thickness as a function of exposure to the electrically charged bath. Thicknesses range from less than 0.0005 in. to 0.003–0.005 in. depending on the material and the specific nature of the process; coating thickness are also measured in weight per square area, effectively providing the coverage volume.

Aluminum coatings provide better heat-resistance because of their higher melting temperature as well as improved reflectivity and appearance. There are two different types applied according to the application. Type 1, a 0.001 in. thick coating used in applications requiring heat-resistance and Type 2, a 0.002 in. thick coating used in applications requiring corrosion-resistance. These coatings are used mostly on smaller parts and fasteners because of the capital investment required for this higher temperature process.

Tin coatings are commonly used in electrical applications involving circuit boards and related hardware as well as in smaller mechanical parts. The use of tin or tin alloys in place of other metals such as cadmium or lead is a result of efforts to minimize or eliminate the use of toxic materials. They are applied most commonly by hot dip or electrodeposition methods to thicknesses ranging from 0.015 to 0.033 in. (0.4 to 1.5 μm). The nontoxic characteristics of tin coatings combined with corrosion resistance, good solderability, ductility, lubricity, and low cost have resulted in their increased demand.

Zinc has been used to protect metal surfaces from atmospheric conditions for many years. The corrosion protection afford by zinc is unique because of its anodic behavior when placed on steel. A small scratch or damaged area, exposing the steel, will be protected by the adjacent zinc particles, because zinc will corrode in preference to the steel. Its relatively low melting temperature and cost also make it an economical alterative to aluminum. Application thicknesses range from 0.0005 in. for electroplating and up to 0.005 in. or more when hot dipped. Zinc spraying can also be performed which produces a rougher, porous surface that is better for painting.

Cadmium coatings have fallen out of favor because of their toxic nature and adverse environmental effects. Although they flourished as a cost effective method of providing corrosion protection for ferrous metals they are highly toxic and should not be filed, ground, or welded because of the toxic byproducts produced during these manufacturing operations.

Electropolishing is a unique process of metal surface preparation that actually removes surface material (reverse plating) rather than applying it to produce a smoother, more uniform surface while also passivating it. The process is typically applied to stainless steel parts but can also be used on selected ferrous metals, such as 1075, 1080, and 1095, as well as nonferrous metals, such as aluminum and copper alloys. Because it is a removal process that is controlled by exposure time, it can be used for deburing of parts of all shapes and sizes while improving their appearance. The benefits of this surface finish is that it produces a hygienically clean surface which is resistant to contamination and related bacteria growth, it eliminates surface corrosion and processing damages, such as forming marks and welding oxides and surface discoloration, as well as other microsurface defects that could contribute to early fatigue failure. This process also produces a microsurface finish that reduces friction, improves adhesion, and overall appearance, providing a finish similar in appearance to chrome or nickel plating.

Surface properties of plain carbon steels can also be altered by the application of thermal spray coatings. These coatings can be applied in thicknesses ranging from 0.0005 to 0.200 in. for the purpose of improving wear-resistance, corrosion-protection, and greater heat tolerance on less expensive materials or those with reduced strength or hardness properties. There are four common thermal spray processes (Table 4.10).

These applied coatings can be used on a wide range of ferrous and nonferrous metals, ceramics, and carbides and are limited in application to line-of-sight surfaces. The materials adhere to the surfaces via mechanical bonding as a result of the high velocity and molten state of the material as it is

**Table 4.10**  Thermal Spray Processes

| Process | Description |
| --- | --- |
| HVOF | High velocity oxygen fueled ($v = M2$) |
| Plasma | Application of high melting point materials |
| Electric arc | For use on materials of high porosity which require high rates of material application |
| Flame | Application of low melting point materials |

applied. Although this bond type can produce significant adhesion, the fact that the interaction is mechanical and not at the atomic level limits the magnitude of the adhesion; hence, the brittle coating is one that does not perform exceptionally well when subjected to impact loads. Coatings applied to softer materials subjected to impact loads also do not perform well because of the difference in material stiffness, resulting in the fracturing of the coating. Because of this, the most successful applications involve sliding wear or material slurries, both of which do not involve high impact forces but rather a high degree of abrasion. For this process to be successful, the surface must be properly prepared and cleaned of all contaminates to allow direct bonding to occur. A quality assurance process usually follows that confirms that the magnitude of the adhesion by the applied coating is within an acceptable range.

Welding is most often used for joining materials but is also used to surface coat or "hard face" low-carbon steel metals in an attempt to reduce surface wear due to abrasion, cavitation, erosion, galling, or impact environments. The materials used for hard facing are, generally, cobalt, nickel, iron, or carbides with iron-based grades, which are the most common applications due to cost considerations. The matrix alloys generally contain up to 35% Cr, up to 30% Mo, and up to 13% W, with smaller amounts of silicon and manganese. These iron grades include pearlitic, austenetic, martensitic, or high-alloy steels applied using MIG welding techniques. Selection of the appropriate material and application thickness depends heavily on the application. Coating thicknesses are generally greater than 0.078–0.118 (2–3 mm) due to the nature of the welding process, but they can be ground back as needed to obtain the desired thickness and finish.

Diffusion treatments are those which involve the movement of atoms within a material. Specifically, when applied to metals, they involve the introduction of the desired atoms at the surface of the material to produce the desired hardness and wear characteristics. The process is a complicated one that follows Flick's law, which mathematically depicts the transportation process involved:

$$J = D \left( \frac{\partial c}{\partial x} \right)_t$$

where $J$ is the flux of matter per unit area perpendicular to the projected surface, $(\partial c / \partial x)$ is the concentration gradient normal to the plane, and $D$ is known as the *diffusion coefficient*.

Application of this process involves exposing the metal surface to either a gas or high-carbon powder graphite to produce a case hardening effect on grades that are normally difficult to harden, or are unhardenable. The earliest form of this process involved packing the plain carbon steel part in a bed of carbon powder and heating it to 815–955°C (1500–1750°F) to initiate the diffusion process which will produce a depth of surface hardness that is proportional to the square root of the time of exposure. This process is known as pack carburizing and produces a hard case over the softer steel, providing improved wear properties while maintaining a significant toughness. Gas carburizing produces similar results by exposing the part to a hydrocarbon-based carrier gas flow, which often is based on natural gas or propane, to produce a case hardened surface to a depth that is a function of process temperature and time.

Additional popular diffusion methods include carbonitriding and gas nitriding. Carbonitriding is a modified form of carburizing rather than a nitriding process. The distinction of this process over gas carburizing is the introduction of ammonia gas into the carrier gas to increase the nitrogen content, which improves surface hardness. Case depths greater than 0.030 in.

can be obtained using similar carburizing process temperatures and times. Gas nitriding is a process in which the metal is exposed exclusively to an ammonia carrier gas to introduce nitrogen into the surface, producing a harder, more wear-resistant surface than that obtained in carburizing. It also improves fatigue life, galling, and high-temperature resistance, while resulting in fewer process-related distortions than observed with carburizing. These characteristics are ideal for machine tool applications. A similar process, ion nitriding, is an extension of gas nitriding, using plasma discharge technology, which results in minimal distortion and excellent surface finish.

For medical devices and those applications requiring improved wear or sterilization, physical vapor deposition (PVD) coatings can provide this capability. These include:

- TiN — titanium nitride
- ZrN — zirconium nitride
- CrN — chromium nitride
- TiAlN — titanium aluminum nitride
- AlTiN — aluminum titanium nitride

The advantages of PVD include: sterilization, improved wear resistance, reduced friction, biocompatibility, chemical barrier, and decorative colors and aesthetics.

### Carburizing Grades

- Improved hardenability compared to straight carbon steels
- Case hardens easier and to a greater depth
- Selected free machining grades

## Common Grades

### Nonhardenable

1020 – A low carbon steel with good all around properties.

### Hardenable

1040 – A hardenable grade of medium carbon steel.
1140 – Free machine grade of 1040.

## Alloy Steels

Carbon steels become alloy steels when any of the elements exceed the specified ranges for carbon steels (i.e., >1.65% manganese and >0.6% silicon and copper). They also are considered alloys if selected additions are made to obtain specific properties or minimum amounts are specified for elements such as aluminum, boron, chromium, cobalt, columbium, molybdenum, nickel, titanium, tungsten, vanadium, and zirconium.

Alloy steels are improvements on carbon steels, formed by adding various elements to the parent material in order to increase its strength, hardness, corrosion-resistance, or machinability. For example, adding up to 0.5% sulfur to a carbon steel improves its machinability without compromising on its strength or corrosion resistance. Adding 0.3 to 4.0% chromium improves strength and corrosion-resistance. Adding molybdenum or vanadium improves both surface characteristic of the material its ability to through harden. These alloys add to material costs when compared with plain carbon steels because of their increased processing times, either at the foundry or during the piece part fabrication, and the added cost of the alloying elements. The effect of other alloying elements must also be considered (Table 4.11).

This classification of alloy steels also encompasses grades that are of low alloying content and have exceptional strength properties. They are known as high-strength low-alloy (HSLA) steels and typically contain less than 1% carbon. They contain less than 5% of any alloying element and are graded on their resulting mechanical properties rather than by a rigorous analysis of their alloy composition, allowing a wide latitude in the chemical makeup of the alloy. Additionally, those that possess a tensile strength greater than 200 ksi are classified as ultrahigh-strength steels. Although this designation is an arbitrary one, it is helpful in providing further clarification of these materials and their important properties.

Generally, the improved properties are a result of a shift in the TTT (Time, Temperature, Transformation) curve to the right, increasing the allowable time for quenching, which is necessary to harden the steel. The combination of material

**Table 4.11** Principal Effects of Alloying Elements in Alloy Steels

|  | Typical ranges in alloy steels (%) | Principal effects |
|---|---|---|
| Aluminum | <2 | Improves nitriding processing while restricting grain growth and promoting oxygen removal |
| Sulfur and phosphorus | <0.5 | These additions reduce toughness (ductility) and weldability while improving machining properties |
| Chromium | 0.3–4 | An increase in hardenability, corrosion resistance and high temperature strength while also improving wear resistance due to the formation of chromium carbides |
| Nickel | 0.3–5 | Can be added to improve either hardenability or ductility, but not at the same time |
| Copper | 0.2–0.5 | Improves corrosion resistance due to the presence of an oxide film |
| Manganese | 0.3–2 | Promotes a reduction in the required hardening temperature while increasing the ability to be hardened |
| Silicon | 0.2–2.5 | Primarily added to control the oxygen content during foundary processing as it relates to the level of impurities |
| Molybdenum | 0.5–1.0 | Primarily added to improve high temperature performance when combined with chromium while also increasing hardenability and grain refinement |
| Vanadium | 0.1–0.3 | Added to improve high temperature hardenability by creating carbides while also producing a finer grain structure |
| Boron | 0.0005–0.003 | Increases hardenability |
| Lead | <0.3 | Improves machinability |
| Nitrogen | 0.1 | Improves strengthening |

chemistry and increase in processing time allows the entire material section to cool in a manner that produces a martensite structure throughout. The SAE Iron and Steel Committee defines full hardness as an alloy with a minimum of 90% martensite. The desired combination of strength and toughness, or percentage hardness, is highly dependent on the processing requirements and end-use. If the material were too hard, it would be susceptible to cracking, while, if it were too soft, it would probably not survive in the intended application. A Jominy hardness test (ASTM A255/A304) can be performed to determine the material's response to heat treating.

Although the primary objective of alloying is to improve hardness, strength and elevated temperature resistance over plain carbon steels. The extent to which this occurs is highly dependent on the material composition and application temperature, but improvements ranging from 20 to 100°F increase in the softening temperature are not uncommon. Comparitively, alloy steels have improved properties over plain carbon steels but at the added expense of both raw material cost and processing time (Figure 4.11).

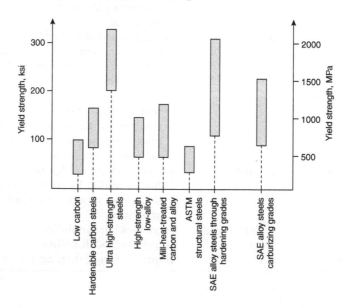

**Figure 4.11**   Comparison of alloy and plain carbon steel material strengths.

## Properties and Designations

The improved properties of alloy steels over plain carbon steels are a result of the change in material microstructure due to the introduction of foreign elements. How these elements are absorbed into the steel is instrumental in defining the resulting properties. There are three basic interactions involved: substitutional solution, interstitial solution, and clustering of like alloy atoms. These interactions result in greater resistance at the slip planes, increasing the material strength. Additional process treatments (as discussed below) further enhance the process to produce a wide variation of material properties.

### Numbering

The generic SAE alloy steel classifications are designated by a four (and sometimes five) digit number. The first two digits designate the type of alloy steel and the nominal alloy content followed by the last two (or three) digits which indicate the nominal carbon content (Table 4.12).

Standard grades are identified as described above while additional designators are included to the number in order to indicate variations to the standard (Table 4.13). As with plain

**Table 4.12**   SAE Ferrous Alloy Designations

| Designation | Primary alloy constituents |
|---|---|
| 13XX | Manganese |
| 2XXX | Nickel |
| 3XXX | Nickel–Chromium |
| 40XX/44XX | Molybdenum |
| 41XX | Chromium–Molybdenum |
| 43XX | Nickel–Chromium–Molybdenum |
| 46XX/48XX | Nickel–Molybdenum |
| 5XXX | Chromium |
| 6XXX | Chromium–Vanadium |
| 7XXX | Tungsten–Chromium |
| 8XXX | Nickel–Chromium–Molybdenum |
| 92XX | Silicon–Manganese |
| 9XXX | Nickel–Chromium–Molybdenum |

**Table 4.13**    SAE Ferrous Alloy Prefix/Suffix Designation

| Designator | Description | Example |
|---|---|---|
| A (prefix) | ASTM designation for ferrous structural quality materials. It is placed in front of an arbitrarily assigned number. To fully define the material, this letter/number combination along with the grade (chemical composition) and the class (desired strength level) must be provided | A516, Grade 55, 55 ksi |
| EX (prefix) | These designate experimental grades as defined by SAE. EX followed by a two digit number | |
| M (prefix) | Designates merchant quality steels | M1010 M1020 |
| B (mid-fix) | Designates the addition of Boron to the base grade. This additive increases the base strength of the material. Boron behaves similarly to carbon in hardening steel 0.0005–0.003% added cost savings — it is cheaper to use boron than some other expensive alloy such as chromium to increase hardness | 15B21 15B35 50B46 50B60 |
| L (mid-fix) | Designates the addition of lead to the base grade. This improves the machining capabilities | 12L14 41L40 |
| Bi (suffix) | A lead-free alternative is using bismuth as an additive which acts an internal lubricant much like lead | 4140Bi |
| CT (suffix) | Fine-grained, calcium– aluminum treated alloy steel. Can improve machining by a factor of three. Machining chips are small rather than long and stringy | 4140CT 8620CT |

**Table 4.13** *(Continued)*

| Designator | Description | Example |
|---|---|---|
| H (suffix) | These steels have hardenability requirements as well as the limits and ranges of chemical composition | Available in most standard alloy grades including 4140 and 4340 |
| X (suffix) | Indicates the addition of niobium, vanadium, nitrogen and other alloying elements<br>Found on callouts for SAE HSLA steels | SAE 942X<br>SAE 945X |
| V (suffix) | Vanadium modified alloy grades | 4335V |

carbon steels, these designators indicate an enhancement to one or several selected material properties, which has been made to produce an improved grade. These improved grades are more costly to produce and procure because of the additional processing involved and their limited availability.

As stated previously, HSLA steels are a class of metals which have <0.25% C, varying amounts of copper and selected alloys for improved strength. These materials are primarily graded based on their mechanical properties rather than a rigorous chemical analysis, resulting in chemical compositions that vary widely. Microalloying is used to improve strength of these grades and involves the addition of small quantities of selected elements. HSLA are typically not heat treatable and produced in hot-finished form for use in structural applications. Their compositions are such that the materials lend themselves to welding and mild corrosion protection when left untreated as a result of a durable oxide surface that forms because of the addition of copper. Severe or aggressive environmental conditions are best served by the use of some form of corrosion protection, such as a paint or thermal spray coating.

### Strength and Hardness

Alloy steels have significant strength and hardness advantages over plain carbon steels because their alloyed structure.

In their hot and cold rolled, nonheat treated forms, alloy steels have tensile and yield strengths in the 100 ksi range, while heat treated (quenched and tempered) alloy steels have tensile and yield strengths above 100 ksi (689 Mpa). In comparison, plain carbon steels have tensile and yield strengths in the 50 ksi range and those that can be heat treated can be strengthened to the 100 ksi range. The range of hardenability is wide for alloy steels with the hot and cold rolled materials ranging from 90 $R_b$ to not greater than 40 $R_c$.

### Heat Treatability

The most notable benefit of alloying plain carbon steel is to improve the material's strength, hardness, and related performance properties by heat treating it. The alloy additions moves the TTT diagram to the right, increasing the allowable quenching time from one second to several seconds which makes it practical to successfully strengthen these materials. Tempering commonly follows the quenching operation to improve toughness with little loss of hardness. The other factors that influence the hardness are the alloy content and the cooling rate used as it affects grain size.

### Fatigue

The fatigue properties for alloys steels are similar to those of plain carbon steel because sufficient carbon is present to establish it. The fatigue limit is proportional to ½ $S_{uts}$ for materials with a tensile strength <200 ksi and, therefore, is typically greater for alloy steels due to the nature of their higher strength. As with plain carbon steels, no fatigue limit is exhibited when in a corrosive environment requiring appropriate design precautions be taken.

### Corrosion

Alloy steels are not as susceptible to most forms of corrosion as plain carbon steels due to the presence of alloying elements. Varying the type and quantity of these elements can significantly alter the corrosion-resistance as well as the mechanical properties of the alloy. Copper is commonly added to improve the corrosion-resistance of many alloy steels. A common corrosion process of alloys and other high-strength steels with hardnesses greater than 30 $R_c$ is hydrogen embrittlement, which is a surface phenomenon similar to the carburizing process that results in surface cracking and premature failure. Plating processes can contribute to this failure mode if the steel is not carefully maintained.

### Machinability

Machinability ratings for alloy steels are generally less than those found for plain carbon steels and typically range from 40 to 80% of the ratings for the 1212 plain carbon steel standard, with the variation being dependent on the alloy composition. Similar microstructure modifiers, such as lead and sulfur, are added to improve the chip cutting ability, but with less profound results due to the other alloying effects. These additions may limit the ability to weld these grades of steel.

### Weldability

Alloy steels having more than 0.4% carbon have a tendency to crack when welded because of the rapid cooling that takes place in the heat affected zone (HAZ) as the adjacent material rapidly conducts the heat away. This quenching process results in undesirable formation of martensite in the weld, producing a brittle weld that may fail immediately after cooling or while in service. To minimize this effect, the materials may be preheated prior to joining, and to eliminate it, the materials should be stress-relieved immediately after welding.

### Unique Properties

The combination of high strength and moderate to good corrosion-resistance is desirable in many design applications, and this is most often found in alloy steels at a reasonable cost. High-strength shafting, such as 4140, is a medium-carbon steel that has many applications in factory automation and transportation systems as it provides high hardenability and good fatigue, abrasion, and impact-resistance.

The numbering system for alloy steels is not an intuitive system that can be easily recalled based on the number involved but rather is one that is best memorized. Because design engineers perform a variety of tasks and use a wide range of materials, they may perform material selections on a limited or infrequent basis, making it difficult to accurately recall this type of information. The following sections provide some basic insight into the specific material designations while the details of specific materials in these classifications can be found in the appendix or other reference sources, such as manufacturer's specifications and those of their distributors.

**13XX:** The major alloying element of this series of alloy steels is manganese (1.6 to 1.9%), the addition of which results in poor machinability and fracture toughness. Mill products including both hot and cold rolled materials in a wide range of hardnesses are available, but none exceed 30 $R_c$. Heat treating can be applied to some grades, such as 1330, but it is limited to the cold rolled materials. Hot rolled materials are used in construction applications, while other grades, such as 1335, are available in cold heading grades for processing into high-strength bolts and cap screws.

**40XX:** This series of alloys includes both manganese (0.7–0.9%) and molybdenum (0.20–0.25%) to produce materials that, when annealed, are soft, with poor machinability. Heat treating typically involves oil quenching or carburizing to produce a material with a low tendency to microcrack because of its soft properties. These grades can be found in applications such as gears or shafting that can benefit from surface-hardening.

**41XX:** This series is produced with chromium (0.5–0.95%) and molybdenum (0.12–0.30%), resulting in an excellent wear- and abrasion-resistant material, which can be nitrided. It can be either oil- or water-hardened, depending on the grade, but is best oil-quenched to limit cracking problems which are more prevalent above 0.4% carbon. This series is also available in 41LXX grades, that include lead for improved machining capabilities but limited weldability, which is considered better than most alloy steels. Applications include shafts, axles, gears, hand tools, sprockets, chains, couplings, and fasteners, with maximum operating temperatures limited to less than 400°F.

**43XX:** This series is similar to the 41XX series as it is alloyed with molybdenum (0.15% max.) and chromium (0.50–0.80%), it also has nickel (1.83% max.) and sometimes may include manganese (0.70–0.90% max.) for added strength. The amount of alloying makes this a high-grade material and an expensive selection. This material is used in applications that require improved properties, such as better hardenability and higher strengths (>200 ksi) than the 41XX grades, and improved high-temperature performance. Corrosion-resistance is improved by the alloying, while machinability is inversely proportional to the hardness. When high accuracy is required, it is best to rough machine in the annealed or normalized state followed by heat treatment and finish machine/grind. The 4320 grade is known for its low and high cycle fatigue response as is the grade 4340, which also heat treats better than 4320. These grades are found in aircraft parts, axials, crank shafts, and gears as well as other safety critical and heavy duty applications. The 4340 grade is the most popular medium-carbon, low-alloy steel, and it can be deep hardened (thick sections) by oil-quenching while maintaining ductility, toughness, and strength.

**46XX:** Case hardening is the most notable use for these grades as they can be routinely surface-hardened to 62 $R_c$ while retaining superior core toughness. These are alloyed with nickel (0.85% max.) and molybdenum (0.25% max.) and machine well prior to carburizing with minimum distortion during post–heat treatments. The grades 4620 and 4640 are commonly used in high-performance applications requiring case-hardening combined with a tough core.

**47XX:** Alloyed with nickel (0.85–1.83%), chromium (0.45% max.), and molybdenum (0.20–0.35%), these grades are carburizing steels with intermediate case-hardenability and medium core-hardenability. During machining, tools tend to wear quickly due to alloying elements and they have slower feed rates. But due to their hardenability characteristics they are used in piston pins and universal crosses. Machining is best carried out in the annealed stage, and after heat treatment only finishing is suitable.

**48XX:** Alloyed with nickel (3.5% max), manganese (0.8% max), and molybdenum (0.25% max), this series is capable of being heat treated to good core strength and ductility. These grades are often used for carburizing to develop a hard, wear-resistant case with excellent fatigue props. This series may be readily machined in the normalized or annealed condition. These are not corrosion-resistant alloys and will rust, hence they require a protective coating for any exposed application. A common grade is 4820, which is used in tractor and automotive gears and pump parts to resist wear, because it has high hardenability in both case and core. The fracture toughness of 4820 is low at the surface but increases toward the core. In hot rolled normalized and annealed condition 4820 has a yield strength of 485 Mpa (70 ksi).

**50XX:** This series is alloyed with chromium (0.28 to 0.5%). Generally boron is added to improve strength and hardenability. This series has a medium hardenability rating, mostly used for springs and hand tools. A common grade is 50B60, a heavy spring steel, which has a yield strength of 1772 Mpa (257 ksi) in quenched and tempered condition along with high hardness (>60 $R_c$). They have poor machinability due to their hardness, as seen with 50100.

**51XX:** This series is alloyed with chromium up to 1.0%. It is similar to 50XX, with boron added for strength improvement. These grades exhibit excellent toughness and high ductility, with a high tensile-yield ratio. Like the 50XX series, 51XX grades are commonly employed in heavy spring applications, primarily in the automotive field for suspension leaf springs. Machining these grades can be very difficult in the "as rolled" condition, requiring that the alloy be annealed prior to machining to obtain maximum speeds and feeds. Weldability is poor due to their high carbon and chromium content. For best results, preheat the section and stress-relieve after welding. Either gas or arc welding methods may be used. These alloys are not known for their corrosion-resistance and have ordinary steel characteristics in corrosive environments. An example is 5140 which is carbo-nitrided for heavy duty gearing. It is susceptible to embrittlement when quenched and tempered. It requires that welding be conducted in an annealed or over tempered condition followed by a heat treatment to counter martensite formation and cold cracking.

**51XXX:** This series is alloyed with chromium up to 1.03%. The heat treatability of this series is good, usually done in oil medium. If distortion is not a major problem the quenching can be done in water but distortion and cracking may occur. It is water-quenched, generally, in cases where the section sizes are large and applications are not critical to minor distortion unlike gear applications. In the heat treated state they have poor machinability as in the case of 51100, which, for better machining, should be annealed or spherodized to prevent slower feed rates and tool wear. Welding can be done by all conventional methods but requires an annealed or preheated state to prevent cracks and martensitic HAZ formation. Like the previous 5XXX series these grades are not good at resisting corrosion and need protection.

**52XXX:** Alloyed with chromium up to 1.45%, this series can be through-hardened and is often used in roller bearings. Heat treatments for through-hardening of the alloys consist of heating to 1500°F followed by an oil-quench. A normalizing heat treatment at 1600°F and slow cooling, to relieve machining stress, may be employed prior to heat treatment. These grades may be formed by all conventional methods, including cold forging or stamping and hot or cold upset forming. These are high-carbon alloys typically used in bearing applications where welding is not applicable or appropriate. Like other 5XXX series, 52XXX grades are not corrosion-resistant. Their applications usually are such that they are used with lubricants that inherently provide a corrosion barrier. An example is of 52100, which is exclusively used for ball bearings. Machinability of 52100 is good by conventional methods. A spherodizing anneal should be used for optimal machining in this case.

**61XX:** This series contains chromium up to 0.95% along with vanadium (0.13 to 0.15%) and is similar to 4130 and 4140 but with higher carbon content. These grades are particularly known for their good shock-resistance and toughness, with abrasion-resistance in their heat treated condition. They are oil-quenched after heating for hardening and are good through-hardening steels. The machinability of this series is fair. Improved machining can be obtained by annealing and spherodizing. Corrosion-resistance is not good and requires post-processing to obtain useful protection. For example, medium-carbon low-alloy 6150, known for its low distortion characteristics and high surface hardness, is difficult to weld, but can be welded when properly preheated. Post-weld stress-relieving is essential to stabilize the finished part. The grade 6150 is commonly employed in heavily stressed machinery parts including shafts, gears, pinions, and also in hand tool components.

**81BXX:** It is alloyed with nickel (0.3%), chromium (0.45%), and Molybdenum (0.12%). As the designation suggests, boron is added for improved strength and hardenability. These are carburizing steels and find their common application in gears and similar applications requiring a hard wear surface and tough core. They have low distortion after quenching and are commonly oil-quenched. Weldability of these alloys is not good, therefore welding them is usually avoided. If welding is essential, it is done with pre-weld and post-weld treatments to avoid cracking. The machinability is reasonably good when in the annealed state and only finishing is done after heat treatment. An example is 81B45 which has a 65% machining rating and finds use in hardened gears and highly stressed aircraft parts.

**86XX:** This series contains nickel (0.55%), chromium (0.5%), and molybdenum (0.2%) with some portions of manganese and silicon. These grades are hardenable chromium, molybdenum, and nickel low-alloy steels often used for carburizing to develop a case-hardened part. This case-hardening will result in good wear characteristics. They are carburized and quenched in oil to case-harden. Relatively thick sections can be hardened by this method. Machining is best done prior to carburizing to as near the final dimension as possible. Finish machining, after heat treatment and/or carburizing, should be limited to prevent the removal of the hardened case. Good forming characteristics exist in the annealed condition. Welding can be done by conventional methods but preheating is recommended. The commonly used general purpose grade is 8620, which is used for gears, shafts, and crankshafts because of its superior case-hardness when carburized. Other examples include 8615 and 8627, which are also used for gears.

**86BXX:** This series has nearly the same alloying as 86XX, but includes boron; which is added for strength. They are regarded as high-alloy construction steels selected for heavy sections or for parts subject to particularly severe service conditions. Through-hardening grades exist but heat treating can be a problem with these, requiring mild quenches to prevent distortion. Typical uses include aircraft or truck parts, or in ordnance materials requiring high strength (>250 ksi). A common grade, 86B45, has very good yield strength in its annealed condition along with high hardness. It is machined prior to quenching and generally not recommended for parts which need welding.

**87XX:** This series is alloyed with nickel 0.55% and molybdenum 0.25% and these grades are known as good carburizing steels with medium machinability. Their processing can be difficult due to microcracks induced during heat treatment, limiting their applications. An example includes 8740, which is frequently used for gears and aircraft parts. It is through-hardenable and may also be nitrided for improved surface conditions. Common problems encountered are hydrogen embrittlement and fatigue limit variations depending on the heat treated state.

**88XX:** This series is alloyed with nickel (0.55%), chromium (0.5%), and molybdenum (0.35%). They are also known as good carburizing steels with medium hardenability. As with previous 8XXX series, their heat treatment is performed with caution to prevent cracks and embrittlement. Machinability rating is approximately 55% in a normalized state and it is not recommended to machine these grades after heat treatment. Due to alloying content, weldability is intermediate, it can be performed by the conventional methods but requires pre- and post-heating treatments. An example of this series is 8822 which is often used for applications requiring a case-hardened alloy steel to provide good wear-resistance as well as toughness.

**9XXX:** This series is alloyed with nickel, chromium, and molybdenum to give good carburizing characteristics and hardenability. These grades generally have high impact and fatigue strength. They form a broad category involving several different combinations of alloying elements to give unique properties. The 93XX and 94XX series are known for their superior case-hardness and good core properties. They are not through-hardening alloys. The 93XX and 98XX series can be nitrided for specific applications, such as gears designed for low contact stresses, spindles, rings, and pins. The 93XX series has good ductility even in the hardened and tempered condition. It can be cold worked by conventional methods as used on the other low-alloy steels. Members of the 95XX series are extensively used in steel castings due to their good tensile and yield strength along with good toughness and ductility. The 95XX grades have good machinability and can be welded more easily than others in this series. Corrosion-resistance of the 9XXX series is a little better than ordinary plain carbon steels but these grades require protection in corrosive environments.

## Fabrication Processes

Alloy steels are produced using the same basic processes as those used to produce plain carbon steels, including the basic open hearth or basic oxygen methods. To obtain improved quality alloy steels, these standard fabrication methods are substituted by the use of the electric furnace process; this is indicated by adding a prefix "E" to the materials designation. Although the quality is improved, the processing cost increases, resulting in a more expensive material that must be justified by the application. It is possible to obtain even higher quality steels by the use of vacuum arc remelting (VAR), which involves a secondary remelting of the alloy to remove additional impurities. The improved, uniform nature of the material results in higher strengths, greater fatigue life, and improved repeatability of foundry processing. These steels are most often selected for aerospace applications or as an upgrade to those other applications requiring improved performance as they allow weight and size to be reduced without compromising on performance.

The production methods and quantities typically demanded of alloy steels lends their foundry processing to smaller scale operations known as "mini-mills", which economically produce short runs and specialty grades at a lower cost and faster turnaround. They also are more adaptable to recycling scrap materials because they can run smaller lots, allowing the furnaces and related equipment to be adjusted to account for the composition of the materials to be recycled.

The finished foundry product forms are typical cast and continuous cast ingots that are processed by hot or cold working to produce the required net shapes such as angles and tees or basic machining stock including squares, rectangles, and rounds.

## Raw and Stock Finishes

Alloy steels are produced using similar processes as those applied to plain carbon steels, namely hot and cold finished forms. Common bar stock shapes including rounds, squares and flats are typically available in both hot and cold finished forms with their selection based on the design requirements due to their variable properties (Table 4.14). Plate stock is also available in several general purpose grades such as ASTM A572 or A588 which provide good strength and welding properties at moderate prices. There are several specialty grades of plate stock that are typically used as well; these are intended for specific design applications involving pressure vessel construction or other public domain or military products requiring stringent quality control.

## Processing Considerations

Alloy steels can be subjected to many of the processes known for plain carbon steels, including the most common thermal processes, strengthening, and surface treatments. In addition to these, alloy steels can be subjected to a chemical treatment known to produce a material that allows for increased machining speeds and tool life.

### Thermal Treatments

The same thermal treatments applied to plain carbon steels can be applied to alloy steels. These include hardening, softening, and conditioning treatments (see section on carbon steels). Although the treatments are nearly the same, there may be some minor process variations necessary to achieve the desire effect. For example, alloy steels generally require higher temperatures for stress-relieving, 100 to 200°F below the transformation temperature to allow grains to reform, but

**Table 4.14**  Common Alloy Steel Raw Stock

| Shape | Form | Material | Comments |
|---|---|---|---|
| Plate | | | |
| General purpose | Plate | A572 | High strength and good weldability at moderate price |
| General purpose | Plate | A588 | Improved corrosion resistance over A572 at similar strength |
| Specialty | Plate | A514 | Construction alloy — alloy steel. grade B, H, F, and Q high strength combined with good workability and weldability and exceptional toughness at low temperatures (to −50°F) |
| Specialty | Plate | A285 (alloy or plain) | Pressure vessel |
| Tubing | | | |
| Mechanical | Round | 4140 | |
| Mechanical | Square | | |
| Hydraulic | Round | | |
| Bar | | | |
| Hot finished | Rounds | ASTM A615 Grades 40 and 50 (are these alloys?) | Reinforcing bars: formed from billets of steel For use in construction of reinforced concrete structures. They are identified by bar number, ranging from #3–#9 |
| | Rounds | Low-carbon: 8620CT, 86L20 (case hardening), E4320 (aircraft quality). Medium-carbon: 41L50, 4340, E4340 (aircraft quality) 6150 (annealed), 4130, 4140CT, 4142CT, | Available as precision ground bars in grades 410, 4142, and 4340 |

| | | |
|---|---|---|
| Rounds | 41L40, 41L42, 41L47, 415OCT, 41L50, 4340, AMS 6470 (aircraftquality) ASTM A193 Grade B7 and B16 | Chrome molybdenum alloy steel — alloy steel grades which are specifically designed for highly stressed applications including bolts, flanges, and valves |
| Squares | Low-carbon: not generally available Medium-carbon: 4140 annealed, 4140CF, 4145, 4150, 8620 | Medium-carbon (4140) and case-hardening (8620) grades |
| Flats | Low-carbon: 8620, E4130 (aircraft quality) Medium-carbon: 4140, 4140CF, 4150 | |
| Hexagons | Medium-carbon: 4140CF, 41L40CF | |
| Cold finished | | |
| Rounds | 4140 | Tubing offers similar performance as the bars stock forms: strength, hardenability, wear resistance, toughness, and ductility |
| Squares Flats Hexagons | 4140, 4150, E4150, 8620 | |
| Structural | | There are generally no typical alloys used for commercial structural applications including shape, such as angles, channels, beams, and tees. |
| | 4130 | Alloy structural shapes are typically made from 4130 for aircraft use |
| | A588, Grade A | High-strength, low-alloy (ASTM A588, grade A) |

they still follow the Larson–Miller equation. Temperatures below this range generally are not effective in removing all the stress. Soak time and cooling rate are subject to the behavior of the individual alloy grades.

There are some notable differences in process results for some specific applications such as quenching. For high-alloy steels with no significant carbon, quenching leaves the material in the softest state rather than a brittle one, as seen in typical plain carbon steels. Hardening characteristics of alloy steels are unique in that some grades through-harden while others do not, behaving more like plain carbon steels. Thickness has an important influence on the hardenability. Thinner sections will harden equally for almost all alloy steels while thicker sections will not. Selection of a grade that is known to harden in thick sections (such as a 4340) should be made for those applications requiring through-hardening. A carbon steel with the same carbon content as an alloy steel will not through-harden with an oil-quench, because the required quenching time is too short. The plain carbon steel may require a quench time of 1 s while the alloy steel will require something less than the 10 s that is achievable with the oil quench. Water must be used to obtain a quench time of less than 5 s, but this results in additional problems including distortion and cracking.

### Strengthening Treatments

The primary strengthening treatment is the choice of alloying elements used in the material formulation which will dictate the extent of the improved strength as well as other important properties such as corrosion-resistance. Varying these alloying elements will correspondingly alter the resulting properties.

Most alloy steels respond favorably to cold working, making it the most commonly applied post-processing strengthening treatment. These include cold heading, cold drawing, cold extrusion, and cold riveting.

### Surface Treatments

The need for surface treatments of alloy steels is significantly less than that of plain carbon steels, because they exhibit greater stability in typical atmospheric conditions (i.e., they do not readily rust). The three general forms of chemical, electrochemical, and physical deposition methods all remain viable candidates for improving the surface performance of an alloy steel when the design environment calls for it. The processes involving coatings, diffusion treatments and electropolishing do not change significantly from those used when applying them to plain carbon steels but may involve changes that take into account the different material properties. For example, carburizing grades (<0.2% carbon) of alloy steels are generally oil-quench hardened rather than by using water because of the shift in the TTT diagram to the right, which provides more time for the cooling process.

### Chemical Treatments

Calcium treatment of alloy steels is possible. Calcium is added during the molten processing stage and acts as both deoxidizer and desulfurizer to produce a higher grade material. It results in a fine-grained alloy steel with superior machining characteristics, allowing as much as a 300% increase in tool life while machining at twice the nontreated material recommended speeds. Although the improved machining benefits without loss of mechanical properties is significant, cost and availability are also a concern as they may limit the practical application of these material grades.

## Common Grades

- 4140 — smaller sections, low price
- 4340 — larger section, higher price and lower availablility
- 4150 — alternate
- 8740 — alternate

## Stainless Steels

Corrosion-resistant steels (CRES) are predominately stainless steels, as they offer the greatest resistance over a wide range of corrosive environments as well as excellent cosmetic properties. For a carbon steel to be classified as a stainless steel it must contain a minimum of 10.5% chromium (Cr) by weight and typically does not exceed a maximum of 30%. The carbon content of these metals typically ranges between

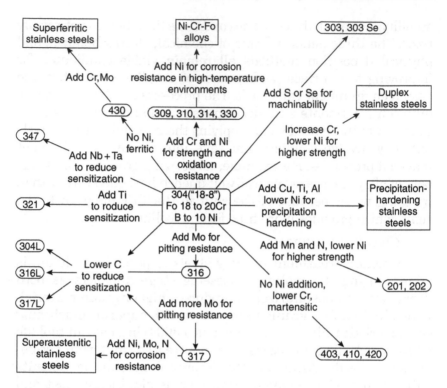

**Figure 4.12** Compositional and property linkages in the stainless steel family of alloys.

*Source*: ASM Metal Handbook: Online Desk Edition, Wrought Stainless Steels: Selection and Application, Figure 1.

0.3 and 1.2%. Other alloys also have an effect on the corrosion-resistance and processing capabilities of stainless steels (Figure 4.12).

Meeting only this requirement is not sufficient to fully meet the classification, as stainless steels also exhibit a natural tendency to produce a passive surface when exposed to ambient conditions. This self-healing property of repairing itself when damaged in the presence of oxygen is unique to stainless steels and is the basis for their resistance to corrosion. The passive surface coating can be compromised when exposed to high concentrations of chlorine, because it excludes oxygen, replacing it with the very corrosive chlorine molecule.

Repair of materials subjected to this condition is possible but difficult as it is necessary to fully abrade the surface while removing the embedded iron particles, allowing the natural passive surface to again form.

In economic terms, stainless steels can compete with higher cost engineering metals and alloys that are based on nickel or titanium, offering a variety of corrosion-resisting properties that are suitable for a wide range of applications. Stainless steels can be manipulated and fabricated using a wide range of commonly available engineering techniques and are fully "recyclable" at the end of their useful life. In addition to their corrosion-resistance, stainless steels also offer other useful properties, depending on their "family" or type. These include the fact that they can be easily cleaned, making them the first choice for applications requiring strict hygiene conditions, such as in hospitals, kitchens, and food processing plants. When the total life cycle costs are considered, stainless steel is often the least expensive material option.

Having a high chromium content is not sufficient for a material to be one that automatically produces its own passive coating and can be classified as a stainless steel. For example, many tool steels have greater than 10.5% Cr by weight but fail to be self-passivating because of their unique combination of carbon and alloying elements.

A direct measure of the effect of chromium on stainless steel can be found in a graph of corrosion rate versus chromium content. The rate of corrosion is significantly reduced after the weight percentage of chromium reaches 10.5% and becomes nearly zero at levels above 15% (Figure 4.3). This general trend of corrosion applies to most classes of stainless steels although within this family of steel, there is variability between grades that should be considered when making specific selections. For example, welding can produce a sensitivity zone that has greater susceptibility to corrosion. Selection of 304L (lower carbon content) for welding reduces the possibility of HAZ corrosion because it is less sensitive to the thermal effects.

There are four basic classes of stainless steels that are categorized by grain microstructure and a fifth, which is noted by its hardening method: austenitic, ferritic, martensitic, duplex

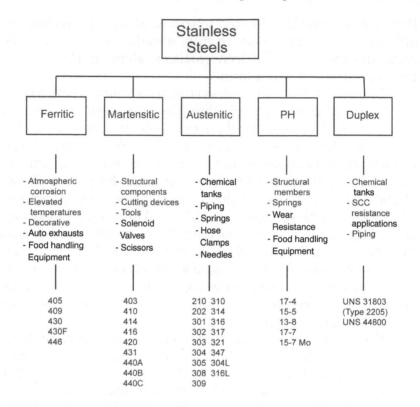

**Figure 4.13** The five families of stainless steels and their common uses and material grades.

alloys, and precipitation hardening (PH) alloys (Figure 4.13). The first three classes describe the material's microstructure, while the fourth class, duplex alloys, is a combination of ferrite–austenite alloys. The PH alloys come in three forms, semi-austenitic, austenitic, and martensitic. These are identified by four basic number series: 2XX, 3XX, 4XX, 5XX, and the PH series suffix (i.e., 17-4PH).

The thermal expansion of austenitic grades (200–300 series) is an important property to consider because this can be 40–50% greater than the expansion of plain carbon steel (Table 4.15). This great a differential in thermal expansion can cause considerable distortion of precision mechanisms or other position sensitive assemblies. The other stainless steel grades

**Table 4.15**  Thermal Expansion Rates of Stainless Steels

| Material | Thermal expansion rate (in/in/°C) |
|---|---|
| Austenitic grades | $7 \times 10^{-6}$ |
| Ferritic and matensitic grades | $10 \times 10^{-6}$ |
| Plain carbon steel | $12 \times 10^{-6}$ |

have expansion rates similar to plain carbon steels and may be better choices when considering materials for designs involving changing thermal environments.

## Austenitic

The austenitic series (2XX and 3XX) is by far the most common class of materials considered when selecting stainless steels because of the superiority of their properties over those of plain carbon or alloy steels with only a moderate cost increase, often worth the expense of the improved properties (Figure 4.14). The major alloying elements include chromium (16 to 26%), nickel (up to 35%), and manganese (up to 20%). The crystalline structure of this series is FCC; its members exhibit excellent toughness, ductility, and formability through a wide range of temperatures, including cryogenic ones and are typically nonmagnetic. Annealed grades of 2XX and 3XX have tensile yield strengths up to 70 and 40 ksi, respectively. They are strengthened by cold working (work hardening) and do not respond to heat treatment as a form of improving strength but do respond to other forms of thermal conditioning. Heavy cold working of some low 3XX series alloys (301 and 304) can result in transforming the austenite to martensite, rendering it magnetic. Alloy 303 represents the optimum in machinability among the austenitic stainless steels. It is primarily used for designs requiring extensive machining or are of high volume and fabricated in automatic screw machines. The machinability rating (compared to B1212) is approximately 78%.

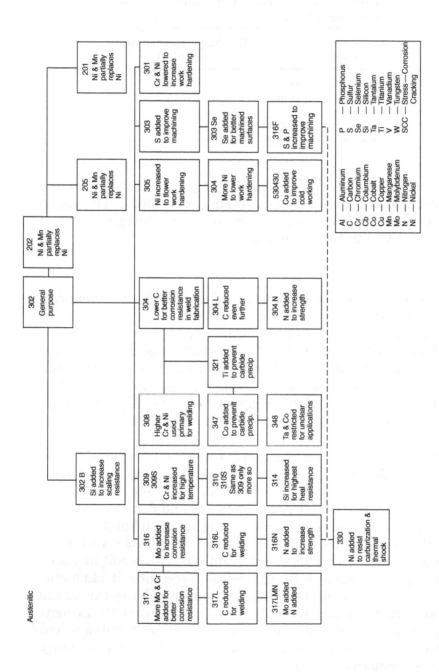

**Figure 4.14** Austenitic stainless steels.

*Source:* ASM Metals Handbook: Properties & Selection: Stainless Steels, Vol. 3, p. 7, Figure 2, ASM International, 1985.

Alloy 303Se (UNS S30323) has selenium rather than sulfur added to improve hot and cold forming characteristics over those of 303. It also provides a smoother machined surface finish and reduced galling in sliding friction applications, such as with fasteners. The machinability rating is also slightly reduced.

Alloying 304 alters its basic properties to produce additional grades in the following manner: molybdenum enhances pitting corrosion-resistance (316 and 317 grades); high chromium grades are for high-temperature environments (309 and 310); titanium and niobium grades prevent intergranular corrosion (321 and 324); reducing carbon and adding sulfur reduces the sensitization effects of 304 and this is noted by the designation "L" (i.e., 304L).

## Ferritic

The ferritic stainless steels are the 4XX series that contain up to 20% Cr and are a BCC structure, resulting in a material that generally has poor toughness and increased notch-sensitivity and is more difficult to machine than austenitic grades (3XX). These grades cannot be hardened by heat treatment (quench hardened) but do respond to thermal conditioning to produce annealed strengths commonly ranging from 40 to 50 ksi. Their chief advantage is their resistance to chloride stress-corrosion cracking, atmospheric corrosion, and oxidation, while remaining a low-cost material. Welding these grades creates a brittle HAZ that is susceptible to sensitization (the loss of the natural ability to create a passive finish), which results in corrosion from welding-induced carbide precipitation, and therefore these grades are not generally preferred for this joining method. Commercial applications of materials of this grade include 405 and 409 which are widely used for automotive exhaust systems because of their good formability and low cost while providing improved corrosion-resistance compared to carbon steels, while not necessarily being the best available stainless steel. Additional examples include 430 and 434 which are used as automotive trim and cooking utensils. These have poor toughness and weldablity, which is in line with their low cost (Figure 4.15).

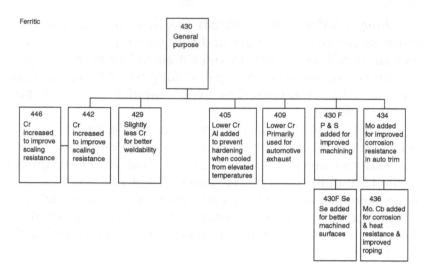

**Figure 4.15**    Ferritic stainless steels.

*Source*: ASM Metals Handbook: Properties & Selection: Stainless Steels, Vol. 3, p. 8, Figure 3, ASM International, 1985.

## Martensitic

The martensitic grades of stainless steels are also the 4XX series, which include up to 18% Cr with low to medium carbon content (up to 1.2%) and a FCC structure. The increased carbon content improves wear- and abrasion-resistance, which can be observed by comparing the poor wear-resistance of 410 (0.1% carbon) to the excellent wear-resistance of 440C (1.1% carbon). Their hardness and strength are also greatly improved and can be as high as 60 $R_c$ and 280 ksi. They can be heat treated like low-alloy steels using quench and tempering techniques. Martensitic stainless steel grades such as 420 are used to make cutlery, gears, shafts, and rollers, while 440 grades are used to make solenoids, shafts, valves scissors, cams, and ball bearings. The 440 grades have improved mechanical properties, such as high tensile strength combined with good fatigue and creep properties, but at the expense of corrosion-resistance. These grades are also magnetic and find use in solenoid valves and other applications requiring a magnetic corrosion-resistant material (Figure 4.16).

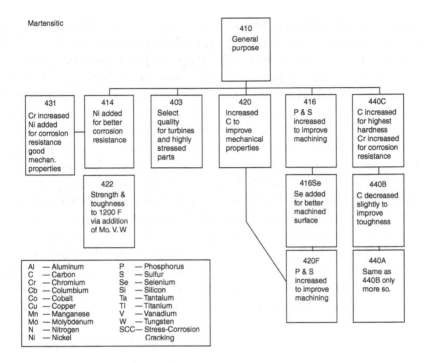

**Figure 4.16** Martensitic stainless steels.

*Source*: ASM Metals Handbook: Properties & Selection: Stainless Steels, Vol. 3, p. 8, Figure 4, ASM International, 1985.

## Precipitation-Hardened (PH)

These materials are uniquely different from the previous three grades because they are hardened when the lattice structure is strained due to the precipitation of excess material phase into the already frozen structure. The material is processed in three steps:

1. Heating to dissolve specialty elements or compounds (similar to dissolving excess salt in hot water);
2. Cooling fast enough to retain the solution effect, suspending the elements in a supersaturated state;
3. Reheating to allow controlled precipitation of submicron forms of the elements.

When the controlled precipitation occurs at room temperature, it is known as natural aging, and when performed at elevated temperatures, it is called artificial aging. The room temperature transition is often incorrectly referred to as "age-hardening." Improper hardening or exposure to elevated temperatures above the aging temperature (approx. 900°F) results in an over-aged condition, which is weaker. This overaged condition is necessary to prevent cracking when rapidly heated during forging. The artificial aging treatment is followed by cooling using oil or air. The condition number indicates the aging temperature employed. Different heat treats will exhibit various corrosion rates and dimensional changes. Dimensional changes occur between Condition A and the treated form ranging from 0.0004/0.0006 to 0.0010/0.0012 per inch, from H900 to H1150. Because of this variation, it is best performed after part fabrication with consideration of the material needed to perform the finish operations, not machined in the heat treated condition.

PH stainless steels can be all three grades: semiaustenitic, austenitic, and martensitic depending on their composition and processing. These steels also include elements such as titanium, aluminum, or copper and generally are low in carbon.

## Duplex

The duplex grades of stainless steels are a combination of ferritic and austenitic grades that typically have equal amounts of these microstructures and can contain up to 30% Cr. They are low in carbon content (<0.03% C). Additional elements that promote austenite formation include nickel, cobalt, manganese, copper, carbon, and nitrogen and those that promote ferrite formation include chromium, silicon, molybdenum, vanadium, aluminum, niobium, titanium, and tungsten. Synergism is observed in tensile yield strength, which can be improved by about 50%, from 45–50 to 80–100 ksi. Duplex alloys have improved chemical resistance compared to 303 and 304 as well as improved stress-corrosion cracking (SCC), resistance, and weldability. Elevated temperature performance is limited to 250°C because of possible embrittlement.

These stainless steels are used in the oil, gas, and petrochemical industries as well as the paper and pulp producing industries because of their resistance to the common caustics present in the related processing activities. The processing environments of these industries include chlorinated solvents and seawater. There is not a conventional numbering designation for duplex alloys but they are defined in the UNS numbering system.

## Proprietary Alloys

Stainless steels are also produced in specifications that are nonstandard, called proprietary grades. These proprietary grades are generally formulated to meet very specific needs and, therefore, are likely to be very expensive or have limited availability. A sample of these types of stainless steel alloys follows:

*High Alloy Content Grades*: These include those with high Cr or Ni content which provides increased wear- and corrosion-resistance. Specific alloys, such as nickel, can provide improved performance in certain environments such as chloride solutions where it affects the stress-corrosion cracking (SCC) behavior. Below 8% it results in poor SCC behavior as seen with 303 and 304 grades (18.8 stainless), while above that value it steadily improves until its effect is nearly eliminated above 15%.

*Nickel-Free Grades*: Specialty grades that exhibit biocompatibility are available for use primarily in the medical device industry where interaction with human tissue is of primary concern. Nickel-free grades are produced to meet stringent industry health standards, while still possessing the intrinsic properties that stainless steels have to offer. They are used in surgical devices, dental implants, jewelry, and food handling.

*Heat-Resistant Grades*: Elevated temperature performance is imperative in many applications that require corrosion-resistance, providing the need for heat-resistant grades. Typical stainless steel grades retain their desired properties up to 800°C. Demanding applications, such as flexible couplings for auto exhaust systems, use 316Ti and 439 grades to obtain high temperature performance above 1000°C.

*High PREN Alloys*: These alloys are specifically formulated to provide pitting resistance and are defined by their PREN number (Pitting Resistant Equivalent Nitrogen content). These alloys are highly resistant to pitting and crevice corrosion with resistance values calculated using an equation. Materials which produce values greater than 50 are considered best.

*Shape memory stainless steels (SMSS)*: A shape memory alloy can return to its original shape when heated above a specified temperature employing a mechanism known as the shape memory effect. It is characterized by a small hysteresis between the starting temperature of transformation (Ms) and its reverse (As) and associated with austinetic (FCC) and martensitic (HCP). Corrosion-resistance is good with chromium alloying between 8 and 10%. Cost and availability of these grades are not inline with mainstream stainless steels but are a growing class of specialty alloys that are gaining recognition. Applications of these materials include pipe couplings on aircraft (F-14 Tomcat) and in fuel injector technologies offering 1 ms response time to heating and cooling, providing the needed high-speed switching of fuel.

## Properties and Designations

AISI describes a chemical composition with a three digit number (Table 4.16). All the manganese/nitrogen austenitic alloys begin with 2 and the nickel containing austenitic alloys begin with 3. However, the three digit numbers that begin with 4 contain both ferritic and martensitic alloys. Precipitation hardened grades begin with 6, or, more commonly, with a PH added to them after the approximate chemistry (like 18.8 PH). Alphabetic designations can be added to the numbers to indicate

**Table 4.16** Chemical Composition Limits (%) of Stainless Steel Alloys

| Alloy | Chromium | Nickel | Carbon | Manganese | Sulfur | Copper | Molybdenum |
|---|---|---|---|---|---|---|---|
| 203 | 16.0–18.0 | 5.0–6.5 | 0.08 | 5.0–6.5 | 0.18–0.35 | 1.75–2.25 | None |
| 303 | 17.0–19.0 | 8.0–10.0 | 0.15 | 2.0 | 0.15 min. | None | None |
| 304 | 18.0–20.0 | 8.0–10.5 | 0.08 | 2.0 | 0.03 | None | None |
| 316 | 16.0–18.0 | 10.0–14.0 | 0.08 | 2.0 | 0.03 | None | 2.0–3.0 |
| 321 | 17.0–19.0 | 9.0–12.0 | 0.08 | 2.0 | 0.03 | None | None |
| 347 | 17.0–19.0 | 9.0–13.0 | 0.08 | 2.0 | 0.03 | None | None |
| 410 | 11.5–13.5 | None | 0.15 | 1.0 | 0.03 | None | None |
| 416 | 12.0–14.0 | None | 0.15 | 1.25 | 0.15 min. | None | None |
| 440C | 16.0–18.0 | 0.75 | 0.95–1.20 | 1.0 | 0.03 | 0.50 | 0.40–0.65 |
| 15–5 PH | 14.0–15.5 | 3.5–5.5 | 0.07 | 1.0 | 0.03 | 2.5–4.5 | None |
| 17–4 PH | 15.0–17.5 | 3.0–5.0 | 0.07 | 1.0 | 0.03 | 3.0–5.0 | None |
| 17–4 PH H1150 | 15.0–17.5 | 3.0–5.0 | 0.07 | 1.0 | 0.03 | 3.0–5.0 | None |

specific composition changes (e.g., L for low-carbon and Se for a grade containing selenium for machinability).

The UNS system keeps the AISI nomenclature in the first three digits of its five digit designations. The last two digits are 00, which designate any modifications. This five digit number has a prefix S for stainless steels. Since the AISI system is the oldest and the most commonly used numbering method, it will be reviewed here with details of each of the basic types and some important grades.

### Strength Properties

Stainless steels encompass a large range of mechanical properties including an elastic modulus range of 196 to 200 GPa (28 to 29 Mpsi) at room temperature. Carbon steels have a modulus of 30 Mpsi. They can be strengthened by precipitation-hardening, solid solution strengthening, cold working, martensite formation, and duplex structure formation. The mechanical properties are highly depend on the alloy composition, fabrication method, and heat treatment employed. Table 4.17 summarizes the range of mechanical properties of some of the common classes available.

### Heat Treatability

Improved properties can be obtained by heat treating stainless steels with consideration for its prior processing and shaping methods. Care must be taken, while heat treating stainless steels, to avoid contamination as it can have adverse effects, such as damage to the naturally passive surface. Stress-relieving is often performed after machining, forming, welding, or casting to relieve the unwanted stresses with a corresponding reduction in yield strength and hardness.

Annealing austenitic and ferritic grades will cause the nucleation and growth of new crystals whose size will depend on the temperature and time of the cycle. However, annealing martensitic grades is done by austenitizing, followed by very slow cooling, such that martensite does not form again. This is known as a full anneal, which results in a ferritic grade with dispersed carbides. The precipitation-hardened grades are solution treated, which allows the essential elements to go into

**Table 4.17** Mechanical Properties of Some of the Common Classes of Stainless Steels

| Class or series | Tensile strength MPa | Yield strength (0.2% offset) MPa | Elongation % | Hardness Rockwell |
|---|---|---|---|---|
| 300 series of austenitic stainless steels | 517–758 | 220–370 | 45–60 | 70–85B |
| Duplex stainless steels | 724–860 | 386–676 | 22–48 | 80–100B |
| 400 series of martensitic stainless steels | 517–1965 | 276–1896 | 2–30 | 27–57C |
| Precipitation-hardening stainless steels | 827–1724 | 517–1689 | 7–16 | 28–49C |

solid solution. Surface-hardening improves hardness and fatigue properties. Methods like induction-hardening, flame-hardening, carburizing, boriding, and nitriding can be used for this purpose. Cleaning after heat treatment involves grit blasting or salt baths.

### Fatigue Properties

Generally, the fatigue life of a stainless steel is similar to that of plain carbon steels. They exhibit a fatigue limit which can be influenced by temperature, heat treatment, hardness, and surface orientation. Also critical to this property is the orientation of the part relative to the processing it has received. If it has been cold worked, it will exhibit different properties in the longitudinal direction against the transverse direction. Because of this property, three different fatigue tests are used to define the material performance: rotating beam, flexural (for sheet stock), and axial-load tests. Each is intended to simulate a typical design application.

*Corrosion*

The corrosion-resistance of stainless steel arises from a "passive," chromium-rich oxide film that forms naturally on the surface of the steel. Although extremely thin, at 1 to 5 nm thick, this protective film is strongly adherent, and chemically stable (i.e., passive) under conditions that provide sufficient oxygen to the surface. The key to the durability of the corrosion-resistance of stainless steels is its self-repairing property (provided there is sufficient oxygen available) that allows the surface to remain passivated. In contrast to other steel types which suffer from "general" corrosion, where large areas of the surface are affected, stainless steels in the "passive state" are normally resistant to this form of attack.

Common modes of corrosion in stainless steels include crevice-corrosion cracking in stagnant liquids and also pitting attack. Intergranular attack is critical, especially in welded parts.

In general, the corrosion and oxidation resistance of stainless steels improves as the chromium content increases. The addition of nickel, to create the austenitic stainless steel grades, strengthens the oxide film and raises their performance in more aggressive conditions. The addition of molybdenum to either the ferritic or austenitic stainless steels improves their pitting corrosion resistance.

Austenitic stainless steels are resistant to a wide range of rural and industrial atmospheres, resulting in their extensive use in architectural, structural, and street furniture applications. Their resistance to attack by acids, alkalis, and other chemicals has led to a wide use in the chemical and process plant industries. However, austenitic stainless steels tend to pit in saltwater, contrary to their perceived corrosion-resistant properties.

The ferritic stainless steels are used in the more mildly corrosive environments, being often used in trim work and somewhat less demanding applications. Martensitic stainless steels have similar corrosion-resistance as the ferritic grades, and the precipitation-hardening stainless steels are similar to the 304 austenitic grades.

Duplex stainless steels are alloys designed to have improved localized corrosion-resistance, specifically to stress-corrosion cracking, crevice, and pitting corrosion.

## Machinability

The primary concern when machining stainless steels is that they do not cut cleanly, often resulting in stringy chips and hotter running tools. They are available in free-machining grades which typically include either sulfur or selenium in their composition to improve machinability. Lead and phosphorus are also added to improve chip breaking. The precipitation-hardened and duplex families do not have free machining grades like UNS or AISI designations but some PH grades are available as commercial brands. Additions to improve machining can have adverse effects on any of the following: toughness, formability, weldability, and corrosion-resistance. Careful consideration of the requirements is necessary when selecting these grades in order that these limitations can be successfully overcome.

Martensitic grades (400 series) are best machined in their annealed state to avoid finish machining operations that result in high tool wear rates, owing to the hardened condition. Austenitic grades are most difficult to machine as they work-harden during machining, especially those grades containing nitrogen. Cold working before machining can improve their machinability. Ferritic grades are easily machinable with improved machinability for those with lower chromium levels. PH grades are best machined in the solution treated condition. Duplex grades are not easily machinable even in the annealed condition, making their selection an important one for those design requiring extensive fabrication.

The Table 4.18 gives an idea of machinability, weldability, and formability rating on the basis of 1 to 10, where 10 is best.

## Weldability

Stainless steels can be welded by common methods, including TIG and MIG methods. Because the thermal conductivities of stainless steels are one-half of those of plain carbon and low-alloy steels, heat dissipation occurs more slowly in the heat affected zone. This effect can be minimized by welding

**Table 4.18**  Comparison of Machinability, Weldability, and Formability

| Grade | Formability | Machinability | Weldability |
|---|---|---|---|
| 303 | 1 | 8[a] | 1 |
| 304 | 8 | 5[a] | 8 |
| 316 | 8 | 5[a] | 8 |
| 416 | 1 | 10 | 1 |
| 430 | 4 | 6 | 2 |
| Duplex grade S31803 | 5 | 6 | 6 |

[a] Improved grades offer higher values.

techniques which limit welding current, include use of chill bars and appropriate joint design. Joint cleanliness is very important when welding stainless steel because of the detrimental effects of oil, grease, and fragments of steel from carbon steel files and brushes on weld integrity. After welding, the work must be thoroughly clean ground, which will initiate the self-passivating mechanism.

Ferritic grades (400 series) with low carbon and nitrogen levels are very weldable, because sensitization is mitigated and grain growth is restrained by the presence of ferrite stabilizers. Other grades of ferritic nature are not preferred welding candidates. Martensitic grades (400 series) require preheating to prevent embrittlement. Austenitic grades (300 series) are much better than the other two families but have a problem with sensitization because of chromium depletion around the weld area. This problem can be overcome by using extra-low-carbon grades and by avoiding the critical range where the particular grade is sensitized.

### Electrical/Thermal

Generally, the thermal conductivity of stainless steels is one-half that of carbon steels and one-sixth that of aluminums, making their selection for thermal management devices less attractive. Similarly, their electrical conductivity is one-third that of carbon steel and greater than one-sixth that of aluminum, limiting their application as an electrical conductor. Thermal expansion rates for austenitic stainless steels are twice that of plain carbon steels while most other stainless steels are similar in expansion as plain carbon steels. Aluminums have greater than two and a half times the thermal expansion as stainless steels, making it critical to allow for thermal differential growth in tolerance sensitive applications.

### Unique Properties

Self-passivation, corrosion-protection, and attractive appearance are properties unique to stainless steels and form the basis for their selection in many applications. Many other properties are similar to carbon steel, including weldability and machinability. Of course, many properties are significantly influenced by the material grade and processing.

Four basic grades are defined by AISI: type 2XX, type 3XX, Type PH, and Duplex grades. Each offer unique properties that can satisfy a wide range of requirements while offering the common property of superior surface finish and appearance.

**Type 2XX (nickel–manganese–nitrogen):** These are austenitic, nonhardening, nonmagnetic steels that are not as common as the type 3XX series, which only has a few grades available.

**Type 203** — A superior machining stainless steel designed specifically for high-speed automatic machining. It allows higher speeds and feeds while providing greater tool life compared to type 303. Corrosion resistance is similar to type 303. The addition of Cu in the range of 1.75 to 2.25% provides the improved machining properties, but at an added cost above 303. Yield strength is 35 ksi with a hardness of 160 Brinell.

**Type 3XX (chromium–nickel):** These are austenitic, nonmagnetic, nonhardening steels that can be hardened only by cold working.

**Type 303** — It is the most commonly used stainless steel with excellent machining properties because sulfur and phosphorus (free-machining) are added. Atmospheric corrosion-resistance is excellent, as it is in most organic and inorganic chemicals, sterilizing solutions, and most foods and related dyes. It is best not to weld this type as the added sulfur can cause cracking and subsequent corrosion (this is true of all free-machining steels). Its properties are similar to type 304. It is ideal for screw machine products. Its yield strength is 35 ksi, and hardness is 160 Brinell.

**Type 304** — This is the most common and widely used grade from which most other stainless steels are derived. It is weldable using all common methods and has better high-temperature properties than 303, while having the same corrosion-resistance. It also has good formability and corrosion-resistance. Its yield strength is 35 ksi and hardness is 170 Brinell.

**Type 304L** — This basic 18-8 steel (18% chromium and 8% nickel) has a lower carbon content than type 304 in order to achieve superior weldability while maintaining the same corrosion-resistance but at a reduced strength level.

**Type 309** — A high temperature stainless steel with corrosion properties that exceed those of type 304.

**Type 310** — Its higher alloy content yields improved high-temperature properties when compared with type 309. It shows an excellent resistance to oxidation.

**Type 316** — Its high nickel and molybdenum contents provide excellent resistance to corrosion and pitting. It has a good weldability and higher strength than type 304 at elevated temperatures. It is ideal for chemical processing and its yield strength is 35 ksi and hardness is 170 Brinell.

**Type 316L** — This type has lower carbon than type 316 for reduced corrosion of welded parts, while providing the same corrosion-resistance as type 316, but at a higher cost.

**Type 317L** — This is similar to type 316 but has a greater molybdenum content, which provides superior high-temperature performance with higher tensile and creep strengths.

**Type 321** — This is similar in composition to type 304 except for the addition of titanium to improve weld performance of those objects subjected to a highly corrosive media. The presence of titanium limits the possibility of carbide precipitation in welded joints. It is used for making welded items that are subjected to severely corrosive conditions. Its yield strength is 35 ksi and hardness is 150 Brinell.

**Type 347** — This is similar to type 321, but with cobalt and tantalum instead of titanium for superior creep strength and greater hardness. It is used for highly stressed welded equipment. Its yield strength is 30 ksi and hardness is 170 Brinell.

**Type 4XX (straight chromium):** The following are martensitic grades which are hardenable and magnetic:

**Type 410** — It is a general purpose, heat treatable stainless steel, for environments that are not severely corrosive. It is used often in shafting, fasteners, bushings, and pumps. It is a popular grade and has applications in sand and investment castings. Its yield strength is 40 ksi and hardness is 155 Brinell (annealed condition).

**Type 416** — This is a free-machine version of type 410 with sulfur added for improved machinability. It is of a superior quality for fasteners. Its yield strength is 40 ksi and hardness is 155 Brinell (annealed condition).

**Type 422** — It has improved strength and toughness at elevated temperatures over type 410.

**Type 440C** — A high carbon content makes this the hardest of all standard stainless steels. An outstanding candidate for heat treating, it also has good abrasion and wear-resistance. Its corrosion-resistance is good, but only after hardening and stress-relieving. It is used for gears, bearings, seats, and valve parts. Its yield strength is 65 ksi and hardness is 230 Brinell (annealed condition).

The following are ferritic, nonhardenable, magnetic grades:

**Type 409** — It is a low-chromium grade referred to as "utility stainless steel" because of its lower cost and wide applicability in less corrosive environments. It is most widely used for exhaust systems in automobiles. The addition of titanium stabilizes this grade and prevents hardening by martensite formation during welding.

**Type 430** — This is a basic 17% Cr ferritic stainless steel which has good corrosion-resistance, especially in nitric acid, but has average mechanical properties, including less than impressive impact strength. If chromium range by weight is 14 to 16%, the weldability, impact resistance, strength, and hardness are improved with some reduction in corrosion. This lower level Cr grade is the bar and plate grade and is given the number 429.

**Type 6XX (precipitation-hardenable grades — Type PH):** PH types are magnetic in the hardened condition but their wear-resistance is not as good as other martensitic grades (400 series).

**Type 15-5 PH** — This is a martenistic grade containing Cr, Ni, and Cu. It has good toughness and very good forgeability compared with 17-4. There are six standard heat treatments employed with the lowest condition number (i.e., H900) being the strongest. It shows superior performance over 17-4 when exposed to salt fog or chloride solutions. It should not be put into service in the annealed condition. It can be machined in both annealed and hardened conditions, but is more often machined in the annealed state and then artificially aged, taking care to account for dimensional change. Its yield strength is 145 ksi and hardness is 332 Brinell (annealed condition).

**Type 17-4 PH (Type 630)** — This is also a martenistic grade containing Cr, Ni, and Cu but contains more chromium than 15-5. It has excellent corrosion-resistance compared to types 302 and 304 and is the most widely used grade among the PH series. It also has six standard heat treatments, the lowest condition number (i.e., H900) is the strongest. Unlike 15-5 it should not be put into service in annealed condition. Machining characteristics are similar to 15-5.

**Type 13-8 PH** — This is a martensitic grade with excellent toughness and corrosion-resistance, especially stress-corrosion resistance which is the best among the hardenable stainless steels. Used in aircraft parts, pins, lock washers, high-performance shafting, or any part requiring stress-corrosion resistance. The highest strength levels are obtained by cold working before precipitation-hardening, as seen in the production of high-strength, corrosion-resistance fasteners. It should not be put into service in annealed condition. Optimum machining grades are available and noted with the suffix "M" (i.e., Condition H1150M). Removal of coatings and oil prior to heat treating is necessary to prevent contamination. Other surface treatments such as nitriding or carburizing may prevent the material from fully responding to the heat treating process. However, it is more difficult to machine than 17-4, requiring 20 to 30% slower tool speeds.

**Type 15-7 PH (Type 632)** — This is a semi-austenitic grade containing 15% Cr, 7% Ni, and 2% Mo. It is generally supplied in the annealed condition with sheet, strip, and wire available in the cold formed condition. Cold working transforms this grade to martensite (from austenite). Heat treated strengths are typically 1.5-2X those of annealed condition. It has better high-temperature performance than that observed by type 17-7 PH.

**Type 17-7 PH (Type 631)** — These are semiaustenitic grades with 17% Cr and 7% Ni. They are austenitic in the annealed condition and martensitic in the hardened condition. They show superior corrosion-resistance when compared with other martensitic grades (15-5, 17-4, 13-8). Sheet, strip, and wire are available in cold formed condition.

**Duplex Series:** These are mixed microstructure grades which contain austenite and ferrite in various ratios and can have yield strengths greater than twice those of austenitic grades such as types 2XX and 3XX.

**Type 2205** — This is one of the "duplex" family of stainless steels with a structure that consists of approximately 50% austenite and 50% ferrite. This duplex structure provides high strength and good resistance to stress-corrosion. It is used for heat exchangers, as well as oil and gas industry equipment. Its yield strength is 65 ksi and hardness is 290 Brinell.

## Manufacturing Process

The processing methods used to produce stainless steel alloys are similar to those of other ferrous metals. The predominant method of processing is the one using an electric arc furnace, which is followed by the argon oxygen decarburizing process (AOD). This processing allows pure metals of unique compositions to be produced by controlling the carbon and nitrogen contents. Lower carbon content improves welding by reducing the potential for sensitization, while the addition of nitrogen improves the steel's mechanical properties.

Generally, the addition of chromium and nickel improves corrosion-resistance at the expense of mechanical properties, including loss of toughness and tensile strength. The addition of the necessary alloys is made during melting and ladle treatments to control nitrogen and carbon content. The resulting microstructure of the material defines its behavior and subsequently its category.

## Raw and Stock Finishes

Stainless steels are available in those forms common to ferrous metals, such as the hot and cold rolled finishes. Common bar stock shapes such as rounds, squares, and flats are typically available in both hot and cold finished forms. Sheet stock are available in many gauges commonly ranging from 7 GA (0.179 in.) to 26 GA (0.018 in.) thick with various surface finishes (Table 4.19). The most common sheet stock available are types 304 and 316, with some gauges available in types 304L and 316L for use in critical welding applications. While other sheet stock are produced, their pricing and availability will be important factors in their selection.

Stainless steel machine stock is most commonly available in the work-hardened grades 303 and 304 for most shapes, types 304L, 316, and 316L are also found but in reduced

**Table 4.19** Common Raw and Stock Finishes

| Material | Sheet | Rounds | Squares | Hexagon | Flatbar | Angles | Beams | Channel | Tees | Tubing | Pipe |
|---|---|---|---|---|---|---|---|---|---|---|---|
| 203 | | X | X | X | | | | | | | |
| 303 | | X | X | X | | | | | | | |
| 304 | X | X | X | X | X | X | X | X | X | X | X |
| 304L | X | X | | | | X | | | | | |
| 309 | X | | | | | | | | | | |
| 310 | | | | | | | | | | | |
| 316 | X | X | X | X | X | X | | | | | X |
| 316L | X | X | | | | X | | | | | |
| 317L | X | | | | | | | | | | |
| 409 | X | X | | | | | | | | | |
| 410 | X | X | | | | | | | | | |
| 416 | | X | X | X | | | | | | | |
| 420 | | X | | | | | | | | | |
| 430 | X | | | | | | | | | | |
| 440C | | X | | | | | | | | | |
| 15-5 | | X | | | | | | | | | |
| 17-4 | | X | | X | | | | | | | |

**Table 4.20**  Standard Stainless Steel Sheet Metal Finishes

| Finish | Description |
|---|---|
| #1 | Hot rolled, annealed, and pickled |
| #2D | Annealed, pickled, and dull cold rolled |
| #2B | Annealed, pickled, and bright cold rolled |
| #3 | Intermediate polish (approx. 100/120 grit) |
| #4 | Standard polish (approx. 150/180 grit) |
| #6 | Dull satin |
| #7 | High lustre polish |
| #8 | Mirror polish (highest luster) |
| BA | Bright annealed. Bright cold rolled and controlled atmosphere annealed to retain highly reflective surface |

availability. The 400 series is also readily available and is a cost-effective material for general machine applications requiring post-machining heat treatment for improved properties. Other shapes and grades are shown in Table 4.20.

## Processing Considerations

Stainless steels generally are more difficult to machine than carbon steels, requiring slower speeds and feeds. Comparisons are based on AISI type 416 free-machining stainless steel because it has the same machining properties as B1112 free-machining carbon steel which is considered the industry standard reference. It is obvious that type 416 would then be the easiest of the stainless steels to machine, followed by types 430F and 303. The machining characteristics of stainless steels are generally not significantly different from those of plain carbon steels or alloy steels and are highly dependent on the specific grade involved, requiring a detailed review of selected materials when considering this aspect of material selection.

Forming of stainless steel sheet stock is commonly required for electronic cabinetry or machine enclosures. An annealed sheet stock allows for the tightest bends of one-half the material thickness while tempered and rolled stock

**Table 4.21**  Bending Properties of Annealed Stainless Steels

| Type | Free Bend | V-Block |
|---|---|---|
| 301, 302, 304, 305, 309, 310, 316, 321, 347 | 180* R = 1/2 T | 135* R = 1/2 T |

*Note*:  R = radius of bend: T = thickness of material. All bends are parallel to direction of rolling.

**Table 4.22**  Bending Characteristic of Tempered Rolled Stainless Steel

| Type | Temper | Gage 0.050 in. (1.27 mm) and under free bend | Gage 0.051 in.–0.187 in. (1.30–4.75 mm) |
|---|---|---|---|
| 301 | 1/4 hard | 180* R = 1/2T | 90* R = T |
| 301 | 1/2 hard | 180* R = T | 90* R = T |
| 301 | 3/4 hard | 180* R = 1½T | — |
| 301 | full hard | 180* R = 2T | — |
| 302 | 1/4 hard | 180* R = 1/2T | 90* R = T |
| 316 | 1/4 hard | 180* R = T | 90*R – T |
| | | V-Block | |
| 301 | 1/4 hard | 135* R = T | 135* R = 1½T |
| 301 | 1/2 hard | 135* R = 2T | 135* R = 2T |
| 301 | 3/4 hard | 135* R = 3T | — |
| 301 | full hard | 135* R = 3T | — |
| 302 | 1/4 hard | 135* R = 2T | 135* R = 2T |
| 316 | 1/4 hard | 135* R = 2½T | 135* R = 3T |

*Note*:  R = radius of bend: T = thickness of material. All bends are parallel to direction of rolling.

is stiffer and requires greater bend allowances, typically ranging from 1 to 3 times material thickness (Tables 4.21 and 4.22).

Thermal expansion is an additional consideration of selecting stainless steels and their related processing characteristics. As a general rule, the 200 and 300 series, austenitic grades, have a higher rate of thermal expansion than do the 400 series, ferritic, or martensitic grades.

### Thermal Treatments

Stainless steels respond to thermal treatment based on their microstructure. The 200 and 300 series are not thermal treated for strength improvement but may be treated for stress relief due to cold working, or annealed for improved properties. Annealing is performed using a water-quench.

The 400 series and PH grades can benefit from thermal treatment. Tempering 400 series to 900°F typically produces the strongest materials, but at the expense of toughness. Higher processing temperatures reduces strength but improves toughness. Common grades that benefit from tempering include 410, 416, 431, and 440°C.

The PH grades are not considered heat treatable but will benefit from thermal treatment, as it accelerates the precipitation-hardening process. Thermal soaking of the material up to temperatures of 1000°F (540°C) can improve strength and hardness properties more rapidly.

### Strengthening Treatments

There are three methods of increasing the strength of stainless steels: thermal conditioning, cold working, and precipitation-hardening. The common methods of thermal treatments were discussed in the "Thermal Treatments" section and are applied to type 400 and PH series materials.

Additional strength can be imparted to the 200 and 300 series stainless steels by cold working, as these grades do not respond to heat treatments. The work-hardened condition is categorized in the hardness grade as discussed previously under the section on carbon steels (i.e., ¼ hard, ½ hard, etc.) and applies to sheet and strip stock. The availability of these different stock grades can be helpful in the design of springs and retaining mechanisms. The ability to cold work stainless steel is affected by the nickel content, and reduces as the content increases above 9%.

Precipitation-hardening provides the grades of highest strength among stainless steels (>220 ksi) and these grades often are used in place of types 410 and 426 in extreme applications. The strengthening mechanism involves the precipitation of alloys which upset the molecular lattice, resulting in one that is resistant to dislocations. The application of heat accelerates this process and is known as artificial aging.

## Surface Treatments

Stainless steels typically do not need to be treated for corrosion protection, as commonly done with plain carbon steels and with many alloy steels, because of their inherent resistance to corrosion. In the past, when iron bearing machine tooling was used, it was necessary to passivate the material after machining in order to eliminate any embedded iron that could initiate corrosion. This is no longer necessary with the advent of ceramic cutting tools, which do not present the same difficulty. If a concern remains, owing to shop handling practices that might embed iron into the surface, requesting passivation as a post-processing operation may be appropriate.

Surface coatings are generally not required on stainless steels because of the inherent corrosion resistance of the material. When additional protection is required, the most common form applied involves physical vapor deposition (PVD) coatings. Applications involving medical devices and those requiring improved wear or sterilization are the best candidates and benefit from both of these process advantages. The deposition methods include the following:

- TiN — titanium nitride
- ZrN — zirconium nitride
- CrN — chromium nitride
- TiAlN — titanium aluminum nitride
- AlTiN — aluminum titanium nitride

The advantages of PVD coatings also include reduced friction, chemical barrier protection, and biocompatibility while also providing decorative colors and aesthetic pleasure.

Surface hardness can also be improved by a treatment which involves diffusion of large quantities of carbon into the surface without the formation of chromium carbides. It is known commercially as Kolsterising. It can be applied to 304 and 304L, 316 and 316L, duplex alloys and high nickel alloys and does not result in a change in size, shape, surface roughness, or color. The surface hardness can be >70 $R_c$ to a depth of 22–33 microns (0.0008–0.0012 in.), providing excellent wear resistance.

Common Grades

- 303
- 304
- 304L
- 316
- 316L
- 440C
- 17-4PH

## Tool Steels

Steels used in the fabrication process of other materials must be very resistant to extreme conditions ranging from intense heat (machining or heat treating equipment) or cold (cryogenic processes) to high impact or wear environments. The materials used for many of these applications are known as tool steels, which are highly alloyed steels processed by electric furnace techniques (Figure 4.17).

Processing may also include secondary remelting which is often necessary to obtain the desired material properties. Carbon content is typically 0.5 to 2.0% with major alloying elements including chromium, molybdenum, tungsten, and vanadium. Tool steels are most often machined in their non-heat treated form to near net geometry, then heat treated to

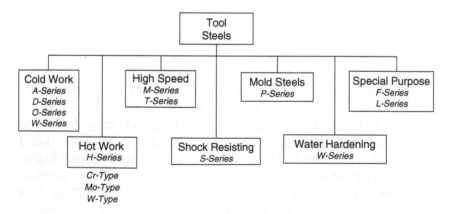

**Figure 4.17**  AISI classifications and designations for tool steel.

the desired condition and finish-ground to their final form. Because of this method of processing hardness can range from 35 $R_c$ for mold tool steels (Type P) to 65 $R_c$ for most others in their hardened material condition. Controlling distortion is critical in processing tools steels to prevent excessive cost in removing or adding material to compensate for volume changes that occur during heat treating.

Tool steels are selected when extreme conditions involving wear, toughness, and resistance to softening at elevated temperatures is required (Figure 4.18). No tool steel provides a solution to all of these environments so they must be carefully selected and processed to obtain the optimum combination. Successful implementation of a tool steel involves: proper tool design, selection of proper tool steel, accuracy of tool fabrication, and proper heat treatment. Tool steels, almost always, are heat treated to obtain the desired material condition for effective results.

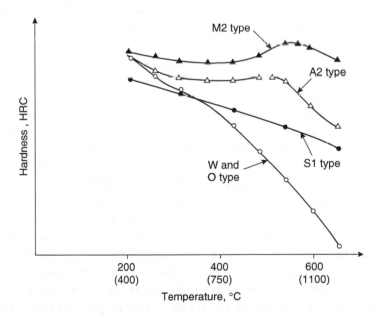

**Figure 4.18** Comparison of different tool materials on various hardness seales.

*Source*: ASM Metal Metals Handbook, Online, Metal Handbook Desk Edition-Tool Steels-Classification and Characteristics, Figure 1.

The corrosion-resistance of most tool steels is fair to poor, meaning these grades are likely to show some form of corrosion even at room temperature conditions unless they are cared for by applying a coat of oil or grease. This is because they are not alloyed, like stainless steels or other CRES materials, to produce a passive surface. For example, chromium is in solid solution to form carbides, not in solid solution to make a passive surface, as is the case with stainless steels.

The cost of tool steels is generally a small portion of the overall tooling cost because the processing costs account for the majority of the expense. Tools steels require rough machining, heat treatment, and final grinding to produce the desired tolerances and resulting functionality.

## Properties and Designations

Tool steels can be specified by different standards but the most commonly used designation is the AISI system in which each type is given a capital letter that denotes a group. Those within each group different grades are given code numbers to identify. Such as A for air hardened, H for hot worked, O for oil hardened etc. A two digit number follows the letter to specify particular grade.

There are seven categories of tool steels as classified by AISI. These categories are based on the predominate characteristic of the tool steel or type of heat treat processing required. The details of these categories and their constituents' series or groups are given in the Table 4.23.

The ASTM standards for tool steels include A600, A681, and A686. ASTM A600 sets forth requirements for both tungsten and molybdenum high-speed steels, A681 is applicable to hot-work, cold-work, shock resisting, special purpose, and mold steels; A686 covers water-hardening tool steels. However, tool steels may be purchased according to a trade name because the buyer may find a better performance with respect to the desired tool steel properties from such a product than a grade specified by AISI or ASTM (Table 4.23).

A listing of other ferrous material properties including tool steels is helpful in understanding their contribution to

**Table 4.23** Classification of Tool Steels into Categories and Subgroups

| Category | AISI Series | Description | AISI grades available | Applications |
|---|---|---|---|---|
| High-speed | M | Molybdenum high-speed steels | M1–M4, M7, M10, M30, M33–M36, M41–M44, M46, M47 M50, M52 | M-Type are used for lathe centers, blanking dies, hot forming dies, lathe cutting tools, drills, taps. T-Types are used for tool bits, milling cutters, taps, reamers, drills, broaches. |
| | M | Intermediate high-speed steels | | |
| | T | Tungsten high-speed steels | T1, T2 T4–T6, T8 T15 | |
| Hot work | H | Chromium hot work steels | H10–H14, H19 | Extrusion dies, forging dies, die casting, hot shear blades, plastic molds, punches and dies for piercing shells, hot press, etc. |
| | H | Tungsten hot work steels | H21–H26 | |
| | H | Molybdenum hot work steels | H42, H43 | |
| Cold work | A | Air-hardening medium alloy | A2–A10 | Gages, blanking, drawing and piercing dies, shears, forming and banding rolls, lathe centers, mandrels, broaches, reamers, taps, threading dies, plastic molds, knurling tools. |
| | D | High C, high Cr | D2–D5, D7 | |
| | O | Oil hardening | O1, O2, O6, O7 | |

**Table 4.23**   (*Continued*)

| Category | AISI Series | Description | AISI grades available | Applications |
|---|---|---|---|---|
| Shock resisting | S | Shock resisting | S1, S2, S5–S7 | Form tools, chisels, punches, cutting blades, springs, trimming, and swaging dies, concrete and rock drills, bolt cutters. |
| Low alloy special purpose | L | Low-alloy, special purpose | L2, L6 | Gages, broaches, drills, taps, threading dies, ball and roller bearings, clutch plates, knurls, files, dies, cutting tools, form tools, knives. |
| Low C mold steel | P | Low-carbon mold steel | P2–P6, P20, P21 | Plastic-molding and zinc die-casting dies. |
| Water-hardening | W | Water-hardening | W1, W2, W5 | Blanking dies, files, drills, taps, countersinks, reamers, jewellery dies, and cold-striking dies. |

**Table 4.24** Comparison of Ferrous Metal Properties with Tool Steels

| Properties | Carbon steels | Alloy steels | Stainless steels | Tool steels |
|---|---|---|---|---|
| Density (1000 kg/m$^3$) | 7.85 | 7.85 | 7.75–8.1 | 7.72–8.0 |
| Elastic modulus (GPa) | 190–210 | 190–210 | 190–210 | 190–210 |
| Poisson's ratio | 0.27–0.3 | 0.27–0.3 | 0.27–0.3 | 0.27–0.3 |
| Thermal Expansion (10$^{-6}$/K) | 11–16.6 | 9.0–15 | 9.0–20.7 | 9.4–15.1 |
| Melting point (°C) | | | 1371–1454 | |
| Thermal conductivity (W/m-K) | 24.3–65.2 | 26–48.6 | 11.2–36.7 | 19.9–48.3 |
| Specific heat (J/kg-K) | 450–2081 | 452–1499 | 420–500 | |
| Electrical resistivity (10$^{-9}$ -m) | 130–1250 | 210–1251 | 75.7–1020 | |
| Tensile strength (MPa) | 276–1882 | 758–1882 | 515–827 | 640–2000 |
| Yield strength MPa) | 186–758 | 366–1793 | 207–552 | 380–440 |
| Percent elongation (%) | 10–32 | 4–31 | 12–40 | 5–25 |
| Hardness (Brinell 3000 kg) | 86–388 | 149–627 | 137–595 | 210–620 |

*Source*: www.efunda.com.

this material family (Table 4.24). Note that tool steels have similar material properties as the other materials in this family.

*Strength Properties*

Tool steels are used primarily for their wear-resistance for cutting, hardness, hardenability, and hot hardness. This is in contrast to the common consideration of tensile strength and yield strength for other carbon and alloy steels. They are used in applications that require them to have strength and toughness properties coupled with good dimensional stability.

Most tool steels have a modulus of elasticity in the range of 210 GPa (30.4 Mpsi) at room temperature giving them suitable dimensional stability during processing. This modulus reduces on increasing the temperature; it changes from a value of 185 GPa (26.8 Mpsi) at 500°F to 150 GPa (21.7 Mpsi) at 1000°F.

Each class and series of the tool steels described above have some set of unique properties which make them suitable for specific machining or die making operation. These high-speed steels, especially the M series, are known for their high hot hardness and wear-resistance but have low toughness. The T series is almost similar to the M series in these properties; their usual working hardness ranges from 63 to 70 $R_c$.

Chromium and tungsten hot worked steels (H series) have half the wear-resistance of high speed steels but with twice the toughness. The hot hardness of the H series is slightly lower than high-speed steels (M series). The usual working hardness of chromium and tungsten hot worked steels is 35 to 58 $R_c$.

For lower wear-resistance, hot hardness and greater toughness, shock-resistant steels (S series) and low-alloy special purpose steels (P series) are used. Both have the highest levels of toughness among tool steels but very low wear-resistance and hot hardness. Their usual working hardness can range from 45 to 58 $R_c$.

In applications which require almost equal levels of toughness, wear-resistance and hot hardness, air-hardened, medium-alloy, cold-worked steels (A series) are used. Their usual working hardness range is from 48 to 66 $R_c$. In general, tool steels with greater toughness and hardening will have lower wear-resistance and hot hardness and vice versa.

### Heat Treatability

The ability to heat treat tool steels is far more crucial than for ordinary steels. The special alloying compositions of tool steels can affect the hardening cycle and tempering characteristics. The alloying additions can move the isothermal transformation to longer or shorter durations, or the shape of curve. Heat treatment instructions supplied by the manufacturer must be followed for satisfactory results.

Tool steels are high performance materials that almost always require heat treatment, before an optimum combination of the desired properties can be obtained. The type of heat treatment varies between grades and will be discussed within each category. The general notation is to use the quenching medium to describe the material. For example, type O tool steel is an "oil-hardened" material that is quenched in oil to obtain unique properties.

High-speed tool steels (M & T series) can both be hardened to at least 63 $R_c$ with fine grain size. They have deep hardening characteristics and should not be normalized. Heat treatment includes preheating from 1350 to 1500°F and hardening at 2100 to 2300°F and can be air-hardened, oil-quenched, or salt bath-quenched.

Hot-work tool steels (H series) can also be hardened in a fashion similar to high-speed steels. Their preheat temperature is the same but hardening temperatures range from 1800 to 2000°F. The exception, in this group, is chromium hot-work tool steels which must be held at the hardening temperature from 15 to 40 min unlike all other grades which only require 2 to 5 min.

The cold-work tool steels (A, D, O, or W series) have a major difference in heating rates while hardening. They cannot be rapidly heated like other grades. All cold-work grades are slowly or very slowly heated to hardening temperatures and can be held at that temperature from 15 to 60 min depending on the grade. Their hardening temperatures range from 1500 to 1800°F.

The mold steels (P series) also have a lower hardening temperature, of about 1500 to 1800°F, but must be heated slowly for best results.

The water-hardening steels (W series) are known for their shallow hardening and high distortion. They also have the highest resistance to decarburization.

### Fatigue Properties

Fatigue properties of tool steels are not considered an important performance criterion. Most failures of tools occur from shock loading or wear. Generally, the fatigue properties of tool steels are equivalent to or better than those of structural steels, but because fatigue is not a common failure method in tools they are typically not considered in the selection process.

## Corrosion

Corrosion characteristics also are not important considerations in the selection of tool steels. Tools and dies are rarely used in corrosive environments. The process of corrosion is usually not due to moisture but to high temperature, since most of the applications involve a higher temperature. Even when applied in a corrosive environment, tool steels have far superior corrosion characteristics than structural steels owing to the high amount of alloying elements such as chromium, nickel, molybdenum, and tungsten.

High-alloying additions give tool steels very good corrosion-resistance. Apart from alloying, most of the tool steels are surface treated for wear-resistance. These platings and coatings are mostly carbides, nitrides, or oxides. Besides providing excellent wear-resistance and preventing adhesion between workpiece and tool, these coatings also provide corrosion-resistance.

## Machinability

Machinability of tool steels is important in applications where a large degree of machining is required to form the tool, or where the numbers of tools to be made is large. It is quite less than carbon or alloy steels. The best among tool steels in this aspect are the water-hardening tool grades. Their machinability is considered about 30% of a free-machining steel grade such as B1112. All tool steels are rated against the water-hardening grades because they are referenced at 100% machinability. A10 is marked at about 90%, P2-6 at 75–90%, H 10-14 at 60–70%, M2 at 40–50%, D2-7 at 30–40% while M15 and T15 are at about 25–30% with regard to machinability, compared to water-hardening tool steels. It should be noted that all the machinability ranges are in the annealed condition because that is the best condition to machine the material (Table 4.25). Hardening treatments are performed after the machining is complete and is followed by finish grinding. Machinability can be improved by altering the composition or including a preliminary heat treatment if the machining required is extensive.

**Table 4.25**  Machinability Ratings of Typical Annealed Tool Steels

| Type | Machinability rating |
|------|---------------------|
| O6 | 125 |
| W1, W2, W5 | 100[a] |
| A10 | 90 |
| P2, P3, P4, P5, P6 | 75–90 |
| P20, P21 | 65–80 |
| L2, L6 | 65–75 |
| SI, S2, S5, S6, S7 | 60–70 |
| H10, H11, H13, H14, H19 | 60–70[b] |
| O1, O2, 07 | 45–60 |
| A2, A3, A4, A6, A8, A9 | 45–60 |
| H21, H22, H24, H25, H26, H42 | 45–55[b] |
| T1 | 40–50 |
| M2 | 40–50 |
| T4 | 35–40 |
| M3 (class 1) | 35–40 |
| D2, D3, D4, D5, D7, A7 | 30–40 |
| T15 | 25–30 |
| M15 | 25–30 |

[a] Equivalent to approximately 30% of the machinability of B1112.
[b] For hardness range 150–200 HB.
*Source*: ASM Metal Metals Handbook: Table 3, p. 18–7, ASM International, 1985.

Grinding is measured as a grinding ratio which is the volume of metal removed per volume of wheel wear. The higher the grindability ratio, the easier it is to grind. High-speed steels have better grinding ratios than other classes.

### Weldability

With few exceptions, all tool steels can be welded by common methods including shielded metal arc, gas tungsten arc, plasma arc, or electron beam processes. Die steels used for blanking, forming, forging, drawing, embossing, coining, or hot and cold trimming can be salvaged or reclaimed using one of these processes. However, they require special pre- and post-weld treatments and precautions to obtain a good quality weld.

Repairing a tool is mostly done by welding as it saves material cost and machining time. Welding can also be used to construct or alter tools, but it is most commonly used to repair them. Weldability can be a crucial factor while picking a tool steel if the tool is large and will need repair. Small tools can be replaced economically without repair. Weldability will largely depend on the composition of the tool steel, but the correct method of welding will also dictate the soundness of a weld. The literature provided by the supplier must be consulted to decide upon the best method of welding and the correct procedure.

In general, tool steels require special pretreatment and precautions to obtain a quality weld. The deep hardening grades with low alloy content are readily weldable. Higher alloy content grades are difficult to weld but following the correct method of welding and parameters will result in a sound weld.

### Surface Treatments

Tool steels are typically not put into service as machined, but are surface treated to prolong tool life by increasing wear-resistance. These coatings are extremely hard and brittle and increase tool life and efficiency. Several forms of surface treatments are common including carburizing and nitriding.

Carburizing is not recommended for high speed tools but is useful for applications such as cold-work dies. Carburizing is done at 1900°F, producing a 0.002 to 0.01 in. deep case.

Nitriding is performed in liquid state at 1000°F, because gas nitriding produces a very hard case, which is not suitable for most applications. Nitriding is found to be very effective in increasing tool life. Aerated nitriding baths and sulfur containing nitriding baths can be used to provide resistance to adhesive wear and resistance to seizing, respectively.

Oxide coatings are also used by immersing the tool in aqueous solutions of sodium hydroxide and sodium nitrate at 285°F or heating in a steam atmosphere of these solutions at 1050°F. Oxide coatings are not as effective as nitriding but do help in reducing friction and adhesion between workpiece and tool.

Hard chromium plating is also utilized to a thickness of 0.0001 to 0.0005 in. These platings are brittle and not suitable for tools that are to be used in shock-loading conditions. Its brittleness may be reduced by tempering below 500°F.

Iron sulfide coatings can be applied electrolytically to prevent seizing. These coatings are applied at 375°F using a bath of sodium and potassium thiocyanate. Because of the low temperatures, the tools can be coated in fully hardened and tempered conditions without affecting hardness.

Other important coatings include tungsten carbide, titanium carbide, and titanium nitride. These coatings can be applied using chemical or physical vapor deposition. Physical vapor deposition processes use lower temperatures, and hence coatings can be applied to fully hardened and tempered tools without affecting their hardness.

## Cold Working Tool Steels

A number of different types of tool steels may be classified for cold work applications. These include high-carbon high-chromium types, air-hardening types, and oil-hardening types. These are meant to be used in places where the operating temperatures are not over 300°C because of their low resistance to softening at elevated temperatures. Hence they do not possess the hot hardness characteristic of the hot working tool steels. However, they have good hardness at room temperatures along with wear-resistance and good finishing which makes them ideal to be used in dies for plastics and lower melting temperature metals like zinc.

### Type O

This grade is an oil-hardened material that has superior cold working properties with high carbon content, and medium to high alloying elements. The common designations are O1, O2, O6, and O7, which have varying amounts of alloying elements. They have high wear-resistance at room temperature. This property deteriorates as the temperature increases, and they lower resistance to softening at elevated temperatures. They are fully hardenable through slow quenching, which results in low distortion and a reduced tendency to crack. A surface hardness of 56–62 $R_c$ can be attained by oil quenching followed by tempering at 175–315°C. Tools made from O type tool steels can be successfully welded if the recommended heat treatments are followed. They are commonly selected for use in dies and punches for blanking, trimming, drawing, flanging, and forming.

### Type A

Type A tool steels are air-hardened and have a high carbon content (0.45 to 2.85%) and high content of alloying elements (Mg, Cr, and Mo), which allows them to achieve high hardenability. These grades can be through-hardened in sections up to 4 in. thick, while exhibiting low distortion and minimal tendency to crack. Their inherent dimensional stability makes them suitable for gauges and precision measuring devices. Their corrosion-resistance is fair while their wear-resistance is high, but they exhibit limited toughness and shock-resistance. Typical applications include knives, large blanking dies, thread roller dies, long punches, rolls, master hubs, trimming and forming dies, plastic molds, precision tools, and gauges.

### Type D

A type D tool steel will typically contain 1.5 to 2.35% C, 12% Cr, and about 1% Mo, the exception being D3. All of the D series exhibit excellent high-temperature wear-resistance and are fully air-hardenable. The high C and Cr content makes them more susceptible to distortion and cracking. They also have high resistance to softening at elevated temperatures. Wear-resistance is high, like type A tool steels, but they exhibit lower toughness and shock resistance. Their corrosion-resistance is reasonably good and comparatively better than type A. They are typically used in blanking dies, thread rolling, gauges, brick molds, shear and slitter knives, and burnishing tools.

### Type W (Water-Hardened)

The water-hardening tool steels are, essentially, plain carbon steels with small amounts of Cr and V and are the least expensive grade of tool steels. The nominal composition of W1, the most popular grade, is 0.95 to 1.05% C, 0.25% Mn, and 0.2% Si. They must be water-quenched to obtain the necessary hardness and will harden with a $\frac{1}{16}$ in. deep hard case and a soft, resilient core. Thin sections are an exception to this behavior, as they will through-harden. The ability of W series steel to meet the needs of many applications is dependent on the depth of the hardened case. Hardenability can be increased by adding vanadium. They have a low resistance to softening at elevated temperatures making them more suitable for cold working operations. They exist in four grades, beginning with grade 1, a highly controlled and certified form, to grade 4, a commercial grade that is not rigorously controlled for hardenability.

W1 has 100% machinability and is the baseline for machinability of the other tool steels. W1 is commonly used for hand-operated metal cutting tools, cold heading, embossing taps, and reamers, as well as cutlery. Heat treatment of W1 is somewhat dependent upon section size or the intricacy of the part. Processing of large sections or intricate shapes requires a slow preheating to 1100°F after which the temperature is increased to 1500°F, followed by a hold of 10 to 30 min, and then the sections are quenched in water or brine. Because type W tool steels are, essentially plain carbon steels they will corrode unless protected. They are used in a wide variety of tools, such as blanking dies, countersinks, drills, files, reamers, taps, and wood working tools.

## Hot Working Tool Steels

These tool steels are formulated to withstand the demanding combination of heat, pressure, and abrasion, associated with manufacturing processes such as punching, shearing, and forming of metals. They have a medium level of carbon, 0.35–0.45%, and Cr, W, Mo, and V contents within a range of 6–25%. There are 13 grades available but the grades H11, H12, and H13 are consumed in large quantities and are, therefore, the easiest to find. Hot working tool steels have quite good hardenability and are readily hardened to above 50 $R_c$. The quenching media is air for smaller sections and oil for larger sections because of the greater mass. The ability to

work a material when it is hot is known as red hardness
(~50 R$_c$) and with this grade of tool steel it can be maintained
after long exposures to temperatures in the range of 500 to
550°C. These hot working tool steels are divided into three
subgroups: chromium (H10–H19), tungsten (H20–H26), and
molybdenum (H42 and H43), and these are discussed in detail
in the following sections:

### Type H (Chromium: H10 to H29)

The chromium based hot working tool steels benefit
from the addition of Cr and some carbide forming elements
which give them good hot hardness. They are also low in
carbon, which promotes good toughness. Greater amounts of
W and Mo increase hot strength but at the expense of
toughness. The Si content can be increased to provide
improved oxidation-resistance up to 800°C. They are deep-
hardening and can be air-hardened in sections up to 6 in.
thick. Due to their low alloy and carbon content they can be
water-cooled during service without cracking. The tensile
strength of chromium-based type H tool steels can be as high
as 300 ksi (2070 Mpa) at room temperatures with good toughness. Their weldability is good with low thermal expansion
and above average resistance to corrosion and oxidation.
Typically, they are used for die casting dies, forging dies, man-
drels, bolt dies, bulldozer dies, forming punches, and hot
shears.

### Type H (Tungsten: H20 to H26)

The tungsten-based hot working tool steels have as
principal alloying elements W, C, Cr, and V. Because of this
higher alloying content they are more resistant to softening at
high temperatures than chromium-based hot work steels.
However, it also makes them more prone to brittleness at working hardnesses of 45 to 55 R$_c$, preventing them from benefiting
from water-cooling during service without cracking. Type H
tungsten tool steels are typically heat treated using oil or hot
salt and, when appropriate, water. The use of air-hardening may
also be appropriate when low distortion is required but at the
expense of increased processing costs. Surface scale becomes

a matter causing concern with these type H tool steels, because their hardening temperatures are higher than the chromium hot work steels, making them more prone to scaling. In many aspects they are like high-speed steels but possess greater toughness. Their breakage can be minimized by preheating to service temperatures. They are used to make hot forging dies and extrusion dies for brass, nickel alloys, and steel.

### Type H (Molybdenum: H42 and H43)

Only two grades are commonly available for the type H molybdenum tool steel. These contain Mo, Cr, V, C, and varying amounts of W, making them similar to tungsten hot work steels with almost identical characteristics and uses. Their high toughness is due to a lower carbon content providing the advantage of lower cost when compared to tungsten hot work steels. Extra precautions and care is required when heat treating, because they can easily decarburize during the process. Their corrosion-resistance is fair, and their uses are similar to tungsten hot work steels.

## High-Speed Tool Steels

These tool steels were developed largely for use in high-speed cutting tool applications where it is critical to maintain material properties at high temperatures. If elevated temperatures are not involved, the use of an oil-hardened (type O) or water-hardened (type W) tool steel would be more appropriate, considering material costs. There are two major types of high-speed tool steels: type M and type T. The type M steels are the most commonly used, constituting >95% of all high-speed steel produced in the United States. Both types are very similar in performance, but the M series has a much lower cost. Use in cold work applications is also possible with some grades of high-speed tool steels. High hardness can be obtained by heat treating and is retained at elevated temperatures produced in common machining applications. This property is known as red or hot hardness. There are more than 25 different high-speed tool steels available, with the following two being the most common among them.

*Type M*

The type M high-speed tool steels contain Mo, W, Cr, V, Co, and C as the principal alloying elements. They provide slightly greater toughness than type T tool steels, while having the same hardness level. They can readily be decarburized and damaged from overheating. Deep hardening is possible, but care must be taken to obtain the desired material conditions because of their sensitivity to the austenitizing temperature and heat treating atmosphere. Full hardness is developed when they are quenched from temperatures of 1175 to 1230°C. Their maximum hardness varies with composition but can be as high as 68 $R_c$, as is possible with M40. Heat treatment is performed in salt baths as well as vacuum furnaces. Welding them also requires that attention be paid to processing conditions; preheating is necessary before welding to prevent cracking and distortion. High-speed tool steels such as M40 are used to make cutting tools for machining modern, very tough, high-strength steels.

*Type T*

Type T tool steels contain W, Cr, V, Co, and C as principal alloying elements and are also characterized by their red hardness and wear-resistance. Deep hardening to 65 $R_c$ or higher in sections up to 3 in. is possible by quenching in oil or salt baths. It is accomplished by quenching from temperatures of 1205 to 1300°C. As with type M, successful welding requires precautions involving pre- and post-weld treatments. Their wear-resistance and hot hardness properties make them suitable for many cutting tool applications, such as drills, reamers, taps, broaches, and milling cutters. Other uses for type T tool steels include dies, punches, precision bearings, and hydraulic pump components.

## Specialty Tool Steels

These tool steel provide unique properties for specific application environments.

*Type S*

Some tool steels are more resistant to shock- or impact-loading than others. Type S tool steels provide greater shock-resistance at the expense of wear-resistance.

The principal alloying elements in the S series are Mn, Si, Cr, W, and Mo, with a common carbon content of about 0.5% in all variants that produces a combination of high strength (323 kpsi), high toughness and low to medium wear-resistance. These properties also lend themselves for consideration in structural applications. Their hardenability varies based on the composition, with a case hardening of 60 $R_c$ and core hardness of 40 to 45 $R_c$ being possible for toughness. To achieve optimum hardness, a relatively high austenitizing temperature is required. Their machinability in the annealed condition is quite good. Welding them requires a slow preheating and a post-weld treatment. Their corrosion-resistance is fair, requiring an oil coating or similar protection in harsh environments. Type S tool steels are used primarily for chisels, rivet sets, punches, driver bits, and other applications requiring high toughness and resistance to shock loading. Use in structural applications is not uncommon because of their high strength.

### Type L

Type L tool steels previously were produced in a wide range of alloy content and physical properties, but now only two grades of this series are commonly available, namely L2 and L6. These grades are oil-hardening fine-grained steels. The nominal composition of L2 is about 0.5% C, 1% Cr, and 0.2% V. The L6 grade contains up to 2% Ni in addition to C, Cr, and V for increased toughness. Although L2 and L6 are both oil-hardening grades, large sections of L2 can be water-quenched. The machinability of L6 is very good, obtaining a rating of 90% of the machinability of the baseline W group water-hardening low-alloy tool steels rating of 100%. L6 may be cold worked by conventional means as with the low alloy steels. Tempering is performed in the range of 350–1000°F to obtain a 45–62 $R_c$ range. The L series provides a good combination of strength and toughness and these grades are typically used in machine tool applications such as bearings, springs, arbors, cams, collets, chuck parts, and rollers.

### Type P

The P series contains Cr and Ni as principal alloying elements which results in very low hardness and low resistance to work hardening in the annealed condition. Among the seven grades available, P2 and P6 are carburizing steels, produced to tool steel quality standards. These factors make it possible to produce mold impressions in a cold state; after the impression is formed the mold is carburized, hardened, and tempered to a surface hardness of about 58 $R_c$. Types P20 and P21 normally are heat treated to 30 to 36 $R_c$, a condition in which they can be readily machined into large intricate dies and molds. Because these steels are prehardened, no subsequent high-temperature heat treatment is required, thereby avoiding distortion and size changes. Type P21 is a precipitation hardening steel containing aluminum that is supplied prehardened to 32 to 36 $R_c$. This steel is preferred for critical finish molds because of its excellent polishability. P series have low resistance to softening at elevated temperatures so they are almost exclusively used in low temperature die casting dies and in molds for injection molding of plastics. Their resistance to softening at elevated temperatures requires them to have adequate cooling when used in these molds.

## Super Hard Tool Materials

Recently, tool materials based on high-speed steels and cemented carbides made by powder metallurgy techniques have been developed. Presently, no formal designation system is in place for these types of tool steels. They are far harder than any existing tool steel and have moderate corrosion-resistance and excellent high-temperature resistance. The tool materials made from cemented carbides typically involve vanadium carbide, tungsten carbide, chromium carbide, and silicon carbide. They possess excellent wear-resistance and high hardness, but exhibit an expected low toughness. Their principal advantages over conventional tool steels are their superior grindability in the hardened condition, faster response to hardening by heat treatment, improved thermal expansion properties (more uniform size change) and the ability to make more highly alloyed steel compositions. The grindability of high-speed steels produced by powder metallurgy may be two or more times better than the same type produced by conventional means. Uses include metal working and wear-resistant applications involving forming and stamping tools.

## Fabrication Processes

Tool steels obtain their performance characteristics in part because they are a highly refined material that is produced in smaller lot sizes, allowing more accurate process control. Typical alloying elements used in their production include cobalt, tungsten, and vanadium, which differentiates them from carbon or alloy steels. Unlike stainless steels, the alloying elements are not added to impart corrosion-resistance but to enhance the material properties that differentiate them from stainless steels. These alloying elements form carbide phases that enhance the material properties. For example, the size of carbides formed and their quantity determine the level of abrasion and wear-resistance, with larger carbides providing greater resistance resulting in the ability to perform cold-working and high-speed tool steel operations.

Raw materials are carefully selected for tool steels to assure high quality and are generally processed in an argon oxygen decarburization (AOD) vessel to achieve the desired composition tolerances. The resulting high purity material has improved mechanical properties including toughness, ductility, and fatigue strength as well as improved processing properties such as uniform machining and the ability to hold a high-quality finish. Electro-slag remelting (ESR) involves post-process remelting of an ingot. Through the remelting process impurities are removed, as the molten metal releases trapped contaminants, and controlled cooling refines the grain structure. In fact, some suppliers perform this process twice to further improve the purity and resulting quality of their product.

The heat treating characteristics of tool steels are significantly different from alloy steels due to their alloy content. The higher alloying content allows them to reach higher hardness by heat treating. The process parameters are not as severe as those necessary to harden alloy steels, resulting in less distortion and cracking. The advantage of these characteristics is a material that has improved processing properties such as uniform machining and the ability to hold a high quality finish.

## Raw and Stock Finishes

Tool steels are not generally found in the same form factors as carbon or alloy steels, instead they are most commonly found in bars or rounds. Hot finished shapes are standard. The most common terminologies used with finished shapes are:

*Decarburization-free (DCF)*: The effects of heat treating (decarburization) are machined from the hot finished product and it is supplied with a machined finish rather than typical hot finish found on carbon steels. This process is also known as prefinishing, where no rough grinding or milling is required, which saves fabrication time. Cost-saving is also realized because the material can be purchased at near net shape, reducing the quantity purchased.

*Standard finish (SF)*: This involves hot finishing without any surface cleanup.

In many applications the service life of a tool steel can be increased by surface treatments such as oxide coatings, plating, carburizing, nitriding, and sulfide treatments. These coatings are not available on standard available stock but are applied by the supplier to make a grade more marketable or at the request of a customer. Regardless, the required surface treatment should be specified. The standard purchased condition is annealed with a DCF finish (Table 4.26).

## Processing Considerations

As a general rule, tool steels are more difficult to machine than plain carbon steel with an average machinability rating being 30% of the standard B1112. Finished machining is typically performed in the annealed condition but recent developments in carbide cutting tools are altering this trend. Carbide tooling can be used in high-speed machining of hardened materials more efficiently than traditional machining

**Table 4.26**  Maximum Sizes Available

| Material | Flat stock (in) (w × t) | Round stock $d_{max}$ (in) |
|---|---|---|
| A2 | 12 × 8 | 28 |
| H13 | 12 × 8 | 28 |
| A6 | 12 × 6 | 16 |
| D2 | 12 × 8 | 30 |
| D3 | 10 × 4 | 5.75 |
| L6 | 10 × 4 | 13 |
| O1 | 12 × 8 | 26 |
| O6 | 12 × 6 | 9 |
| S5 | 8 × 4 | 24 |
| S7 | 12 × 8 | 26 |
| M2 | None | 8.125 |
| M4 | None | 4.5 |

steels. For example, the machining of H13 tool steel with a hardness of 53 $R_c$ is more economical when it is done in the hardened state than in the annealed condition, followed by heat treating and grinding to final dimensions.

Processing tool steels is difficult but predictable when in their hardened condition. Similar to most carbon steels, the ability to harden it is generally proportional to the carbon content and related chemistry which is optimized for tool steels. The heat treating processes are well understood and defined by thermal expansion graphs depicting the material size change for a range of tempering temperatures (Figure 4.19).

The ability to weld tool steels is an important characteristic because of the realities of the manufacturing environment which involves tool damage and breakage. Welding is a much more cost-effective solution than remaking the entire mold or even a portion of it. This property is a matter of concern for large sections, because of their replacement cost. Use of deep hardening tool steels such as type D are the best candidates for repair by welding when controlled pre- and post-processing is used.

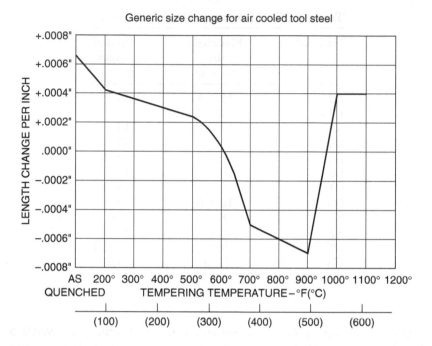

**Figure 4.19**   Generic air hardened tool steel size change over its typical thermal conditioning range.

### Thermal Treatments

Tool steels are typically subjected to the same thermal treatments as those performed on plain carbon steels, including the hardening, softening, and conditioning treatments reviewed in that section. It is the chemistry of tool steels that produces a significantly different outcome when compared to plain carbon steels with specific considerations for distortion and decarburization.

Material distortion is a concern with tool steels because of the nature of most of their applications. Tool steels are used in close tolerance designs because of their stability and performance characteristics. Distortion can be controlled by following manufacture's data on distortion versus tempering temperature and quenching techniques involving controlled heating and cooling. Some applications (such as cutting hobs for gears) are used in the unground condition, requiring close control over their heat treat to produce the desired dimensions.

All high-carbon tool steels are subject to some degree of surface decarburization and require that precautions be taken to control this condition. A certain amount of carbon is oxidized from the outside skin as a result of the rolling, forming, and annealing processes used to form the steel. Because of this, it is customary to remove the surface scale (or bark) to a depth that depends on the cross-sectional shape. For example, a bar of dimensions $1 \times 1/2$ would require that a minimum of 0.031 be removed from the thickness dimension (per side) to remove any decarburized material. If this surface scale is not removed, surface cracking may result during quenching owing to the internal stresses.

### Strengthening Treatments

The primary strengthening treatment of tool steels is heat treatment. Unlike plain carbon, alloy, or stainless steels, tool steels do not benefit from work-hardening to improve strength. Controlled aspects of heat treat processing, including careful consideration of the processing time, temperature, and atmosphere are the primary considerations to obtain a good quality material for use in critical engineering applications.

### Surface Treatments

Generally, tool steels are not materials that improve greatly from surface treatments, because the material itself is designed to provide the required performance without the additional processing. Certain applications can benefit from the use of surface treatments which can be applied to extend tool life by imparting additional properties that improve performance for specific applications.

Oxide coatings and sulfide treatments prevent or reduce adhesion of the cutting tool to the workpiece resulting in improved life of the tool. Application of an oxide coating by bath or steam methods produces an adhesion-resistant surface that can double tool life when working with softer materials such as copper or plain carbon steels (excluding free machining grades). Sulfide coatings produce similar results and are applied using a low temperature (190°C; 375°F) electrolytic bath.

Chrome plating can be applied to reduce adhesion and increase tool life. A plating thickness of 0.0001–0.0005 in. (0.0025–0.0125 mm) is common for this application. Aspects of concern for this approach include high cost and the possibility of inducing hydrogen embrittlement as a result of the plating process, which could lead to early tool failure.

Liquid nitriding provides a similar improvement, while also increasing surface hardness, providing a light case hardness effect to the tool. Use of gas nitriding produces a brittle case that lacks the required toughness for most applications. This process will often result in a two-to-threefold increase in tool life and in some cases up to a fivefold increase, making it a popular method. Common uses include drills, taps, and reamers.

Carburizing is used for nonimpact cold working applications to increase wear-resistance. The surface case of steels treated in this manner are brittle and do not respond well to elevated temperatures, restricting their use with tool steels to cold working applications that do not involve high impact loads, such as shearing dies.

## Common Grades

Tool steel selection is very application dependent, requiring consideration for the intended life of the part and its environment (Table 4.27). The tool maker determines how the part will be made as well as how it will be used. This is not altogether different from the selection method for applications other than

**Table 4.27**   Common Grades of Tool Steels

| Application | Tool steel |
|---|---|
| Single-point tool | M2, M4 or T4, T5 |
| Drills | M1, M2, M7, M10 |
| Reamers | M1, M2, M7, M10 |
| Threaded taps | M1, M2, M7, M10 |
| End mills | M1, M2, M7, M10 |
| Cold shearing | D2, A2, S2, S5 |
| Hot shearing | H10 |
| Blanking dies | O1, A2, D2, D4 |

production tooling ones, but is generally predetermined because of the extensive application data for the materials in very specific use environments.

## Cast Iron Materials

Cast irons are ferrous metals that are similar to steel but with a much greater (grade dependent) carbon content; they are alloyed with carbon and silicon to a degree that exceeds the solubility limit, resulting in the excess carbon precipitating out in the form of carbon flakes or nodules. The carbon content of cast iron ranges from 1.8 to 4.0% depending on the type of iron, with white iron typically having the least amount of carbon and ductile iron the highest. Cast irons are graded based on their total carbon content, which includes both that which is in solution and that which is in a solid form, resulting in a composite material consisting of precipitated graphite in a solid matrix. Comparatively, carbon steels contain up to 2% carbon because of their solubility but does not remain in solution at greater levels.

The term cast iron refers to a large family of ferrous alloys which are multicomponent systems that solidify as eutectics. In addition to their high carbon content, cast irons also contain appreciable silicon, usually from 1 to 3%, and thus they are actually iron–carbon–silicon alloys. The high carbon content and the silicon in cast irons make them excellent casting alloys, because their melting temperatures are appreciably lower than those of steels. Molten iron is more fluid than molten steel and less reactive with molding materials. Formation of lower density graphite in the iron during solidification reduces the change in volume of the metal from liquid to solid and makes production of more complex castings possible. Cast irons, however, do not have sufficient ductility to be rolled or forged.

Cast irons are available in four basic forms (Figure 4.20). The most commonly used of the four are the gray and ductile irons because of their availability and desirable properties. The properties of cast irons are controlled by four foundry variables which include the base chemistry, the inoculation or the presence of growth sites, solidification rate, and cooling

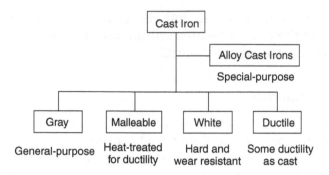

**Figure 4.20**   Cast iron.

rate. The manipulation of these variables is important to producing the desired ferrite-to-pearlite ratio which determines the mechanical properties. Ferrite is softer and lower in strength compared with pearlite, and altering this ratio produces the many useful properties of cast irons, including increased mechanical damping for softer grades (Figure 4.21). Increasing the rate of cooling can be useful for producing a material that has higher wear resistance or surface hardness due to the more rapid cooling of the surface, effectively quenching it to produce the desired material condition. Course flaked graphite in a ferrite matrix has a thermal conductivity of one and a half times that of low carbon steel.

Cast irons are attractive design materials, owing to several of their unique physical properties such as mechanical damping, compressive strength, and modulus. Mechanical damping is inversely proportional to tensile strength and dependent on the amount of precipitated carbon that is present in the matrix. Higher amounts of carbon result in higher damping but lower tensile strength. The compressive strength of cast irons can be three to four times that of its tensile strength and are dependent on the amount of graphite in the matrix. When a cast iron is in tension, the presence of graphite acts as a stress riser, weakening the material considerably. When in compression, the graphite has little influence over the compressive properties. This behavior is responsible

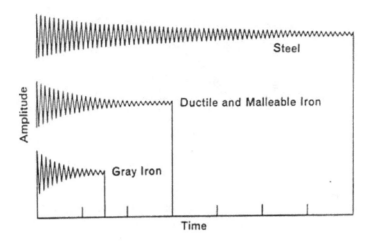

**Figure 4.21** Damping characteristics of gray cast iron compared to steel and maleable iron.

for cast irons being selected for machine tool bases and other large supporting structures that must support large compressive loads while subjected to varying vibration loads. The modulus of cast irons varies widely between the different grades as it is dependent on composition and processing; this can produce significantly different material structures (Figure 4.22). The resulting stress–strain curve does not obey Hooke's law and shows lack of linear proportionality, requiring that the tangent or the secant modulus be defined.

## Properties and Designations

Depending primarily on composition, cooling rate, and melt treatment, cast irons can solidify according to the thermodynamically metastable Fe–Fe$_3$C system or the stable Fe–Graphite system. How the liquid iron solidifies determines the resulting material properties. When the metastable path is followed, the carbon-rich phase in the eutectic is the iron carbide; when the stable solidification path is followed, the carbon-rich phase is graphite. The formation of stable or metastable eutectic is a function of many factors including the nucleation potential of the liquid, chemical composition, and cooling rate.

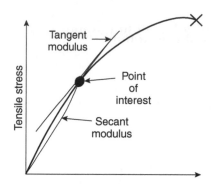

**Figure 4.22**   Stress-strain curve for gray cast iron with tangent modulus and secant modulus obtained at a particular point of interest.

The first two factors determine the graphitization potential of the iron. A high graphitization potential will result in irons with graphite as the carbon-rich phase, while a low graphitization potential will result in irons with iron carbide.

These two basic types of eutectics (stable austenite–graphite or the metastable austenite–iron carbide, $Fe_3C$) exhibit wide differences in their mechanical properties, such as strength, hardness, toughness, and ductility. Widening the scope of the metallurgical processing of cast iron to produce a desired mix of these properties involves the manipulation of the type, amount, and morphology of the eutectic. The addition of selected minor elements is also vital to the successful production of iron. For example, nucleating agents, called inoculants, are used in the production of gray iron to control the graphite type and size. Trace amounts of bismuth and tellurium are used in the production of malleable iron, and the presence of a few hundredths of a percent of magnesium causes the formation of the spherulitic graphite in ductile iron.

It is difficult to designate cast irons by chemical composition because of the similarities between the grades. Historically, the first classification of cast iron was based on its fracture characteristics. Two types of iron were initially recognized:

- *White iron*: exhibits a white, crystalline fracture surface because fracture occurs along the iron carbide plates; it is the result of metastable solidification ($Fe_3C$ eutectic).

- *Gray iron*: exhibits a gray fracture surface because fracture occurs along the graphite plates (flakes); it is the result of stable solidification (graphite eutectic).

The various types of cast irons are more appropriately classified by their microstructure, which is based on the form and shape in which the majority of carbon occurs in the iron. The designation includes the matrix as well as the graphite shape. Each of these types may be moderately alloyed or heat treated without changing its basic classification.

This system provides for four basic types of cast iron: white iron, malleable iron, gray iron, and ductile iron.

As indicated earlier, white iron and gray iron derive their names from the appearance of their respective fracture surfaces. Ductile iron derives its name from the fact that in the cast form it exhibits measurable ductility. In contrast, neither white nor gray iron exhibits significant ductility in the standard tensile test. Malleable iron is cast as white iron and then heat treated to impart ductility. In addition to these four basic types there are other special forms of cast iron (Table 4.28).

Physical properties of the more common classes of cast irons can be found in Table 4.29.

### Numbering
Classifications of cast iron were originally based on their fracture characteristics, and then, because of advances in metallography, are based on microstructure. There are four basic types of cast irons: white, malleable, gray, and ductile. In addition to these four basic types there are other commercially important types such as mottled and compacted graphite cast iron. Numbering of the various grades of cast irons differ for each class. Gray cast irons are numbered according to ASTM A48 which uses tensile strength (in ksi) as the number, that is, class 20 means the grade have a minimum tensile strength of 20 ksi. Ductile irons use a similar system for designation by using tensile strength (in ksi), yield strength (in ksi), and elongation (in percentage). It is convenient to use this system as it readily gives user an idea of the material

**Table 4.28**   Special Forms of Cast Iron

| Cast iron grade | Description |
| --- | --- |
| Chilled iron | A white iron that has been produced by cooling very rapidly through the solidification temperature range to produce a hard surfaced material with a soft core |
| Mottled iron | A cast material that solidifies at a rate inter mediate between those of chilled and gray iron and which exhibits microstructure and fracture features of both types |
| Compacted graphite cast iron (vermicular iron) | A cast material characterized by graphite that is interconnected within eutectic cells as is the flake of graphite iron in gray iron. However, in compacted graphite iron it is coarser and more rounded (i.e., its structure is inter mediate between the structure of gray and ductile iron) |
| High-alloy cast irons | A cast material produced both in flake and spheroidal graphite configurations. Generally containing over 3% of added alloy, can also be individually classified as white, gray, or ductile iron, but the high-alloy irons are classified commercially as a separate group. Used primarily for applications requiring corrosion resistance or a combination of strength and oxidation resistance |

**Table 4.29**   Important Classes of Cast Irons

| Commercial designation | Carbon-rich phase | Matrix | Fracture |
| --- | --- | --- | --- |
| Gray iron | Lamellar graphite | Pearlite | Gray |
| Ductile iron | Spheroidal graphite | Ferrite, Pearlite, austenite | Silver-gray |
| Compacted graphite iron | Vermicular graphite | Ferrite, Pearlite | Gray |
| White iron | $Fe_3C$ | Pearlite, martensite | White |
| Mottled iron | Lamellar graphite $+Fe_3C$ | Pearlite | Mottled |
| Malleable iron | Temper graphite | Ferrite, Pearlite | Silver-gray |
| Austempered ductile iron | Spheroidal graphite | Bainite | Silver-gray |

properties. For example, grade 65-45-12 is used in applications requiring high ductility and impact resistance. ASTM, ASM, and SAE have different classifications for special purpose grades in all classes of cast irons.

### Strength Properties

The strength properties of cast irons are most significantly affected by the shape and size of the graphite they contain. Nodular shaped graphite, as found in ductile irons, will provide greater strength than the flake graphite, found in gray iron. But the size of the nodules does not influence the strength nearly as much as variation in flake size. The mechanical properties of cast irons vary widely, with ultimate tensile strengths ranging from 25 to 100 ksi. Compressive strengths of cast irons are not highly influenced by graphite shape or size and can be as high as 150 ksi. Their moduli of elasticity also vary based on material type and range from 20 to 30 Mpsi. Cast irons are also known for their hardness, which can be altered by heat treatment, with common values ranging from 130 to 330 BHN.

Among other mechanical properties cast irons are well known for their vibration dampening capabilities fostering their use in engine blocks and other automotive components. The presence of soft graphite in the matrix provides the damping characteristics by absorbing acoustic and mechanical energy. Gray irons provide ten times the damping of low-carbon steels while ductile irons provide three times as much damping. Cast irons have poor impact resistance and impact loading applications are to be avoided. Nodular cast iron provides the best impact strength while the flake graphite forms provide the least impact strength.

The presence of graphite in cast iron improves wear-resistance as it acts as a lubricant, reducing sliding friction. Use of cast irons in applications requiring wear-resistance is one of the most common. Hardness is important for wear-resistance, and cast irons require suitable heat treatment to maximize the wear properties. Alloy cast irons containing chromium and nickel have higher hardness, therefore, give better wear-resistance but with the penalty of higher material and manufacturing costs. Apart from affecting mechanical properties, the shape and form of graphite also affects other properties like electrical resistivity, magnetic properties, and corrosion-resistance. The selection of a proper microstructure is crucial to the properties desired for the designed application.

### Heat Treatability

Cast irons are commonly heat treated to improve their wear resistance; this is easily achieved because of their favorable response to the treatment. However, malleable cast irons are not heat treated, as they convert to white cast irons upon heat treatment. The thermal expansion and conductivity of cast irons are typically very close to those of steels providing less distortion and greater flexibility in heat treatment cycles. The matrix structure can also be controlled by heat treatment as it is possible to have more than one kind of matrix in a cast iron. Hardening treatments yield a martensitic matrix. Other possible matrices are ferritic, pearlitic, and ledeburite. Hardness can be used as a measure of strength, wear, and machinability only within a given grade, not across grades because although the hardness may be similar, the other material properties will be significantly different. Common hardness ranges are from 130 to 330 BHN.

### Fatigue Properties

While the higher carbon content of cast irons provides several advantages of these materials over others, (improved wear-resistance and vibration damping), it is also responsible for degradation of ductility and fracture toughness. The carbon, usually in form of graphite, serves to nucleate fatigue and fracture processes at lower strains. Since the free graphite in cast iron acts as an inherent notch, the fatigue properties are greatly influenced by its quantity, size, and shape. The flake form of graphite lowers fatigue strength because of increased stress concentrations. The nodular form has better fatigue strength because of fewer stress concentration. Endurance ratios range from 0.35 to 0.60 depending upon the grade of iron. Other causes of stress concentrations can be slag inclusions and porosity induced in the casting process. The continuous casting process reduces these factors and improves fatigue properties.

### Corrosion Properties

The corrosion-resistance of cast irons is generally better than that of unalloyed steels. The presence of slag (3%) which is a silicate does not corrode in most environments. Corrosion-resistance varies from grade to grade because of differing composition. The graphite in cast iron is not attacked by the environment as much as the matrix. The most common method is to use alloy grades of cast iron in corrosive environments. Higher silicon grades have better corrosion-resistance, but they are brittle and difficult to manufacture. Nickel, chromium, and molybdenum are other common additions which improves corrosion-resistance. Unlike mechanical properties, the shape of graphite does not have a significant effect on corrosion-resistance. Ni-resist grades are a common solution to corrosive environments. Also high-nickel austenitic grades, high-chromium, and high-silicon grades offer excellent corrosion-resistance in acidic and alkali environments.

### Machinability

Cast irons vary greatly in machinability, ranging from white irons, which are extremely difficult to machine, to ferritic cast irons, which are among the easiest metals to machine. Between the two extremes are grades of gray, ductile, compacted graphite, and malleable cast irons, with varying degrees of machinability. This property is related directly to the microstructure of cast iron as the presence of graphite provides free machining and the amount of ferrite in matrix determines the ease of machining. A martinsitic or fine pearlitic matrix makes the machining difficult. Gray and ductile irons have better machining characteristics as they provide long tool life and better surface finish. Malleable iron is also easily machinable and said to have 25% better machining than that of free-cutting steel. The most difficult to machine are the white and high-silicon grades of cast irons. They require heavy tooling and rigid setups. Alloying elements can also affect machining characteristics. Phosphorus has an adverse effect on machining while tin, in small amounts, in gray irons increases machinability. Dimensional stability is generally good with cast irons, but, as with all materials, the amount of residual stress can have a significant impact on the part stability, often leading to the need for stress-relieving heat treatment.

The best surface finishes are obtained with flaked size graphite; smaller flakes are preferred as they allow for frequent chip breaks. Larger flakes and nodules can result in tearing of the material by the cutting tool, producing a poor finish. Machined RMS finishes of 32 or better are common in cast irons.

### Weldability

Due to the wide range of compositions and microstructures found in cast irons, their weldability ranges from easy to very difficult. Welding procedures must be suited to the type of cast iron to be welded. Cast iron is commonly welded in order to repair defects or damage in the castings and not as a method of manufacture. White cast iron is generally considered unweldable, while ductile iron is generally easier to weld than gray iron. The ferritic and pearlitic matrices cause a heat affected zone which is primarily martensitic. It requires further heat treatment to reduce brittleness and cracking. Reducing the size of HAZ can also be done by using small heat input, smaller diameter electrodes, use of low melting point rods and preheating. Base metal cleaning is also crucial to get a sound weld. Most castings have a carburized or oxidized surface from the casting process which should be cleaned. Any traces of sand, rust, oil, paint grease, or moisture should be removed prior to the welding process.

Cast irons are produced in four major forms: white, gray, malleable and ductile. There are also several speciality grades that offer unique formulations to address specific design environments.

### White Cast Iron

White cast iron is formed when much of the carbon remains in solution, resulting in massive carbide formation within the matrix instead of graphite upon solidification. It has a pearlitic matrix with large amounts of iron carbide present, and is called white cast iron as its fracture produces a white crystalline surface. In order to retain much of the carbon in the solution the carbon and silicon content must be kept relatively low (i.e., 2.5–3% C and 0.5–1.5% Si) and it must be subjected to a high rate of cooling to promote rapid solidification.

White cast iron is known for its excellent wear and abrasion-resistance resulting from a matrix structure containing large amounts of iron carbides. Quenched hardness up to 600 HB is possible along with high compressive strengths and good property retention at temperatures up to 500°C. Varying hardness levels can be produced in selected areas of a casting by promoting localized rapid solidification of the iron during processing. The resulting white iron that occurs at the surface of the casting is known as chilled iron and is produced by making that portion of the mold, where the white iron is desired, out of a material, such as iron or graphite, that can extract heat very rapidly. These sections of the mold are called chills. White iron does not have the easy castability of other irons because its solidification temperature is generally higher, and it solidifies with carbon in its combined form as iron carbide. Although white cast iron can be welded, welding is not recommended.

### Gray Cast Iron

Gray cast irons contain 2.5–4% C and 1–3% Si as well as 0.1–1.2% Mg to produce the desired microstructure. A substantial portion of the carbon precipitates out of solution to form flakes, which are exposed during fracture to give the fractured surface a gray appearance. The microstructure is a matrix of pearlite with graphite flakes dispersed throughout. Foundry practices can be manipulated to produce the necessary pattern of flakes to enhance desired properties. A simple and convenient classification of the gray irons is found in ASTM

**Table 4.30**   Typical Properties of Gray Iron Grades (ASTM A48)

| ASTM class | Tensile strength (Mpa) | Minimum section thickness (mm) | Compressive strength (Mpa) | Torsional shear strength (Mpa) |
|---|---|---|---|---|
| 20 | 152 | 3.2 | 572 | 179 |
| 25 | 179 | 6.4 | 669 | 220 |
| 30 | 214 | 9.5 | 752 | 276 |
| 35 | 252 | 9.5 | 855 | 334 |
| 40 | 293 | 15.9 | 965 | 393 |
| 50 | 362 | 19.0 | 1130 | 503 |
| 60 | 431 | 25.4 | 1293 | 610 |

*Source*: ASM Metals Handbook, Table 5, p. 5–6, ASM International, 1985.

specification A48 which classifies various grades of gray iron in terms of their, tensile strengths (Table 4.30). Class 20 indicates a tensile strength of 20 ksi. For gray cast iron the class label ranges from 20 to 60. Low-strength grades are common due to their good resistance to thermal shock and related cracking. The material properties for classes 20 through 60 including strength, modulus of elasticity, wear-resistance, and ability to be machined for finishing increases while resistance to thermal shock, damping capacity, and ability to be cast in thin sections decrease. Verification is obtained via test specimens; although the test specimen performance may not be identical to the casting, it represents a level of consistency that is necessary to ensure that the material is good. When fatigue strength is a consideration, it is commonly taken at 40% of the known tensile strength.

In addition, ASTM A436 covers austenitic gray iron castings that are characterized by uniformly distributed graphite flakes, some carbides and the presence of sufficient alloy content to produce an austenitic structure.

SAE defines gray iron on the basis of its specification J431 in which grades are denoted by GXXXX (i.e., G followed by four digits). The SAE designations are more specific than those of A48 (Table 4.31). For example, an iron intended for heavy sections, such as grade G3500, is specified to have

**Table 4.31** Typical Uses of SAE Grades of Gray Cast Irons

| SAE grade | Applications |
| --- | --- |
| G1800 | Used where strength is not a primary concern |
| G2500 | Small cylinder blocks, cylinder heads, pistons, clutch plates |
| G3000 | Diesel cylinder blocks |
| G3500 | Diesel engine blocks, tractor cylinder blocks and heads, heavy flywheels, heavy gear boxes |
| G4000 | Diesel engine castings, liners, cylinders and pistons, camshafts |

higher strength and hardness in the standard test bar than does grade G2500, which is intended for light-section castings.

Gray irons are cast for a wide variety of applications because of their good fluidity within the mold and low shrinkage. Successful production is dependent on the fluidity of the molten metal and cooling rate which is a function of section thickness. The "as cast" hardness of gray cast iron ranges from 156 to 302 HB and can be increased to 550 HB (62 $R_c$). Hardening is achieved by rapid air cooling or quenching followed by tempering at 150 to 650°C to improve toughness and stress-relieving. The hardenability of an unalloyed gray iron is about equal to that of a low alloy steel. Hardenability can be increased by the addition of Cr, Mo, or Ni.

Gray cast iron is not recommended for applications requiring high impact-resistance. By comparison, it has considerably lower impact strength than ductile iron, malleable iron, or a cast carbon steel. The presence of carbon flakes improves their machining properties, making them superior to those of most other cast irons of equivalent hardness, and virtually all steels. Wear-resistance is excellent, assisted by the flaked graphite which acts as a dry lubricant, allowing gray cast iron to be frequently used where resistance to wear is required. However, different grades exhibit varying degrees of wear-resistance. In addition, gray cast irons also possess excellent vibration damping properties due to the graphite present.

Gray iron does not obey Hooke's law because Poisson's ratio and modulus of elasticity change as a function of applied stress due to physical properties of the matrix. The presence of graphite flakes prevent the material from conforming as it is loaded in tension and actually results in an expansion of the material. The reverse is true for compression as Poisson's ratio increases when compressive stresses are greater than 25 ksi.

The modulus can vary considerably because of both section size and chemical composition, while also reducing linearly with increasing temperature. It is higher in compression than tension, resulting in a shift of the axis of symmetry that provides greater factor of safety when in bending.

The modulus in tension is usually determined, arbitrarily, as the slope of the line connecting the origin of the stress-strain curve with the point corresponding to one-fourth the tensile strength (secant modulus). It is also possible to use the slope of the curve near the origin (tangent modulus). The secant modulus is a conservative value suitable for most engineering work, as design loads are seldom as high as one-fourth the tensile strength (Figure 4.24). However, in the design of precision equipment where design stresses are very low, the use of the tangent modulus may represent the actual condition more accurately.

### Ductile Cast Iron

This cast iron is similar to gray cast iron because the graphite is precipitated out of solution, but is different in that it results in spherical particles rather than flakes. It is produced using similar materials as white or gray irons but of somewhat higher purity, which reduces its brittleness. It is also known as nodular or spheroidal graphite cast iron. This is done so using inoculants such as magnesium or cerium in finite quantities during a process called nodulizing to produce desired results. Ferrosilicon is also added to control the nodule size. A typical composition will contain 3–4% C and 2–3% Si, with nodulizers to produce graphite in the form of nodules rather than flakes. This matrix has greater tensile strength and higher toughness than gray or malleable irons and does not require heat treatment to produce nodules as is required for malleable irons. The unique combination of strength and ductility make it desirable for structural applications.

Ductile iron has good strength and toughness combined with good machinability, which makes it desirable for many applications. Cost of machining ductile iron is similar to that for gray cast irons, since they have similar hardness and are similar to cast mild steel, making their manufacturing cost-competitive. Ductile iron has unique dampening characteristics that make it desirable in making quiet running gears and machinery bases that need to absorb vibrations to be most effective. It also has a low coefficient of friction because of the graphite, which increases its effectiveness in sliding applications such as gears of machine ways. The automotive and agriculture industry are the prime users of ductile iron, for motor casing and gear trains. It can also be used for pipes because its ductile properties prevent facture. Its corrosion-resistance is also similar to that of gray iron.

ASTM A536 employs a system for grade designation of ductile irons based on their physical properties; this system uses numerical placeholders for tensile strength (ksi), yield strength (ksi), and elongation (percentage). This system makes it easy to specify nonstandard grades that meet the general requirements of ASTM A536. For example, grade 80-60-03 translates to 80 ksi minimum tensile strength, 60 ksi yield strength, and 3% elongation. This grade is commonly used in applications where ductility is not critical. Grades 65-45-12 and 60-40-18 provide increased elongation and are used in applications that require high ductility and impact resistance. These are available in both ferritic and martensitic grades (Table 4.32).

Heat treatments for ductile irons are the same as those for gray cast irons, producing significant improvements in their mechanical properties; these include annealing, stress-relieving, normalizing, and quenching. A process known as austempering is also applied to significantly increase both tensile strength and toughness and is defined by ASTM A897 to include five different grades (Table 4.33). Austempering utilizes a high-temperature salt quench-tank that causes austenite to transform to acicular

**Table 4.32**  Material Property Summary Contrasting the Two of the most Common Cast Irons: Gray and Ductile

| Material Property | Ductile iron | | | Gray Iron | |
|---|---|---|---|---|---|
| | 65-45-12 | 80-55-06 | 100-70-02 | G2 | G1 |
| Tensile strength (psi) | 65,000 | 80,000 | 100,000 | 40,000 | 20,000 |
| Yield strength (psi) | 45,000 | 55,000 | 70,000 | – | – |
| Elongation (%) | 12 | 6 | 2 | – | – |
| Fatigue strength (psi) | 40,000 | 40,000 | 40,000 | 20,000 | 10,000 |
| Shear strength (psi) | 54,000 | 72,000 | 90,000 | 58,000 | 29,000 |
| Compressive strength (psi) | 110,000 | 115,000 | 120,000 | 120,000 | 90,000 |
| Torsion strength (psi) | 54,000 | 72,000 | 90,000 | 60,000 | 30,000 |
| Density (lbs/in$^3$) | 0.260 | 0.260 | 0.260 | 0.260 | 0.260 |
| Hardness (BHN) | 180 | 229 | 279 | 241 | 160 |
| Modulus of elasticity (Mpsi) | 23 | 25 | 25 | 20 | 20 |
| Relative wear resistance | Fair | Good | Excellent | Excellent | Fair |
| Coefficient of thermal expansion (in/°F) (range: 70–212°F) | $6.4 \times 10^{-6}$ | $6.4 \times 10^{-6}$ | $6.0 \times 10^{-6}$ | $5.5 \times 10^{-6}$ | $5.5 \times 10^{-6}$ |
| Thermal conductivity, (Btu/hr/ft°F) | 18.68 | 18.68 | 18.68 | 30.84 | 30.84 |
| Relative damping capacity | Low | Low | Low | High | High |
| Heat-treat response (R$_c$) | 55–60 | 55–60 | 55–60 | 55–60 | 55–60 |

ferrite and high-carbon austenite, allowing strengths comparable to carbon steels to be obtained (Table 4.34).

Hardenability of these materials is good but their heat transfer capability is limited because graphite is in nodular form rather than flake form, reducing the heat transfer capabilities. Generally, ductile iron has greater hardenablility than eutectoid steel with a comparable composition.

As with welding of other cast irons, welding of ductile iron requires that special precautions be taken in order to obtain optimum properties in the weld metal and the adjacent

**Table 4.33**  Grades of Ductile Cast Iron as Defined in ASTM A897

| Grade | Tensile (min) ksi | Yield (min) ksi | Elongation % | Impact ft.lbf | Hardness HB |
|---|---|---|---|---|---|
| 125-80-10 | 125 | 80 | 10 | 75 | 269–321 |
| 150-100-7 | 150 | 100 | 7 | 60 | 302–363 |
| 175-125-4 | 175 | 125 | 4 | 45 | 341–444 |
| 200-155-1 | 200 | 155 | 1 | 25 | 388–477 |
| 230-185 | 230 | 185 | n/a | n/a | 444–555 |

heat affected zone. The main objective is to avoid the formation of cementite in the matrix material, which makes the welded region brittle and prone to fracture. In ductile iron, it is desirable to retain the nodular form of graphite. However, the formation of martensite and fine pearlite can be removed by tempering.

### Malleable Iron

This form of iron contains 2–3% C and 1–1.8% Si, most of its carbon is in the form of irregularly shaped graphite nodules instead of flakes. Malleable iron is produced by first casting it as a white iron and then heat treating it to convert the iron carbide into irregularly shaped nodules of graphite. This form of graphite is known as tempered carbon, because it is formed in the solid state during heat treatment. This heat treatment consists of two stages: graphitization and cooling. During graphitization, white cast iron is heated up to 940°C and held from 3 to 40 h depending upon the structure desired; cooling can transform the austenite into different kinds of matrices and can take up to 30 h. The only heat treatment typically applied is quench hardening because it is extensively heat treated when manufactured so the product form is already nearly optimal. Flame and induction hardening can also be employed when appropriate.

Generically, their properties are similar to gray cast irons, as malleable iron possesses considerable ductility, toughness, and high impact resistance owing to its combination of nodular graphite and a low carbon metallic matrix. Most grades can

**Table 4.34  Mechanical Properties of Austempered Ductile Iron with Cast Irons and Cast Steels**

| | Austempered ductile iron | Cast Iron | | | Steel | |
|---|---|---|---|---|---|---|
| | | Gray (A48) | Malleable (A47) | Ductile (80-55-06) | Cast (A27) | Forged (AISI 862) |
| Hardness, Brinell | 269–550 | 140–301 | 110 | 192 | 130 | 341–388 |
| Ultimate strength ($10^3$ psi) | 125–230 | 20–55 | 50 | 81 | 63 | 167–188 |
| Yield strength ($10^3$ psi) | 96–185 | | 32.5 | 53 | 35 | 120–149 |
| Elongation (%) | 10–<1 | | 10 | | 30 | 12–14 |

bend without breaking because of their high ductility, which can reach 20%. Consequently, malleable and ductile irons are suitable for same applications with the choice depending upon the desired cost. The modulus of elasticity is in the range of 25 to 27 Mpsi (nearly as stiff as steel).

Malleable iron also exhibits high resistance to corrosion, excellent machinability, good magnetic permeability, and low magnetic retention, making it useful for magnetic clutches and brakes. Its wear-resistance is very good at higher hardness values because, as with other cast irons, the presence of solid graphite significantly reduces the surface friction. Its fatigue strength is typically 60% of maximum yield with a compressive strength four times the yield strength. Its good fatigue strength and damping capacity makes malleable iron useful for long service in highly stressed parts.

There are two common types of maleable iron: black heart and white heart. Black heart malleable iron is the only type produced in North America and is the most widely used variety throughout the world. White heart is an older type and is essentially decarburized throughout in an extended heat treatment of white iron. Malleable iron can be made with a ferritic matrix or pearlitic–martensitic matrix.

The hardness of ferritic malleable iron is 110-156 HB with high strength. Cu is added (about 1%) to increase its corrosion-resistance. The pearlitic–martensitic malleable irons have hardnesses in the range of 250–320 HB after heat treatment. The pearlitic–martensitic type has good fracture toughness with excellent wear-resistance.

Welding is not recommended for this material as it produces a brittle white iron in the weld zone thus losing ductility. If welding is necessary an annealing treatment is required afterwards. Silver brazing and tin lead soldering can be satisfactorily used. Alloy designations are defined by ASTM A47 using a five digit number (XXXYY) with XXX identifying the minimum yield strength in ksi (e.g., 24.5 ksi) and YY standing for the percentage elongation (e.g., 10%). This example would produce a designation of 24510. Common examples of ferritic malleable iron are 32510 and 35018. Examples of pearlitic–martensitic malleable irons according to ASTM

A220 include 40010, 45008, and 80002. Applications in auto-motive parts include frames, suspensions, wheels, differential carriers, differential cases, bearing caps, and steering gear housings.

### Ni-Resist Cast iron

Austenitic cast irons containing 18–36% Ni, up to 7% Cu and 1.75–4% Cr are used for both heat-resistant and corrosion-resistant applications. Known as Ni-resist, this type of cast iron contains an austenitic matrix with about 10% of alloy carbides. The austenitic structure is suitable for corrosive environments in sour well oils, salts, saltwater acids, and alkalies. The iron is relatively low modulus and chrome is added to produce carbides that help improve wear resistance. Ni-resist is denser than gray or ductile irons and will have a higher coefficient of thermal expansion. Ni-resist parts maintain relatively high impact properties even at low temperatures. The alloy–carbide network remains stable at elevated temperatures, providing good wear properties in moving parts that operate at temperatures up to 1300°F.

Ni-resist iron exhibits good resistance to high temperature scaling and growth up to 815°C in most oxidizing atmospheres, good performance in steam service up to 530°C and can handle sour gases and liquids up to 400°C. In the presence of appreciable amounts of sulfur the maximum temperature of use is 540°C. Austenitic irons have the advantage of having considerably greater toughness and thermal shock-resistance than the other heat-resistant alloy irons, although their strength is rather low.

Ni-resist is manufactured to produce a material conforming to ASTM A436, Type I. Continuous cast Ni-resist grades contain type VII graphite type A flakes, size 4 to 6, as defined in ASTM A247. The matrix is austenite with alloy carbide. The austenitic matrix provides excellent corrosion-resistance and is nonmagnetic. The carbide network provides resistance to wear but does not adversely affect machinability. The austenitic structure is achieved by adding a specific combination of nickel and copper and will remain stable in temperatures up to −300°F.

## Alloy Cast Iron

Alloy cast irons are considered to be those casting alloys based on the iron–carbon–silicon system that contain one or more alloying elements intentionally added to enhance some specific property. The addition of small amounts of elements to control the shape of graphite is termed inoculation rather than alloying. Alloying elements, including silicon when it exceeds 3%, are usually added to increase the strength, hardness, and hardenability or corrosion-resistance of a basic iron. In gray and ductile irons, small amounts of Cr, Mo, and Ni are used primarily to achieve high strength or to ensure the attainment of a specified minimum strength in heavy sections. Otherwise, alloying elements are added exclusively to enhance resistance to abrasive wear or chemical corrosion or to extend service life at elevated temperature.

## Fabrication Processes

Cast irons are generally formed either in discrete parts using molds or by continuous casting of a generic shape of constant cross-section. Mold casting has been successfully employed for centuries with little change in a process that involves the pouring of molten steel into a fixed shaped cavity. There are six commonly used methods:

- Sand casting — as implied by the name, this process involves the use of sand as a means of handling the molten iron as it cools into a desired form. A wooden pattern is used to first define the shape in the sand, then it is removed and the cavity filled with molten iron. This is an inexpensive method for limited volume applications.
- Permanent mold casting — similar to a sand casting but done using permanent mold made from a water-cooled steel mold. This is a costly method that is best suited for high volume applications.
- Die casting — a process similar to plastic injection molding involving the pressurized injection of molten metal into a mold. This is expensive, requiring a large number of parts to amortize the high tooling cost.

- Shell casting — this process involves first making a casing or shell of the pattern (or actual part), splitting it to remove it, reassembling the pieces and finally placing shell in sand. The sand supports the shell while the molten metal is poured into it. After cooling, the part is removed by breaking the shell to expose the completed part. This process is labor intensive but is a good one for intricate parts that are produced in low volumes.
- Investment casting — this casting process produces similar results as shell casting but is aimed at higher volumes. A mold is built to form a wax pattern which is then removed, coated with a ceramic material, heated to remove the wax, and then placed in sand. The molten metal is then poured inside, allowed to cool, and then the ceramic shell is removed.
- Centrifugal casting — this process involves producing a part by using a rotating drum with the mold being the inner diameter of the drum. It is rotated while molten metal is poured inside, forming the desired part which is removed when cooled by splitting the mold. This is most commonly used to form pipes.

Continuous casting is a relatively newer process which was fully developed after the World War II. As the name states, the process involves continuously pouring molten metal from the bottom of the crucible on to a water-cooled mold, forming a skin that allows it to be further handled down the line. The process inherently produces a higher quality casting than that obtained in molds, because the material is drawn from the bottom of the crucible, away from the slag and other impurities that float on the surface. It is also subjected to differential cooling results which produces varied cross-sectional material properties, with the outer region generally being made up of a finer graphite structure than the core to give a combined surface hardness and overall toughness. Two forms of graphite are typically produced, flakes (present in gray iron) and nodules (present in ductile iron). The solubility limit at which these form involve many factors that are not easy or economical to control, requiring additions, known as inoculants to be added which force the graphite out of solution and make it possible to control the size and shape of the graphite particles.

The casting process requires careful consideration of three parameters of the cast part that will often determine the method of casting selected. These are part size, required tolerances and surface finish. Generally, the larger the part, the more expensive the tooling and handling equipment. Those processes which require hard tooling such as die casting or investment casting, are not used to produce large parts (>30 lbs.), while sand casting has no such limitation. The trade-off here is the quality of the part and the surface finish. Larger parts can have internal cavities caused by shrinkage while sand casting does not have the improved surface finish afforded by hard tooled processes. ASTM and the American Foundrymen's Society provide a set of standards for casting process.

## Raw and Stock Finishes

Cast iron materials are most often supplied in their net shape in an as cast condition and requiring post processing, including machining, surface treatments and less often, heat treating. The size of the casting can have significant effects on the quality of the part, which include reduction in tensile strength and porosity. Generally, tensile strength decreases with increasing bar diameters because of the difference in solidification and cooling rates. For example, the graphite flake size at the center of a 20 in. diameter bar will be larger than the flakes in a 2 in. diameter bar on the outside edge, resulting in lower tensile strength. Concern for quality may require the preparation of test bars, but because of this behavior, separately cast test bars are not useful in predicting the strength in castings or in continuous cast bar stock. The test bar must come from an actual casting to be of any engineering value.

Net shape castings are designed to the user's specification and produced in quantities based on setup and processing times. This method allows efficient use of material to produce a structure that is highly efficient based on strength-to-weight criteria.

Round stock is available in as cast and cold finished forms from the continuous casting machines. The cold finished form is from a continuous casting which is machined to a finished dimension, removing undesirable surface effects. They can vary in size from 0.625 to 20.00 in. in diameter. Tube stock is typically only available in cold finished forms with the ID and OD machined to remove effects of cooling. These are known as trepanned forms. Common stock sizes range from 1½ ID × 2½ OD to 7 ID X 9 OD with IDs being more variable as required by the user as long as sufficient wall thickness is maintained.

Rectangular and square stock are generally available in as cast forms in sizes ranging from ¾ × 1½ in. to 14 × 21 in and square sizes range from 1½ to 12¼ in., respectively.

## Processing Considerations

The selection of cast iron is typically driven by the unique material properties and the cost savings afforded by the net shape casting process. If part quantities are sufficiently high to justify the nonrecurring costs of the mold, then significant processing time can be saved by working with the net shape part. Processing of the part most often includes machining with some applications requiring heat treating, but this is not common.

### Thermal Treatments

Cast irons are generally not heat treated because the desired properties are obtained in the as cast condition. When it is required, there are four types, none of which change the size of the graphite in the matrix.

- Stress-relieving (800 to 1200°F) — it is generally not required on as cast rounds, squares, and rectangles. It can be required on net shape castings depending on the complexity of the casting and its mass. No change in the matrix structure occurs nor is there any change in mechanical properties. Hold times of 1 h at a temperature of 1000°F will generally be sufficient to relieve most stresses.
- Annealing — It is a softening process used to improve machinability and is accomplished by controlling the

amount of ferrite in the matrix; the more ferrite, softer the material, resulting in better machining. Only the matrix is changed for the base cast material, leaving the graphite and any alloy carbides basically unchanged. The annealing process is generally classi- fied into low- medium- and high-temperature processes with a corresponding change in soak time as necessary to obtain the desired results. Times are generally defined as time at temperature per inch of cross-section.

- Hardening — A process of heating to 1650°F and hold- ing as required by part thickness, followed immedi- ately by an oil-quench, and tempering at temperatures above 400°F are common. The as cast surface has many stress risers due to its rough surface finish which can result in cracking during the quenching operation. Because of the potential for cracking and ease of machining, it is best to carryout heat treating after the machining operation. Surface hardening is also common and it is typically performed by flame- and induction-hardening.
- Austempering — often applied to ductile iron to produce a material with excellent wear and impact prop- erties. It involves placing the iron in a high-temperature salt quench-tank that assists in the transformation of an austenite microstructure to a ferrite and austenite matrix with excellent mechanical properties.

The best method of evaluating heat treatment results is by test- ing the hardness of the finished part. Brinell tests are preferred for cast iron because it uses a 10 mm ball to create the surface indentation, which reduces the effects of the soft graphite affecting the results. Rockwell tests are acceptable, if used and interpreted correctly, but suffer a distinct disadvantage arising from the use of a pointed indenter or small ball (1.6 mm diame- ter). The average value of several Rockwell tests is preferred in order to avoid the negative effect of graphite on the results. A more conclusive method for evaluating the success of the heat treatment is to prepare a sample for microstructure evaluation and examine the etched structure at 100 × magnification.

The primary effect of heat treating of cast irons is material growth resulting from the decomposition of carbides from their pearlitic structure. The amount of growth depends on the chemistry and microstructure of the material. Since growth occurs as a result of the decomposition of carbide, irons having a higher pearlite content will grow more when heat treated than ferritic grades.

### Strengthening Treatments

The structure and processing of cast irons do not lend themselves to secondary operations such as work-hardening to increase strength. The thermal treatments described above are the most common methods of strengthening.

### Surface Treatments

Surface treatments, such as carburizing or nitriding, that are applied to other ferrous metals are not used in the case of cast irons because their compositions do not allow them to respond to these treatments. Removal of surface material is performed in order to eliminate (or reduce) stress risers in as cast forms, which may cause cracking during quenching, and also to obtain uniform material properties. As discussed above, the differential cooling that results in a casting can produce a brittle surface and tough core that is often not desirable in as cast forms.

## Common Grades

Cast irons are typically available in different graphite shapes and matrix structures. The graphite shapes include flake (ASTM Type VII, A 4–5), fine flake (ASTM Type VII, D 6–8), and Nodules (ASTM Type I & II).

The nonalloyed matrix structure for common cast irons will range from all ferrite to all pearlite. Ferrite is softer and of lower strength when compared to pearlite, with a wide range of properties that can be made available by varying the ferrite-to-pearlite ratio. The matrix structures available include ferrite, ferrite and pearlite mixture, and pearlite, with the ferrite being the easiest to machine and the pearlite the hardest.

Gray iron and ductile iron are the most common grades in use. Partially ferritic to high-pearlitic grades are available. Ductile iron grades that are commonly selected include 65-45-12, 80-55-06, and 100-70-02.

## SELECTION CONSIDERATIONS

Selection of ferrous metals can be a significant challenge for a design engineer because of the wide range of material options available. Selection may not driven by physical properties, such as thermal or magnetic behavior, but rather by mechanical properties and related material characteristics, such as strength, toughness, and hardenability. In addition, designs using these forms of material are typically very cost-sensitive as they are commonly placed into service in consumer products and industrial applications. Because of the last requirement, the selection process is typically a progressive one, beginning with the least expensive material family, carbon steels.

Carbon steels are a family of low cost, easily machinable materials that are available in wide range of compositions that allow variations in strength and toughness. These materials, as with all ferrous metals, offer fatigue-resistant performance. The significant limitation of carbon steels is their poor corrosion-resistance properties, which commonly causes them to require secondary processes such powder coating or oxide treatments for protection from corrosive attack.

Alloy steels should be considered next in applications requiring high strength as they offer that property as well as greater corrosion resistance than carbon steels. If a high level of corrosion-resistance is important, then stainless steels should be considered. Alloy steels are selected based on strength, hardness, and thickness of the section to be hardened.

Stainless steels are selected, first, for their corrosion-resistance and, second, for their strength. These materials generally do not require a secondary process as they have a naturally passivating mechanism that protects them from most forms of corrosion. They are harder to machine and form but easier to weld than carbon steels.

Tool steels are typically used to cut, shape, and form other materials rather than for placement in the end product. The demand on their performance is unique and often greater than most other ferrous metal applications, requiring a highly refined material with consistent properties. The chosen composition and the subsequent heat treatment they require depend on the type of tool work required. The resulting environments require material properties that provide a range of performance, including wear-resistance and toughness as a function of hardness and red hardness, or the ability to retain material properties at elevated temperatures. Wear-resistance implies a hard material that is not tough, requiring a compromise between the two for any given application. The ability to attain the required level of hardness is a function of the heat treatment method: air-, oil-, or water-hardened. Air-hardened grades are selected when the tool requires extreme accuracy and safety or if deep hardening is required. Water-hardened grades are selected when size change and warpage do not matter, as they provide good wear-resistance and toughness. Oil-hardened grades are for those applications requiring close tolerances of size and shape and for those forms which may crack when quenched in water. Those of the red hard category are subjected to heating above 300°F during service while requiring varying degrees for wear-resistance and toughness. These are selected based on the desired elevated temperatures material properties.

As the name implies, cast irons are primarily produced in net shape castings that require machining and finishing operations to make them complete. Gray iron is most common because of its economical processing costs while also providing desirable damping characteristics. Cast irons are best used in compressive applications because the nature of the material structure is to crack when under tension but not compression. This asymmetric loading property can be an advantage when employed properly as seen in their use in machine tool frames and other forms of structures intended to support large compressive loads. Selection of cast irons are driven by the their functional requirements, net shape factors and material properties.

## Performance Objectives

When selecting a ferrous material a short list of candidates can be obtained by establishing the important objectives of the selection based on the design function and related constraints. The most common, primary performance requirements of any ferrous material are corrosion-resistance and strength. For example, if the material must not corrode, plain carbon steels can be immediately eliminated, because they do so unless post-processing is employed to prevent it. Similarly, if a structural member is required to support a specific load it must be sufficiently strong not only to support the load but also to provide a reasonable safety factor.

Secondary performance requirements are those that enhance product performance, influence product life, involve processing methods, or influence product cost. These include physical properties, such as thermal and electrical characteristics; mechanical properties, such as hardness, stiffness and toughness; and processing properties, such as heat treatability and weldability. The specific requirements of each design will determine which of these has priority in the material selection process and will undoubtedly influence the product success.

Carbon steels are most often selected for the high performance-to-cost ratio, ease of processing and because they are readily available. Alloy steels are commonly selected for their high strength and heat treatability. Stainless steels are selected, primarily, for their corrosion-resistance and also for their appearance, and lastly, for their mechanical properties. Concern over their stress corrosion cracking (SCC) also limits their use.

Tool steels are selected for their high hardness and wear properties as well as high strength properties. Material grades are available for a wide range of operating temperatures including room temperature cold working to red-hot temperatures.

Cast irons are commonly selected for their combination of low cost, good compressive strength and high damping characteristics. The ratio of ferrite (soft form of steel) and pearlite (hard form of steel) has a significant effect on the resulting properties.

Consideration of secondary selection criteria will also be required to make successful material selections. For example, a comparison of thermal conductivities of ferrous metals and with those of nonferrous metals provides insight into the available ranges of this property (Figure 4.23). It should be noted that even with the same material family there can be significant variations of thermal conductivity as stainless steels are considerably less conductive than carbon steels.

Conversely, electrical resistivity has a behavior which is the inverse of that of thermal conductivity (Figure 4.24). In reviewing this type of property data in a graphical format it becomes readily apparent as to why aluminum is chosen as a conductor over other ferrous metals.

Another important secondary property of ferrous metals is their thermal expansion characteristics (Figure 4.25). Ferrous metals generally have the lowest thermal expansion of all metals making them a stable material for those applications requiring high precision over a wide temperature range.

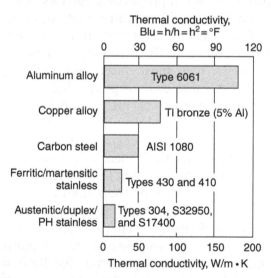

**Figure 4.23** Comparison of thermal conductivity for carbon steel, copper alloy, aluminum, and stainless steels. From ASM Online.

*Source*: ASM International: Metals Handbook Desk Edition, Wrought Stainless Steels: Selection and Application, Physical and Mechanical Properties of Stainless Steels, Figure 3.

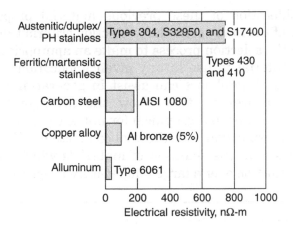

**Figure 4.24** Comparison of electrical resistivity for carbon steel, copper alloy, aluminum, and stainless steels. From ASM Online.

*Source*: ASM International: Metals Handbook Desk Edition, Wrought Stainless Steels: Selection and Application, Physical and Mechanical Properties of Stainless Steels, Figure 5.

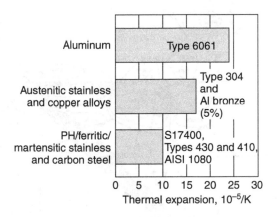

**Figure 4.25** Comparison of thermal expansion for carbon steel, copper alloy, aluminum, and stainless steels. From ASM Online.

*Source*: ASM International: Metals Handbook Desk Edition, Wrought Stainless Steels: Selection and Application, Physical and Mechanical Properties of Stainless Steels, Figure 4.

Consideration of these previous material properties are only a few of the many secondary selection criteria that must be applied to the selection process to make an appropriate decision on what material to use in any given application. Computer-based material databases can assist in generating these comparative measures as they allow the arrangement of the material properties in a graphical format to compare, for example, thermal expansion with thermal conductivity. The resulting plot can assist in the selection of materials which exhibit low thermal distortion over a large temperature range (Figure 4.26).

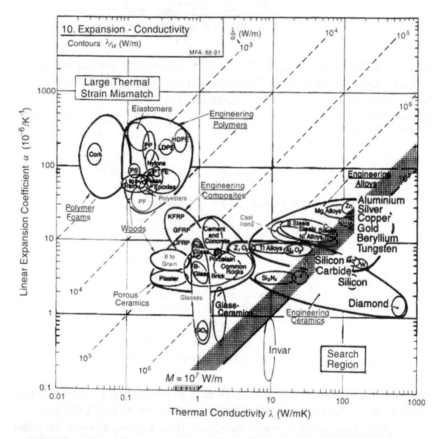

**Figure 4.26** Comparison of thermal expansion for carbon steel, copper alloy, aluminum, and stainless steels. From ASM Online.

*Source*: Reprinted with permission from Materials Selection in Mechanical Design, Michael F. Ashby, p. 153, Fig. 6.40, 1999, with permission from Elsevier.

## Selection Methodologies

Selecting a ferrous metal can be difficult considering the number of materials that are available. At the risk of oversimplifying the process, the following figure provides a selection flow process to help guide one quickly toward the most appropriate material family (Figure 4.27). The selection is most often based on the required corrosion-resistance and tensile strength.

Once these two requirements are determined, secondary performance objectives can then be considered and applied to the selection process. Within the subset of a particular material

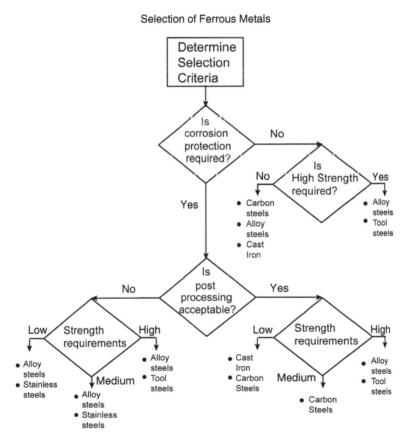

**Figure 4.27**   Ferrous metals selection flow chart.

family, further refinement will be required to satisfy the performance objectives. At this point in the process, it is not unusual for materials to be selected based on cost and availability rather than purely on their performance merits, requiring careful consideration of the necessary compromises. Use of material indices and an electronic database can assist in this process.

# 5

## Nonferrous Metals

There are many useful metals that do not contain iron as their primary element, which can be used for structural as well as nonstructural purposes. These nonferrous metals are used in many everyday applications, such as beverage cans, compact disks (which have an aluminum layer), copper nickel coins, and jewelry made of gold, platinum, and silver. The corrosion behavior of copper limits its practical alloying level in gold, requiring the gold always be greater than 14 karat (i.e., the copper content must be less than 31.2%) because if it is less, the Cu content tends to dominate, resulting in a green oxide finish.

Fatigue properties of nonferrous metals significantly differ from those of ferrous metals as they have no fatigue limit as typically represented by an $S–N$ curve. An exception to this rule is titanium, which exhibits steel-like fatigue resistance, making it an exceptional metal of high specific strength and stiffness. The corrosion-resistance of nonferrous metals is also a desirable attribute, supporting their selection. Unlike ferrous metals, resistance to corrosion is a generic property for materials such as copper or titanium, while others require some form of surface modification to prevent corrosion. But, even when

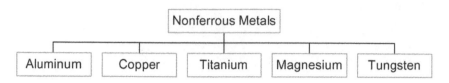

**Figure 5.1** Common nonferrous metals for use in engineering applications.

left unprotected, the corrosion rates are generally slower than those of ferrous metals. The actual corrosion rates are highly dependent on the application environment.

The processing of nonferrous metals can also be quite different from that of ferrous metals, making understanding their properties a greater challenge and not a direct extension of the procedures commonly used for ferrous metals. For example, the hardness of copper is commonly qualified as one-fourth hard, one-half hard, or full hard (indicating the extent of cold work) rather than being quantified in comparable terms on a known scale. Similarly, when quenching a nonferrous metal, it becomes soft rather than hard as with ferrous metals.

The structure of this chapter is similar to others in this book and includes detailed reviews of the primary nonferrous materials. A detailed review of each material and its specific numbering system is provided, to help orientate and focus the selection process as well as delineate the specifics of the material family from similar nonferrous and ferrous metals. Gaining an understanding of the crossover considerations is critical to the proper selection of these familiar metals (Figure 5.1).

## CLASSIFICATIONS

### Aluminum Alloys

Aluminum is the second most plentiful metallic element on earth and became an important metal in industry in the late 19th century. Its density and modulus of elasticity are a third of carbon steel; it has a comparable strength while retaining good ductility at subzero temperatures. Aluminum is also highly conductive, both electrically and thermally, making it a material that has many uses in the design of both electrical and mechanical systems. The most significant limitation of aluminum in the design of structures is its lack of fatigue strength (Figure 5.2). Nonferrous metals do not have an infinite fatigue life threshold, as exhibited by ferrous metals, but rather, one that declines steadily throughout the life of the material, regardless of the magnitude of loading.

The corrosion-resistance of untreated aluminum is poor. If left exposed to ambient conditions for an extended period of time, a white oxide powder will begin to form on its surface, which is a result of the surface corrosion. The aluminum oxide forms at a rate proportional to the severity of the environment. Because of this natural tendency to corrode (unlike stainless steels, which have a naturally passive surface) aluminums require some form of surface preparation to preserve the base material.

Aluminum alloys and some other nonferrous metals are unique in that they are nonsparking and nonmagnetic, making them ideal for hazardous environments involving explosive gases and solvents. The lack of iron content is responsible for this behavior.

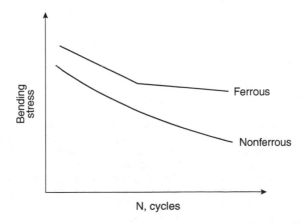

**Figure 5.2** Fatigue limit comparison of ferrous and nonferrous metals.

One of the most desirable characteristics of aluminum is its versatility in processing. It can be machined, cast, rolled, stamped, drawn, stretched, hammered, forged, or extruded into a wide variety of shapes, making it an important part of transportation, aviation, power supply, packaging, and other such industries.

Aluminum alloys can be joined in a wide variety of methods, including fusion and resistance welding, brazing, soldering, adhesive bonding, and mechanical methods. The choice of method will depend upon material thickness, type of joint, quality requirements, and cost. Tungsten electrode methods of welding are commonly used, but regardless of the method the above-mentioned characteristics of aluminum alloys should be kept in mind in order to obtain a clean joint.

### Properties and Designations

Aluminum alloys are divided into two major categories: casting alloys and wrought alloys, with further classification of each category on the basis of the process used to develop a particular property (Figure 5.3). The system was developed by the

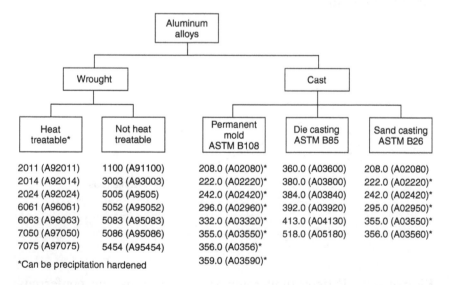

**Figure 5.3**   Aluminum alloys and their common uses.

Aluminum Association and is the most widely used in the United States.

## Numbering

The wrought alloys use a four digit system where the first number indicates the major alloying element, though some other elements may be present in small quantities. The following scheme is for wrought alloys:

- 1XXX: pure aluminum (greater than 99%)
- 2XXX: copper
- 3XXX: manganese
- 4XXX: silicon
- 5XXX: magnesium
- 6XXX: zinc
- 7XXX: other elements (tin, lithium etc.)
- 9XXX: unused series (reserved for future use)

The cast alloys use a three digit system followed by a decimal value. The decimal 0 pertains to casting alloy limits. Decimal 1 and 2 concern ingot compositions, which, after melting and processing, should result in chemistries conforming to casting specification limits. The following scheme is for cast alloys:

- 1XX.X: pure aluminum (greater than 99%)
- 2XX.X: copper
- 3XX.X: silicon added with copper/magnesium
- 4XX.X: silicon
- 5XX.X: magnesium
- 6XX.X: unused series (reserved for future use)
- 7XX.X: zinc added with copper/magnesium
- 8XX.X: tin
- 9XX.X: unused series

As mentioned above, each series can be further divided into subcategories on the basis of the process used. The following symbols are used by the Aluminum Association for describing fabrication stages (Table 5.1).

The heat treated stage has numerous subdivisions that define specific material conditions (Table 5.2). No heat

**Table 5.1**   Fabrication Stage Designation by the Aluminum Association

| Symbol | Description |
|---|---|
| F | Rough stage of fabrication. The symbol is applied to work-hardened products without control of cold working rate and heat treatments. Hence, mechanical properties cannot be defined in these products |
| O | Annealed stage. Only valid in work-hardened products, it is the most ductile stage with low strength |
| H | Work-hardened stage. Represents those alloys whose strength is increased by cold working. It is subdivided into H1, H2, and H3 |
| W | Tempered nonstabilized stage. An unstable metallurgical stage for alloys which are tempered and aged after solution |
| T | Heat treated stage. Represents heat treated products. It is subdivided into T1, T3–T12. Each one represents a particular combination of heat treatment |

**Table 5.2**   Aluminum Heat Treated Subdivision

Subdivisions of T Temperature

| | |
|---|---|
| T1 | Cooled from an elevated temperature shaping process and naturally aged |
| T2 | Annealed |
| T3 | Solution heat treated and cold worked |
| T4 | Solution head treated and naturally aged |
| T5 | Cooled from an elevated temperature shaping process and artificially aged |
| T6 | Solution heat treated and artificially aged |
| T7 | Solution heat treated and stabilized |
| T8 | Solution heat treated, cold worked, and artificially aged |
| T9 | Solution heat treated, artificially aged, and cold worked |
| T10 | Cooled from and elevated temprature shaping process, artificially aged, and cold worked |

Additional digits are used to designate stress relieving:
| | |
|---|---|
| T51 | Stress relieving by stretching |
| T52 | Stress relieving by compressing |
| | T510 designates products that receive no further straightening after stretching, and T511 designates products that receive minor straightening in order to comply with standard tolerances |

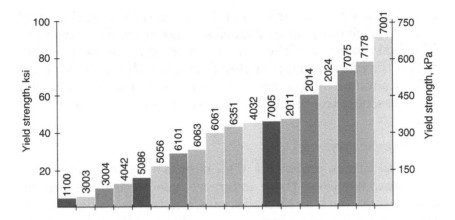

**Figure 5.4** Approximate relative yield strengths for various aluminum alloys

treatment is designated by "−T0" indicating the fully annealed stage. Structural applications never use a material in a −T0 condition, but it may be used in a decorative or some other nonstructural application.

### Strength Properties

Wrought alloys are characterized by their ability to withstand severe physical deformation when worked hot or cold. Cold working increases the strength of all wrought alloys. However, with heat treatable alloys, improved results are obtained either by heat treatment or by combination of heat treatment and cold working. The strongest aluminum alloys are among the most heat treatable. Figure 5.4 shows a comparison of the physical properties of a range of heat treated aluminum alloys.

### Heat Treatability

The property changes accomplished by heat treating aluminum alloys are a result of the solution and precipitation of hardening phases, as determined by solid solubilities. Generally, the solubilities of alloying elements in aluminum are much greater at elevated temperatures. Therefore, by heating and quenching rapidly, such elements become super-saturated in solution, much higher than in the equilibrium conditions, and usually precipitate out as particles of intermetallic compounds. The final properties of the alloy depend on the size and distribution of the precipitated particles. Since pure aluminum and certain aluminum alloys are noted for their low strength and hardness, heat treatment is essential for most of the aluminum alloys.

### Fatigue Properties

The fatigue properties of aluminum alloys are not among the best, however, their extensive use in fatigue critical applications, such as aircrafts, cars, and trucks, indicates that proper selection of an appropriate alloy and the joining techniques can considerably increase fatigue life. Fatigue is an important consideration in weldments and joints as they are critical for stress concentrations. It has been proven that a clean joint, which is properly machined and ground, always has a greater fatigue life. Techniques that use remelting like TIG are also helpful.

### Corrosion

Aluminum and aluminum alloys rapidly develop a surface layer of aluminum oxide, a form of corrosion, when exposed to ambient conditions. Once this layer is in place, they have a strong resistance to corrosion because this corrosive skin protects aluminum from most chemicals, weathering conditions, and even many acids. There are exceptions, such as alkaline substances, which are known to penetrate the protective skin and corrode the metal. Regardless of the application, aluminum needs a protective coating to prevent a long-term building up of the corrosive layer and this is typically achieved through anodizing or the deposition of a chemical film, both of which improve the alloy's appearance.

## Fabrication Processes

### Machinability

Most aluminum alloys have excellent machinability. The ease of machining varies depending on the type of heat treatment rendered. The hardened condition results in the lowest

**Table 5.3**  Machinability ratings for
various aluminum alloys and corresponding heat treated conditions

| Material | Machinability |
| --- | --- |
| 2011-T3 | 1 |
| 4032-T6 | 2 |
| 2014-T6 | 2 |
| 2024-T6 | 2 |
| 7075-T6 | 2 |
| 7178-T6 | 2 |
| 7001-T6 | 2 |
| 6101-T6 | 3 |
| 6063-T6 | 3 |
| 6061-T6 | 3 |
| 1100-H14 | 4 |
| 3004-H32 | 4 |
| 5086-H32 | 4 |
| 3003-H12 | 5 |

machinability while the annealed and tempered conditions
result in good machinability. The micro-constituents present
in aluminum alloys have a great effect on machining because
the abrasive intermetallic particles will make machining
more difficult and are most often associated with heat treatment. On the other hand, constituents that are soft and
nonabrasive are beneficial, as they assist in chip breaking.
These constituents can be intentionally added to produce
high-strength free-cutting alloys.

ASM International rates aluminum machinablity in the
1–2 range on a scale of 1–5 with 1 being the best and 5 the
worst. The 2XXX series is the best among wrought alloys for
machining purposes, when in the tempered condition, because
of its high copper content. For all others, a suitable heat
treated condition and optimal tool design and machine settings are required for satisfactory control over chip and finish
(Table 5.3). High-speed steel tools are used for machining all
alloys but for high-silicon aluminum alloys such as 4032,
carbide tools must be used for best results.

Chemical milling is the preferred method of removing less than ⅛ in. of aluminum from large intricate surfaces. Alkaline solutions with a sodium hydroxide base are popular etching solutions for chemical milling of aluminum. The baths are operated at elevated temperatures, while requiring strict control of the formulations for best results. The complicated nature of this process makes it an expensive method for material removal and one that is generally only cost-effective on parts requiring high strength-to-weight ratios such as aircraft components.

### Weldability

Pure aluminum melts at 1220°F while aluminum alloys melt across a range from 900 to 1220°F, depending upon their alloy content. There is no color change in aluminum when it is heated to welding or brazing temperatures, which makes it difficult to judge when the metal is near its melting point. Its high thermal conductivity, when compared to steel, necessitates high heat input for fusion welding and preheating of heavy sections to assist in the welding process. The high electrical conductivity of aluminum requires higher currents and shorter weld time in resistance welding, as compared to steel, as well as precise control of welding variables.

Aluminum and its alloys rapidly form an oxide coating when exposed to air. This coating must be broken up, during welding or brazing, to obtain a quality weld that is free of contaminants. The oxide film may be removed by fluxes, by action of a welding arc in an inert gas atmosphere or by mechanical or chemical means. It is more important with aluminums than with ferrous metals that the surface be free of contaminants, because they generally will not float to the surface of the weld pool, but rather, sink into the molten weld. When welding ferrous metals, the contaminants and flux float to the surface, leaving a quality weld free of contaminants.

### Forgeability

Aluminum alloys can be forged into a wide variety of shapes and types of forgings. Aluminum alloy forgings, particularly closed die forgings, are usually produced to more highly refined final forging configurations than hot forged carbon and alloy steels. The forging pressure for a given shape varies widely and will depend upon the composition of alloy, the forging process, the strain rate, lubrication conditions, and die temperatures.

Although, as a class, aluminum alloys are generally considered to be more difficult to forge than carbon steels and some alloy steels. The ability to forge aluminum alloys benefits from conventional forging methods in which dies are preheated to 1000°F. Because of their forging properties and ease of obtaining a good surface finish aluminum forgings are very popular for making intricate parts and shapes.

## Raw and Stock Finishes

Aluminum and its alloys can be cast or formed by using virtually all known processes. The standardized products available include sheet, plate, foil, rod, bar, wire, tube, pipe, and structural forms. Some forgings, extruded shapes, impact castings, stampings, powder metallurgy parts, and metal matrix composites are also available for specific applications. The sheet plate and foil constitute about 50% of the available products, while ingots make up nearly 20% and extruded products form approximately 17% of what is available. The remaining 13% consists of rod, bar, wire, forgings, and powders.

## Processing Considerations

There are many processing methods to improve the mechanical and corrosion properties of aluminums. Solution heat treatment and precipitation-hardening are the most common methods of improving the mechanical properties, while surface treatments are required to protect aluminums as they corrode in ambient conditions.

### Thermal Treatments

The most common heat treatment applied to aluminum alloys is solution treating, which takes advantage of the precipitation-hardening reaction (Figure 5.5). It involves soaking the alloy at a temperature ranging from 240 to 375°F for 5 to 48 h to achieve a nearly homogeneous solid solution, and then

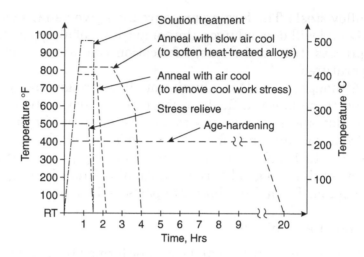

**Figure 5.5**   Thermal treatments for aluminum alloys.

quenching it fast enough to retain the solute in solution. The optimum hardening cycle will vary from alloy to alloy. Annealing is done for stress-relieving or whenever a softer product is desired.

### Hardening

Hardening is mostly done by precipitation-hardening, as discussed above. However, cast alloys will behave differently during hardening when compared to wrought alloys because chills, gates, and mold type affect different areas of the casting resulting in different behaviors during hardening. Because of these concerns, castings are typically not heat treated.

### Surface Treatments

The most common surface treatments for aluminum include anodizing and chemical film coating. As a general rule, aluminum must be coated to prevent it from corroding, which can be detrimental to both its structural strength and appearance. Although design requirements may include other coatings, these are generally applied as a base coating to provide the necessary adhesion for other coatings, such as powder coatings and painting. Wrought forms respond best to this process, while castings or die cast parts may not, depending on the Si or Cu content. Generally, Si >5% will turn the casting black and will not produce a uniform coating because of the variable distribution of Si within the casting, while >3% copper will also have this effect. Si-rich alloys and castings include 319, 333, and 328, and Cu-rich alloys include 2011, 2017, and 2024. Anodizing of castings should be thoroughly reviewed with the experts at the anodizing shop and tested to obtain the most appropriate finish for the material selected.

Anodizing produces an aluminum oxide surface finish that is integral to the material and nonconductive. It forms by the conversion of the surface aluminum to a ceramic coating that gradually gives way to the base aluminum as depth increases, resulting in no distinct boundary between the two materials. The surface finish is porous and requires a secondary operation when a smooth, high-quality finish is required for the application. It is not uncommon to grind away much of the buildup to improve the finish.

There are three principal types of anodizing: chromic (chromic acid), sulfuric (sulfuric acid), and hard coat process (modified sulfuric acid process). The chromic and sulfuric acid processes are performed at room temperature and result in typical buildup of 0.0002–0.0007 in. from the base material with greater penetration. Hard coating is performed at lower temperatures (0–10°C) and higher current densities to form a buildup of 0.002–0.003 in. and similar penetration (Figure 5.6). Because anodizing is a self-limiting process, greater buildup is not possible without changing to the standard process.

The effect of the anodized buildup needs to be considered when designing with anodized aluminum. Primary considerations include the following:

- Masking threaded holes
- Blind holes
- Sharp corners
- Surface roughness and porosity

Typical chromic or sulfuric anodizing does not result in sufficient buildup on threaded features to require masking during anodizing. Application of hard coat is not suggested on

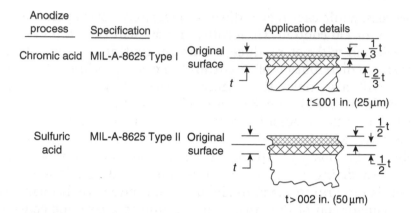

**Figure 5.6** Anodize coating thickness. Thinner coatings are produced by chromic acid processing and result in greater surface penetration than growth while hard anodizing is produced by sulfuric acid processing and produces a uniform growth and penetration.

threaded features because the buildup will result in interference fits, so they must be masked.

Blind holes present a challenge to this process and should be avoided when possible. Although the process involves an agitated bath that promotes constant circulation in the tank, it cannot overcome the stagnation effects of a blind hole, and results in limited surface conversion. Blind holes also make flushing of the part difficult, leaving tank residues in the hole, which prevents proper sealing of the aluminum surface.

Use of sharp corners on anodized parts results in thinner coatings and brittle corners as there is insufficient material in the region to form the necessary surface buildup or penetration. What material remains is a brittle aluminum oxide that can be easily fractured, exposing the base aluminum for corrosion. The best method to prevent this is to use generous radii on all corners (maximum allowable for the design) to allow uniform buildup and penetration.

The nature of an anodized surface is that of many peaks and valleys, with magnitudes directly related to the type of anodizing process. Basic sulfuric and chromic acid processes do not produce extreme surface roughness greater than the pre-anodized finish while hard coating produces the roughest surface of the three methods and can increase roughness two or three times the pre-anodized finish.

| Suitable for hardcoating | Less suitable |
|---|---|
| 5052 | 2024 |
| 6061 | 355 |
| 1100 | |
| 3003 | |

Consideration of the end use also plays a part in the desired processing. Decorative coatings are those that are less than 0.0004 in. thick and are applied for use indoors or in nonreactive environments. Similar coatings that must withstand outdoor exposure are applied in thicknesses ranging from 0.0004 to 0.0007 in. to provide environmental resistance. Coatings of greater thickness are applied using the hard coating process and range from 0.0007 to 0.002 in. and are suited for extreme environmental resistance or mechanical wear applications. The latter, typically, include pneumatic or hydraulic sealing surfaces or carefully designed applications for rolling or sliding wear that require post-finishing of the surface by grinding. The grinding operation removes much of the buildup, to produce a uniform surface, while retaining the more densely packed base material that has less porosity. Surface finishes of less than 10 $\mu$ in. are possible.

The surface is also porous and must be sealed after the conversion process. This is commonly done using boiling water or slightly acidic dichromate solution. Modifications to this sealing process include the addition of teflon powder, which tends to fill the surface voids and produce a finish that has greater corrosion-resistance and improved lubricity. The use of dyes is also common to obtain the required surface color without the need for an additional painting or powder coating process.

Aluminum can also be finished using chemical conversion methods that produce a surface film by reacting with the base metal. They differ from anodized coatings in that they are formed by chemical reaction with chromate, phosphate, or oxide compounds rather than with an electrochemical reaction. Prior to applying either of these methods it is not uncommon to include a chemical etching or brightening process to improve the surface finish. This is done using an acid bath to remove surface scales or small amounts of surface roughness. Chemical conversion coating results in providing an electrically conductive surface finish, which protects the base material from corrosion while improving surface adhesion for additional processing, such as painting or powder coating.

Electroplating is mainly done to enhance the alloy's appearance. Bright chromium is a common form of plating that alters the color of the surface to an attractive blue and does not readily tarnish. Silver is also plated to improve electrical contacts. Tin and copper plating provide surfaces to which joints may be soldered by conventional soldering methods.

A substantial volume of aluminum is painted for both decorative and protective purposes. The most important applications, in such cases, are in residential siding and mobile homes, others are in cans, electronic enclosures, highway trailers, canopies, and sports goods. The surface is first cleaned and anodized to improve adhesion, and then painted. Usually, this all is integrated in one continuous process.

## Commonly Used Aluminums

### 1100

This is a commercially pure aluminum that is not heat-treatable. It is soft and ductile and has excellent workability. It is ideal for applications involving intricate forming because it work-hardens more slowly than other alloys. It has the best weldability of almost all aluminum alloys. It has excellent corrosion-resistance and is widely used in the chemical and food processing industries. It is also used in giftware as it responds well to surface modifications.

### 2011

This is generally considered the free-machining grade with excellent mechanical properties. This alloy includes bismuth and tin as the free-machining elemental constituents that function as discontinuities in the aluminum alloy matrix, rather than low-melting point compounds. It is widely used for screw machine products, automotive parts, hardware, fasteners, and any parts that require extensive machining.

### 2519

This has been developed for improved resistance to ballistic penetration and lack of susceptibility to SCC. It is a modified version of 2219, with reduced Cu content and magnesium increased in small amounts. This material is used in the Advanced Amphibious Assault Vehicle (AAAV).

### 3003

This is, essentially, commercially pure aluminum with the addition of manganese, which increases the tensile strength some 20% over the 1100 grade. 3003 has similar characteristics as 1100 as well as greater yield strength. It has excellent corrosion-resistance and workability, and may be deep drawn or spun, welded or brazed. It is not heat-treatable. Its applications range from cooking utensils, decorative trim, awnings, siding, and storage tanks to chemical equipment.

### 5052

This alloy has the highest strength among the more common grades that are not heat-treatable. Its fatigue strength is higher than most aluminum alloys. In addition, this grade has particularly good resistance to marine atmosphere and saltwater-corrosion. It has excellent workability. It is used in a wide variety of applications: from aircraft components to home appliances; marine and transportation industry parts; heavy duty cooking utensils; and equipment for bulk processing of food.

### 6061

This is the most versatile of the heat-treatable aluminum alloys. It offers a range of good mechanical properties and good corrosion-resistance. It can be fabricated by most of the commonly used techniques and has good workability in the annealed condition. Fairly severe forming operations may be accomplished in the T4 condition. The full T6 properties may be obtained by artificial aging. It has good weldability. This grade is used for a wide variety of products and applications ranging from truck bodies and frames to screw machine parts and structural components.

### 7075

This is a high strength grade and is most often used for aircraft structures and parts requiring a high strength. It is also available in aclad, which combines the strength properties with the coating of commercially pure aluminum for corrosion-resistance. It can also be used in machine parts and ordnance.

## Copper and Copper Alloys

Copper and copper alloys constitute one of the most important groups of commercial metals. Their excellent thermal and electrical conductivities, corrosion-resistance, bearing qualities, good strength and fatigue resistance, and ease of fabrication make them a popular choice in a wide variety of applications. Furthermore, they are easily joined by soldering and brazing, and are available in a variety of textures and colors for decorative purposes. The two most notable families of copper alloys are brass and bronze. In addition, pure copper and beryllium copper are important families (Figure 5.7).

Wrought alloys of copper are produced in a variety of different methods, such as annealed, cold worked, hardened by heat treatments, or stress-relieved. There are four main families of wrought copper: copper, brasses, bronzes, and beryllium

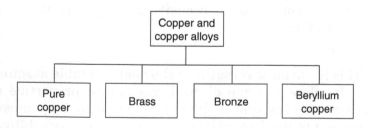

**Figure 5.7**  Copper and copper alloys.

copper. Table 5.4 gives the UNS designations of copper alloys along with their generic names.

Commercially pure copper is used extensively in electric supplies and related products, while some brasses, bronzes, and cupro-nickels are used in applications such as radiators, heat exchangers, and heat absorbers and also in pipes, valves, and other equipment for carrying liquids.

Pure copper has a reddish gold color, which quickly oxides to a dull green. Copper alloys can exhibit a wide range of color because of the influence of alloying on the composition. Since copper often contains natural impurities, or is alloyed with more than one element, it is difficult to say what effect each alloying element has on the color. Electrolytic tough pitch copper contains silver and often trace amounts of iron and sulfur. It has a soft pink color. Gilding copper has a reddish brown color and contains zinc, iron, and lead. Brass is often used as an ornamental metal, since it has an appearance very similar to that of gold and is much less expensive. Brasses contain varying amounts of zinc, iron, and lead and can vary from reddish to greenish to brownish gold. Nickel silver, which

**Table 5.4**   UNS Designations for Copper with Generic Names

| UNS numbers | Types | Generic names |
| --- | --- | --- |
| C10000–C19999: | Wrought | Coppers, high-copper alloys |
| C20000–C49999: | Wrought | Brasses |
| C50000–C59999: | Wrought | Phosphor bronzes |
| C60600–C64200: | Wrought | Aluminum bronzes |
| C64700–C66100: | Wrought | Silicon bronzes |
| C66400–C69800: | Wrought | Brasses |
| C70000–C79999: | Wrought | Copper nickels, nickel silvers |
| C80000–C82800: | Cast | Coppers, high-copper alloys |
| C83300–C85800: | Cast | Brasses |
| C86100–C86800: | Cast | Manganese bronzes |
| C87200–C87900: | Cast | Silicon bronzes and brasses |
| C90200–C94800: | Cast | Tin bronzes |
| C95200–C95800: | Cast | Aluminum bronzes |
| C96200–C97800: | Cast | Copper nickels, nickel silvers |
| C98200–C98800: | Cast | Leaded copper |
| C99300–C99750: | Cast | Special alloys |

contains nickel, zinc, iron, lead, and manganese, can have a grayish white to silvery appearance.

Cast copper alloys generally have a greater range of alloying elements than wrought alloys because of the nature of the casting process. In all classes of copper alloys, certain alloy compositions for wrought products have counterparts among the cast alloys, enabling the designer to make an initial alloy selection before deciding on the manufacturing process. This approach affords the designer a wider range of selection for the cast process than the wrought processes. Cast copper alloys provide greater corrosion-resistance than wrought grades as well as increased thermal and electrical resistivities.

Cast brasses consist of copper–zinc–tin alloys (red, semi-red, and yellow) while cast bronzes consist of manganese bronzes (high-strength yellow brasses), leaded manganese bronze alloys (leaded high-strength yellow brasses), and copper–zinc–silicon alloys (silicon brasses and bronzes). These are low-cost, low-to-moderate strength, general purpose casting alloys with good machinability and corrosion-resistance. Typical applications include marine hardware and automotive cooling systems made from die casting or permanent molds.

There are four main families in the cast bronze alloys: copper–tin, copper–tin–lead, copper–tin–nickel, and copper–aluminum alloys. The composition of these four families are defined as follows:

- *Copper–Tin*: Alloys with 1 to 2% tin which provides improved corrosion-resistance while retaining strength. Uses include electrical contacts, connectors, terminals, and fuse clips (ref. C40000).
- *Copper–Tin–Lead*: The addition of lead improves the machinability without significantly compromising the basic properties of copper–tin alloys (ref. C30000).
- *Copper–Tin–Nickel*: The addition of nickel improves the strength of the copper–tin alloy as well as the corrosion-resistance. Their formability is also good (ref. C70000).
- *Copper–Aluminum*: These alloys contain between 2 and 3% aluminum for added strength while still maintaining

formability. Uses include bushings, wear plates, and hydraulic valve parts (ref. C60000).

## Properties and Designations

Copper is one of the oldest materials, having been employed for centuries because of its abundance and easy formability. Its electrical and thermal properties are integral in the design and functioning of most technical products developed today. Copper's high resistance to corrosion is also an important factor in its selection. Wrought alloys are commonly available in various cold worked conditions (cold worked tempers) with strength and fatigue resistance a function of this condition.

Copper and copper alloys were originally identified by a three digit numbering system as defined by the Copper Development Association (CDA) and ranged from a designation of 100 to 700. This numbering system was altered to a five digit system with a prefix "C" and made part of the ASTM and SAE Unified Numbering System (UNS); the designation ranges from C1000 to C7000 for wrought alloys and C8000 to C9000 for cast alloys.

### Numbering

Copper alloys are commonly classified using six families namely: coppers, dilute copper alloys, brasses, bronzes, copper nickels, and nickel silvers. The UNS system provides a five digit number with a prefix C and is subdivided into cast and wrought alloys.

### Strength Properties

The tensile strength of copper alloys can vary widely depending on the specific alloy and its material condition (Figure 5.8). The most common form of strengthening these alloys is cold working to a specific hardness or temper, as discussed in the section Heat Treatability.

Solid solution strengthening of copper is also a common strengthening method. Copper by itself is relatively soft compared with other metals, as are tin and zinc. Tin is more effective in strengthening copper than zinc, but is also more expensive and has a greater detrimental effect on the electrical

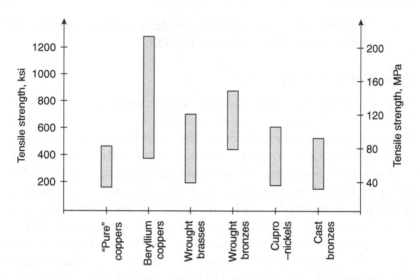

**Figure 5.8**   Tensile strength range for selected copper alloys.

and thermal conductivities than zinc. Aluminum (known as aluminum bronze), manganese, nickel, and silicon can also be added to copper to improve the strength.

Another method of strengthening copper is precipitation-hardening. This process is often used for copper alloys containing beryllium, chromium, nickel, or zirconium. Precipitation-hardening offers distinct advantages as it allows fabrication using a relatively soft solution-annealed form of the quenched metal followed by hardening. The aging process can be performed using relatively inexpensive and unsophisticated furnaces. Often, the heat treatment can be performed in air, at moderate furnace temperatures and with little or no controlled cooling. Many combinations of conductivity, ductility, impact-resistance, hardness, and strength can be obtained by varying the treatment times and temperature. Table 5.5 provides mechanical properties of various wrought and cast copper alloys.

### Heat Treatability

Copper-based materials are generally not heat treated to improve strength or hardness, but, instead, are work-hardened. Annealing is done to return work-hardened material to its soft

**Table 5.5** Mechanical Properties of Selected Wrought and Cast Copper Alloys

| UNS number | Type | Temper | Tensile strength (MPa) | Yield strength (MPa) | Shear strength (MPa) | Elongation (%) |
|---|---|---|---|---|---|---|
| C10100 | Wrought | H04 | 345 | 310 | 195 | 12 |
| C10300 | Wrought | H04 | 345 | 310 | 170 | 12 |
| C11000 | Wrought | M20 | 220 | 69 | 150 | 50 |
| C11600 | Wrought | H01 | 260 | 205 | 170 | 35 |
| C14300 | Wrought | H04 | 310 | 275 | — | 14 |
| C16200 | Wrought | OS025 | 250 | 83 | — | 57 |
| C17500 | Wrought | H04 | 450 | 380 | — | 15 |
| C19400 | Wrought | H02 | 400 | 315 | — | 18 |
| C21000 | Wrought | OS035 | 240 | 76 | 195 | 45 |
| C33500 | Wrought | H04 | 510 | 415 | 295 | 8 |
| C40800 | Wrought | H08 | 545 | 455 | 330 | 3 |
| C51000 | Wrought | H06 | 635 | 550 | — | 6 |
| C61400 | Wrought | O60 | 565 | 310 | 310 | 40 |
| C74500 | Wrought | H06 | 655 | 525 | 405 | 3 |
| C81800 | Cast | As cast | 345 | 140 | — | 20 |
| C82200 | Cast | As cast | 345 | 170 | — | 20 |
| C82500 | Cast | As cast | 515 | 275 | — | 15 |
| C82800 | Cast | As cast | 550 | 345 | — | 10 |

state. Severely deformed metal can be annealed to a more finer and uniform grain size than lightly deformed metal. Grain size control can be done by both cold working and annealing. In addition to grain size control heat treatment is done for homogenization, stress-relieving, solutionizing, precipitation-hardening, quench-hardening, and tempering. Temper designations for copper alloys are described in ASTM B601. This standard assigns alphanumeric code to each descriptive temper designation (Table 5.6).

### Fatigue Properties

Compared to most structural materials, copper alloys are not commonly applied in designs involving cyclic stress. The most common use of copper alloys under dynamic loading is in rotating electrical machinery used for power generation, which uses very dilute copper alloys or pure copper in an annealed state as conductors or other noncyclic applications.

**Table 5.6**   ASTM B601 Temper Codes for Coppers and Copper Alloys

| Copper temper | ASTM code | Cold worked/stress relieved tempers | ASTM code |
|---|---|---|---|
| ⅛ Hard | H00 | H01 temper and stress relieved | HR01 |
| ¼ Hard | H01 | H02 temper and stress relieved | HR02 |
| ½ Hard | H02 | H04 temper and stress relieved | HR04 |
| ¾ Hard | H03 | H08 temper and stress relieved | HR08 |
| Hard | H04 | H10 temper and stress relieved | HR10 |
| Extra hard | H06 | As finned | HR20 |
| Spring | H08 | Drawn and stress relieved | HR50 |
| Extra spring | H10 | | |
| Special spring | H12 | Cold rolled/order strengthened tempers | ASTM code |
| Ultra spring | H13 | H04 temper and order heat treated | HT04 |
| Super spring | H14 | H08 temper and order heat treated | HT08 |
| Extruded and drawn | H50 | | |
| Pierced and drawn | H52 | As manufactured tempers | |
| Light drawn, light cold rolled | H55 | As sand cast | M01 |
| Drawn general purpose | H58 | As centrifugal cast | M02 |
| Cold heading and forming | H60 | As plaster cast | M03 |
| Rivet | H63 | As pressure die cast | M04 |
| Screw | H64 | As permanent mold cast | M05 |
| Bolt | H66 | As investment cast | M06 |
| Bending | H70 | As continuous cast | M07 |
| Hard drawn | H80 | As hot forged and air cooled | M10 |
| Medium hard drawn | H85 | As forged and quenched | M11 |
| Electrical wire | | | |
| Hard drawn electrical wire | H86 | As hot rolled | M20 |
| As finned | H90 | As hot extruded | M30 |
| | | As hot pierced | M40 |
| | | As hot pierced and re-rolled | M45 |

## Electrical and Thermal Properties

Copper and copper alloys are excellent electrical and thermal conductors and are used more widely for this property than any other metal. The best way to *increase* the electrical and thermal conductivities of copper is to *decrease* the alloys or impurities. The existence of impurities and all alloying elements, except for silver, will decrease the electrical and thermal conductivity of copper. Additionally, as the amount of the impurity or alloying element increases, the electrical conductivity decreases. Cadmium has the least effect on copper's electrical conductivity followed by increasing effects from zinc, tin, nickel, aluminum, manganese, and silicon, with the greatest effect from phosphorus. Although different mechanisms contribute to the thermal conductivity, the addition of increasing amounts of elements or impurities also produces a drop in thermal conductivity. Phosphorus is often used to deoxidize copper, which increases the hardness and strength, but severely affects the conductivity. Silicon is used in place of phosphorus to deoxidize copper when it is important to preserve its electrical or thermal conductivity.

### Corrosion

Copper, in its pure form, has good resistance to severe corrosion. It is known to exhibit a surface oxide, much like aluminum, that limits further corrosion but cannot prevent it. Many of the alloys in copper, such as nickel and tin, are more resistant to corrosion than pure copper. Alloy additions, such as aluminum or beryllium, fail to assist in providing the protective oxide films that they form in their pure state and may actually increase the alloy's susceptibility to hydrogen embrittlement or stress-corrosion-cracking. Stress-corrosion-cracking occurs in brasses exposed to ammonia or amines. In copper alloys containing zinc, the removal or de-alloying of the zinc from the copper is also another form of corrosion. In this case, the more active metal is selectively removed from the alloy, leaving a weak noble metal behind. Overall, copper alloys show excellent corrosion-resistance in a wide range of environments.

### Machinability

Copper and copper alloys are usually divided into three groups, with respect to machinability:

1. Free-cutting alloys. These include an appreciable amount of lead, sulfur, or tellurium, which have been added in order to improve the machining characteristics.
2. Moderately machinable alloys. Chiefly the non-leaded brasses containing 60 to 85% copper and the leaded nickel silvers.
3. Alloys difficult to machine. These including non-leaded coppers, low-zinc brasses and nickel silvers, the phosphor bronzes, aluminum bronzes, cupronickel, and beryllium copper.

Group 1 produces short brittle chips and is well suited for rapid machining. Group 2 produces fairly open coil or closely wound helix chips and Group 3 alloys produce long continuous and strong chips, making rapid machining difficult. The machining ratings for copper alloys are based on that of free-cutting brass (61.5% Cu, 35.5% Zn, 3% Pb), which is taken as 100.

### Weldability

Welding of copper alloys is performed commonly by gas, arc, and resistance methods. The mechanical fasteners necessary for joining are easily attached to these alloys by being welded, brazed, or soldered.

## Fabrication Processes

Copper alloys are capable of being shaped to any form by all common fabricating processes. They can be rolled, stamped, drawn, forged, and formed. Due to their rapid stamping capability they are used around the world for coins. Cold working is easily done because of its ability to be heavily deformed without tearing or fracturing and with the added benefit of increased yield and tensile strength. Hot working permits greater changes in shape than cold working and can be used to work them to nearly finish size.

## Raw and Stock Finishes

Copper and copper alloys are available in many fabricated forms, much like aluminum alloys. These include squares, rectangles, rounds, tubes, and hexagons but typically not structural shapes or flanged products (Figure 5.9). The solid shapes are available in different compositions and cold worked hardnesses, as discussed in the properties and designations section.

## Processing Considerations

Copper alloys are similar to aluminums in that they are affected by thermal treatments and require some form of corrosion-protection, even for ambient conditions, to avoid surface discoloration. Thermal treatments generally soften copper alloys, except for quenching treatments, which harden them. Surface

**Copper**
    Rounds
        C110 — Electrolytic Tough Pitch
    Bus Bar
        C110 — Electrolytic Tough Pitch
    Squares
        C110 — Electrolytic Tough Pitch

**Brass**
    Rounds
        C353 — Leaded Brass
        C360 — Free-Cutting
        C464, C485 — Naval Brass
        C694 — Silicon Brass
    Squares
        C360 — Free-Cutting
    Hexagons
        C353 — Leaded Brass
        C360 — Free-Cutting
        C694 — Silicon Brass
    Rectangles
        C360 — Free-Cutting

**Bronze**
    Rounds
        C674 — Manganese Bronze
        C932 — Bearing Bronze
        C954 — Aluminum Bronze
    Tube Hollows
        C932 — Bearing Bronze
        C954 — Aluminum Bronze
    Squares
        C954 — Aluminum Bronze
    Rectangles
        C954 — Aluminum Bronze

**Figure 5.9**   Common stock shapes for copper alloys.

treatments vary depending on the required finish, which can be functional or decorative.

*Thermal Treatments*

Thermal treatments applied to copper and its alloys include homogenizing, annealing, stress-relieving, solution treating, precipitation-hardening and quench-hardening, and tempering. Annealing is applied to wrought copper products, during or after mill processing, or cast products and is intended to increase the ductility and/or toughness. Stress-relieving is usually done on cast parts and wrought products to relieve the internal stresses in the material or parts without appreciably affecting their properties. Care should be taken with alloys that contain oxygen, such as tough pitch copper (copper containing from 0.02 to 0.04% O, obtained by refining copper in a reverberatory furnace) because the hydrogen in the furnace causes embrittlement when it combines with the oxygen in copper.

### Hardening

Copper alloys that are hardened by heat treatment are of two general types: those that are softened by high-temperature quenching and hardened by lower temperature precipitation heat treatments, and those that are hardened by quenching from high temperatures through martensitic reactions. Alloys that harden during low- to intermediate-temperature treatments following solution-quenching include precipitation-hardening and spinodal-hardening types. Spinodal-hardening alloys are copper–nickel-based, with additions of chromium or tin. These alloys offer excellent dimensional stability throughout the hardening process because the process does not involve a phase change (precipitation).

Quench-hardening alloys include aluminum bronzes, nickel aluminum bronzes and a few special copper zinc alloys. Usually quench-hardened alloys are tempered to improve toughness and ductility and reduce hardness in a manner similar to that of alloy steels.

### Surface Treatments

Copper alloys are known for their outstanding corrosion-resistance. Surface treatments applied to copper alloys are mostly for cleaning, polishing, and decorative purposes. Various colors and lusters are available, as well as different polished or buffed surface textures. Copper alloys can be plated, coated with organic substances, or chemically colored to further extend the variety of available finishes. The selection of surface treatment for copper and copper alloys will depend on application requirements for appearance and corrosion-resistance. These treatments can range from a simple low cost chemical process that provides a simple uniform surface to expensive electroplates that provide maximum corrosion-resistance.

Surface preparation includes pickling with sulfuric acid and sand or glass blasting. Chemical and electrochemical cleaning can be done where lubricating oils, grease, oxides, or abrasives are on the surface. Commonly electroplated metals that are applied to copper alloys include silver, chromium, cadmium, gold, nickel, tin, tin–copper alloys, and tin–lead alloys. In addition to electroplating, coloring and organic coating are also common.

## Commonly Used Copper Alloys

*Pure copper (C10100)*: This is widely used, because of its high electrical conductivity, in wires, connectors, and motors. Roughly, one-half of all mined copper is used in the form of electrical wires and cables.

*Yellow brass (C27000)*: This is used widely for electrical contacts along with C50500, which is mainly used for connector springs and thin sockets. For PCB circuits C26000 and C51000 are also used.

*Cartridge brass or 70–30 brass (C26000)*: This brass is 70% copper and 30% brass. This is commonly used in piping, tubing, valves, heat transfer equipment, heavy industrial equipment, fasteners, and ordnances.

*Leaded free-cutting brass (C36000)*: This is perhaps the most widely used of all copper alloys. It is used in fittings, fasteners, screw machine products, and other transport applications.

*90–10 copper nickel (C70600)*: This is an alloy that needs no surface protection, and hence, is increasingly used in brake and hydraulic suspension systems for automotives. The same alloy is also used in cooling systems of cars. It is also used as cladding for ship hulls and legs of oil platforms in sea.

*70–30 Cu Ni (C71500) and 90–10 Cu Ni*: Used widely in condenser tubing, especially where seawater or environments are involved.

*C71300*: Coinage is also a big area of application for copper alloys and is used worldwide as a good coinage grade. However, it is interesting to note that the 10 cent and 25 cent coins in the United States have a pure copper (C11000) core and clad with C71300.

*Cu-5Sn-5Pb-5Zn alloy (C83600):* This is the most common
cast general purpose alloy used for valves and plumbing
hardware. Contains lead particles dispersed around the
single-phase matrix and offers good machinability, with
moderate levels of corrosion-resistance, tensile strength
(240 MPa, or 35 ksi), ductility, and conductivity (15%
IACS).

*C84400:* This is widely used for cast plumbing system
components.

## Titanium

Pure titanium is a silver-colored metal, which has grown as a
metal of strategic importance in the last 50 years. Its density is
approximately 55% of steel with similar strength. Titanium is a
durable, biocompatible metal, which is commonly used in artificial joints, dental implants, and surgical equipment. It also finds
use in aerospace and some automotive applications because its
operating temperature limit is nearly 500°C. Apart from its use
as a structural metal it is also added in small quantities to steels
and other alloys to increase hardness and strength by the formation of carbides and oxides. Titanium can exist in two
allotropic forms: alpha and beta. Its mechanical properties are
closely related to these allotropic phases, with the beta phase
being much stronger but more brittle than the alpha phase.
Hence, titanium alloys are commonly classified as alpha, beta,
and alpha–beta alloys. Titanium alloys have attractive engineering properties, which include a desirable combination of
moderate weight and high strength, property retention at elevated temperatures, and good corrosion-resistance. These properties provide high values of specific strength, $S_y/\rho$, which is
desirable for transportation systems.

Apart from the commercially pure forms of titanium,
there are three principal types of alloys: alpha, beta, and
alpha–beta alloys which are available in wrought and cast
forms (Figure 5.10). In recent years some also have become
available in powder forms. The system of designations for titanium alloys vary from one standard to another; however, the
most prevalent and commonly used system is to name the alloy

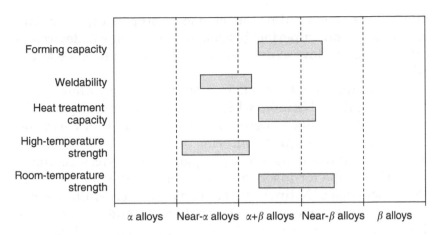

**Figure 5.10** Performance range for various titanium compositions.

by its composition. For instance, Ti-4Al-3V, which means its major alloying elements are 4% aluminum and 3% vanadium.

There are five grades of what are known as commercially pure or unalloyed titanium, ASTM Grades 1 through 4, and 7. Each grade has a different impurity content, with Grade 1 being the most pure. Tensile strengths vary from 172 GPa for Grade 1 to 483 GPa for Grade 4.

Titanium carbide is an important product of titanium and is made by reacting titanium dioxide and carbon black at temperatures above 1800°C. It is compacted with cobalt or nickel for use in cutting tools and for heat-resistant parts and it is lighter weight and less costly than tungsten carbide, but it is more brittle in cutting tools.

One of the primary uses of titanium is as titanium oxide in the form of a white pigment. It is also widely used as titanium carbide for hard facings and for cutting tools. Primarily because of their high strength-to-weight ratio (specific strength), titanium and its alloys are widely used for aircraft structures requiring greater heat-resistance than aluminum alloys. Owing to their exceptional corrosion-resistance they are also used for chemical processing, desalination, power generation equipment, marine hardware, valve and pump parts, and prosthetic devices. Alloy Ti-6Al-4V is widely used in medical applications due to its outstanding mechanical properties, corrosion-resistance, and superior biocompatibility. Many devices in orthopedic, cardiovascular, and dental implantations use this alloy.

Titanium is also found in a shape memory alloy (SMA) material called Nitinol, which is a titanium alloyed with nickel that exhibits superelastic behavior. It is a corrosion-resistant, biocompatible material that has a shape memory property, making it useful for implantable devices requiring an initial shape for insertion and a final shape once in place. The properties of Nitinol rely on its dynamic crystalline structure, which is sensitive to external stress and temperature. The alloy has three defined temperature phases that influence its behavior:

> *Austenite Phase*: This temperature is above the transition temperature and varies depending upon the exact composition of the Nitinol alloy; commercial alloys usually have transitional temperatures between 70 and 130°C (158 to 266°F). The yield strength with which the material tries to return to its original shape is considerable; 35,000 to 70,000 psi. Crystalline structure is cubic.
>
> *Martensitic Phase*: In this low-temperature phase the crystal structure is needle-like, with the crystals aligned. The alloy may be bent or formed easily using a deformation pressure of 10,000 to 20,000 psi. Bending transforms the crystalline structure of the alloy by producing internal stresses.
>
> *Annealing Phase*: In this high-temperature phase the alloy will reorient its (cubic) crystalline structure to "remember" its present shape. The annealing phase for Nitinol wire is in the range of 540°C.

When at room temperature Nitinol is in the martensitic phase and can be deformed as required. When the new shape is heated above its transitional temperature (austenite phase), the crystalline structure changes from needle-like to cubic. The resulting cubic structure does not fit into the same space as the needle-like structures formed when the alloy was bent.

The internal stresses are relieved by material returning to its "remembered" crystalline cubic shape. The crystalline structure changes still occur, even if the alloy has not been deformed but no movement will be observed. Nitinol typically generates a shape resuming force of 22,000 psi. In wire form Nitinol contracts by 8 to 10% of its length but for longer life (greater than 1,000,000 cycles), the contraction should be restricted to only 6%.

SMAs have been in development since 1932 and by 1961 that they were used for other than laboratory experiments. Dr. William Beuhler, working at the U.S. Naval laboratory, discovered an alloy of nickel and titanium that produced SMA behavior. The new alloy was called Nitinol (pronounced night-in-all), representing its elemental components and place of origin. The "Ni" and "Ti" are the atomic symbols for nickel and titanium while the "NOL" represents the first letter of place it was discovered, the Naval Ordinance Laboratory.

The ratio of nickel to titanium in Nitinol is, typically, nearly equal and is not allowed to vary by much, as even the smallest change in the ratio of the two compounds has a dramatic effect on the transition temperature of the resulting alloy. For example, a 1% change in the ratio varies the transition temperature from $-100$ to $+100°C$. Nitinol can be found in medical devices including orthodontics, where it can apply a constant force in the shape of a spring and in catheters, which can be shaped as required during surgery. Post-processing typically involves electropolishing to improve surface finish and remove sharp edges, which are a result of the forming process.

## Properties and Designations

Titanium alloys are unique in the aspect that they have been developed and put to use during the last 50 years as opposed to old metals such as copper, gold, and iron which have been in use for hundreds, if not thousands, of years. The lack of a universal numbering system is a confirmation of the maturity of this material. The alloy designation methods were determined by scientists and people developing the alloys who chose to classify them by their microstructure rather than their composition or properties. Titanium alloys are known for their lower density and good strength as well as for their use in biomedical field and aerospace industry. This section will highlight their numbering and some of the important properties of titanium alloys.

### Numbering

Alpha alloys: These are the titanium alpha alloys commonly contain aluminum and tin, which improve the material's properties by solid-solution strengthening. Other alloying elements may include zirconium, molybdenum, and, less commonly, nitrogen, vanadium, columbium, tantalum, or silicon. Though they are generally not capable of being strengthened by heat treatment, they are more creep-resistant than the other two types at elevated temperatures. The absence of a ductile-to-brittle transition, a problem with beta alloys, makes them suitable for cryogenic applications, too. They are more weldable but difficult to forge. One example is Ti-5Al-2Sn, which is widely used in the annealed or recrystallized condition to eliminate residual stresses caused by mechanical working.

Beta Alloys: Titanium beta alloys are noted for their hardenability, good cold formability in the solution-treated condition and high strength after aging. On the other hand, they have a slightly greater density than other alloy titanium types and exhibit a ductile-to-brittle transition. They are also the least creep-resistant of the alloys. Main alloys of this type include Ti-13V-11Cr-3Al, Ti-8Mo-8V-2Fe-3Al, and Ti-11.5Mo-6Zr-4.5Sn.

Alpha–Beta alloys: Titanium alpha–beta alloys can be strengthened by solution heat treatment and aging. Aging temperature ranges from 480 to 650°C. Ti-6Al-4V is the principal alloy, with its production alone having accounted for about one-half of all titanium alloy production. This alloy exhibits properties similar to plain carbon steel but with the added properties of titanium, which include good formability, corrosion-resistance, and elevated temperature properties. Its uses range from aircraft and aircraft turbine parts to chemical processing equipment, marine hardware, and prosthetic devices. The alloy is also the principal alloy used for super plastically formed and simultaneously diffusion-bonded parts. Other important alloys of this type include Ti-6Al-6V-2Sn, Ti-7Al-4Mo and Ti-3Al-2.5V.

Gamma alloys: Titanium aluminides based on $Ti_3Al$ and TiAl offer high temperature performance to 650°C (1200°F) and a density 50% that of nickel-based high-temperature alloys. They also offer high specific strength and high temperature properties such as strength retention, creep and stress rupture and fatigue-resistance. This alloy finds uses in aircraft engine components, airframe components, automotive valves and turbochargers.

### Strength Properties

The mechanical properties of a specific titanium alloy will depend on the extent of mechanical working and heat treatment. Table 5.7 provides the mechanical properties of some common alloys of each type.

### Heat Treatability

Thermomechanical treatment (TMT) is commonly used to manipulate the microstructure and properties of titanium alloys. The effects of TMT on the properties of titanium alloys are especially pronounced in the case of fracture resistance. The

**Table 5.7** Mechanical Properties of Selected Titanium Alloys

| Alloy | Condition | Tensile strength (MPa) | Yield strength (MPa) | Elongation (%) | Hardness (Rock well) | Impact strength (J) |
|---|---|---|---|---|---|---|
| 99.5 Ti | Annealed | 331 | 241 | 30 | 120 HB | 40 |
| Ti-5Al-2.5Sn | Annealed | 862 | 807 | 16 | 36 HC | 26 |
| Ti-8Al-1Mo-1V | Duplex annealed | 1000 | 951 | 15 | 35 HC | 32 |
| Ti-6Al-4V | Annealed | 993 | 924 | 14 | 36 HC | 19 |
| Ti-6Al-4V | Solution + age | 1172 | 1103 | 10 | 41 HC | 24 |
| Ti-13v-1Cr-3Al | Solution + age | 1220 | 1172 | 8 | 40 HC | 11 |

influence of microstructure on toughness and tensile ductility tends to be an inverse relationship. The particular processing route selected depends heavily on the intended application. For example, fracture critical structures perform most efficiently when fabricated from materials processed to achieve high fracture toughness. TMT is usually accomplished by hot rolling or hot forging. Other working operations such as extrusion or spinning also can be used for this purpose.

### Fatigue Properties

Unlike steels or aluminum alloys, titanium alloys are generally free of defects such as inclusions or porosity because of the double and triple vacuum arc melting procedures employed in producing them. The resulting material structure provides good fatigue properties and almost no crack initiation due to inclusions or porosity. However, fatigue behavior in titanium alloys is very sensitive to surface preparation, which can be more important than microstructural effects. In contrast to working with ferrous metals, inducing compressive stresses in the surface actually reduces fatigue life rather than increasing it. Similarly, the titanium surface finish is more sensitive to surface machining effects than that of ferrous metals, requiring care in producing the final finish. In unalloyed titanium, fatigue life is also influenced by grain size, interstitial contents, and degree of cold work. Decreasing grain size will increase fatigue life. In alpha alloys fatigue life depends on grain size, degree of age-hardening, and oxygen content of the alloy. Age-hardening in alpha alloys makes the crack grow faster, and, hence, reduces fatigue life. The grain size effect is the same as unalloyed titanium, the finer grain size will give longer fatigue life. For near-beta and beta alloys the microstructure plays a significant role in the fatigue limit as does the shape and size of the grains. Endurance limits for most titanium alloys are observed to be $10^7$ cycles or more.

### Corrosion

The presence of a thin, tough oxide surface film provides excellent resistance to atmospheric and sea environments as well as a wide range of chemicals, including chlorine and organics containing chlorides. What makes this possible is a mechanism similar to what occurs with stainless steel, the formation of a stable, self healing surface oxide. Commercially pure titanium is the most commonly used titanium alloy for corrosion applications, especially when high strength is not a requirement, because it is relatively inexpensive. Titanium is near the cathodic end of the galvanic series, allowing it to perform the function of a noble metal, but it may react pyrophorically in certain media. Its corrosion-resistance is commonly improved by the application of an anodizing finish, surface coatings or alloying.

Explosive reactions can occur with fuming nitric acid containing less than 2% water or more than 6% nitrogen dioxide and on impact with liquid oxygen. Increasing the water content above 2% removes the concern. Pyrophoric reactions also can occur in anhydrous liquid or gaseous chlorine, liquid bromine, hot gaseous fluorine, and oxygen-enriched atmospheres.

### Machinability

The machining characteristics of titanium vary greatly and depend upon the alloy composition, heat treatment employed, and resulting hardness. Generally, titanium is more difficult to machine than carbon steels due to its reactive nature. This property can result in poor cutting characteristics if inappropriate speeds and feeds are used during processing, reducing the cutting effectiveness by welding itself to the tool and alternately creating a hardened layer due to heat generated during the chip formation. Low cutting speeds combined with high feed rates limit temperature extremes, while effectively getting below the hardened surface layer. Pure titanium and alpha alloys require lower contact pressures than beta alloys but are still more difficult to machine than plain carbon steels. Rigid setups are required to limit deflection because of the low modulus of titanium.

Titanium reacts rapidly at high temperatures with oxygen, nitrogen, and constituents in cutting tools. The high strength of the alloy requires high contact pressures, which produces high tool-tip temperatures. This combination of chemical activity and heat contributes to seizing, galling, and abrasion and to pyrophoric behavior of small particles of titanium. In addition, titanium has relatively poor thermal conductivity, exacerbating the temperature effects at the tool-tip. The net effect is that machining of titanium requires careful selection of tools, speed, coolant, and atmosphere to get the desired results.

### Weldability

Titanium tends to oxidize rapidly when heated in air above 1200°F (650°C). At elevated temperatures, it has the property of dissolving discrete amounts of its own oxide into solution. For these reasons, the welding of titanium requires the use of a protective shielding, such as an inert gas atmosphere, to prevent contamination and embrittlement from oxygen and nitrogen. Titanium's relatively low coefficient of thermal expansion and conductivity minimize the possibility of distortion due to welding.

## Fabrication Processes

Fabricating titanium is relatively difficult because of its susceptibility to hydrogen, oxygen, and nitrogen impurities, which cause embrittlement. Elevated temperature processing, including welding, must be performed under special conditions that avoid diffusion of gases into the metal.

Forming is more difficult with titanium than with aluminum or iron-based materials. Heat is usually required in most forming operations to reduce the "springback" of the material, improving the accuracy of forming. Casting can be performed, but requires molds made from something other than sand, which is used with ferrous metals, because of the reactive nature of titanium. Special molds using sand combined with organic or graphite binders are typically used. Apart from this consideration, conventional casting methods and mold design principles can be applied.

Superplastic forming can also be applied to those titaniums that exhibit a high strain rate sensitivity. Ti-6-4 with a beta volume of 20% exhibits this characteristic at 870°C.

## Raw and Stock Finishes

Commercially pure titanium and many of the titanium alloys are now available in common wrought mill forms, such as plate, sheet, tubing, wire, extrusions, and forgings. Ingots are produced in sizes ranging from 26 to 36 in. in diameter, weighing 8,000 to 15,000 lb. Critical components such as aircraft turbines and similar high-speed rotating machinery use triple-melted material while most other applications can use double-melted processing.

Castings can also be produced with titanium and some alloys using investment casting and graphite mold casting as the principal methods. Because of titanium's highly reactive nature in the presence of such gases as oxygen, the casting must be done in a vacuum furnace.

## Processing Considerations

The processing of titanium is similar to aluminums in the types of thermal and surface treatments applied. Solution and aging are common treatments to improve its mechanical properties, while anodizing is used to improve corrosion-protection.

### Thermal Treatments

The thermal treatments applied to titanium and its alloys include stress-relieving, annealing (duplex, recrystallizing, and beta), solution heat treating, and aging. These heat treatments are not applicable to all titanium alloys. For example, the alpha and near-alpha alloys can be stress-relieved and annealed without resulting in high strength, as they do not respond to any type of heat treatment. Conversely, the commercial beta alloys always contain a metastable beta phase, which allows strengthening during aging. The beta alloys provide great potential for age-hardening and frequently utilize the stability of their beta phase to provide large section hardenability. The alpha–beta alloys exhibit heat treatment characteristics between that of the alpha class and the beta class, which prevent high levels of strengthening.

At the elevated temperatures required for heat treatments, titanium is susceptible to hydrogen contamination (resulting in hydrogen embrittlement). It is also chemically active and will oxidize in air. Both these factors necessitate a thorough purging of the environment with an oxidizing atmosphere during heat treating of titanium alloys to prevent hydrogen embrittlement.

### Hardening

The most common method of hardening titanium alloys is age-hardening. The beta alloys have greater potential to be hardened over a wide range. Induction-hardening is also employed but care must be taken to prevent oxidation, which can be accomplished by use of a controlled atmosphere. Specially developed plasma nitriding processes are also available for titanium alloys and these are commonly employed for parts used in motorsports.

### Surface Treatments

Surface treatments for titanium alloys include cleaning, finishing, and coating processes. Cleaning and finishing processes are necessary before any coating can be applied, in order to remove scales, grease, and any other foreign particles that may have an adverse affect on the coating process. For heavy sections, grinding and grit blasting processes are preferred to remove scales. For tubes, bars, and sheet products, descaling salt baths are used, which employ an oxidizing salt at a temperature of 750 to 900°F. After the parts are submerged in this bath for 5 to 20 min, they are placed in a sulfuric acid bath for a minute and then rinsed with water. Removal of grease and similar compounds can be done using vapor degreasing, emulsion and solvent cleaning, or alkaline cleaning.

After proper treatment of the surface, any coating process can be employed. The processes commonly used for coatings include anodizing, physical vapor deposition, plasma spraying, platinum or copper coatings and selected wear-resistant coatings. The purpose of coatings on titanium alloys can be either be to increase the wear-resistance or for decoration. Copper plating can increase the lubricity of the surface, allowing it to be drawn into wires. Decorative coatings are usually applied by anodizing in slightly acidified solutions of phosphoric or sulfuric acid.

## Commonly Used Titanium

*Ti–6Al–4V* is used in a wide range of applications including armor plating, medical implants, and golf clubs. It provides the best mix of damage tolerance properties and strength at high temperature, providing good structural strength up to 315°C (600°F) along with outstanding corrosion-resistance and biocompatibility properties. This alloy accounts for half of the U.S. titanium market.

*Ti–5Al–2.5Sn* is a common grade used in biomedical as well as aerospace applications. It exhibits good high-temperature creep properties, and is weldable. It is commonly used in fuselage panels, landing gear beams, wing boxes, frames, flap tracks, hot air ducts, and undercarriage components. It is also used in forging applications.

*Ti–10V–2Fe–3Al* provides a combination of high strength and high toughness, superior to that of any other commercial titanium alloy. It is used in air frames and any aerospace application where high strength at temperatures up to 315°C are involved. It is also used for forgings.

## Magnesium

Magnesium is the eighth most abundant element and constitutes about 2% of the Earth's crust; it is the third most plentiful element dissolved in seawater. Magnesium is probably more widely distributed in nature than any other metal. It can be obtained from ores, such as dolomite and magnesite, from underground brines, from waste liquors of potash, and from seawater; with about 10 million pounds of magnesium in 1 cubic mile of seawater, there is no danger of a dwindling supply.

Magnesium is the world's lightest structural metal, weighing only two-thirds as much as aluminum. It is a silvery white material that does not possess sufficient strength in its pure state for structural uses but when alloyed with zinc, aluminum, and manganese it produces an alloy having the highest specific strength (strength/density ratio). However, some complicated and poisonous surface treatments (chromate treatment, etc.) are necessary when using magnesium alloys for structural metals, owing to their poor corrosion-resistance.

Magnesium's principal use is as an alloying addition to aluminum, to produce aluminum–magnesium alloys, used mainly for beverage cans. Magnesium compounds such as magnesium oxide are commonly used as refractory material in furnace linings for producing iron and steel, nonferrous metals, glass, and cement. Magnesium oxide compounds also are used in agricultural, chemical, and construction industries. Magnesium alloys are used as structural components of automobiles and machinery. Its low corrosion-resistance limits its use in aerospace applications.

It is important to note that magnesium may exhibit fire hazards of an unpredictable nature. When in large sections, its high thermal conductivity makes it difficult to ignite and prevents its burning; it will not burn until the melting point of approximately 1200°F is reached. However, magnesium dust and fine chips are easily ignited; precautions must be taken to avoid this. If they are ignited, they should be extinguished immediately with an extinguishing powder, such as powdered soapstone, clean dry un-rusted cast iron chips, or graphite powder.

## Properties and Designations

### Numbering

Magnesium alloys use a system of identification developed by ASTM and SAE in which the first part of the designation refers to the two principal alloying elements, given by two letters, and the second part, given by two numbers, indicates their percentages in round numbers. If the alloy is heat treated, it is followed by a dash and a letter and in some cases a number indicating the temper. Table 5.8 provides symbols for this system of designation.

The alloys are classified into two major groups, wrought alloys and cast alloys. The mechanical properties of wrought alloys depend upon the fabrication process used and the dimensions of the product. Table 5.9 provides mechanical properties of some magnesium alloys.

### Strength Properties

The specific strength of magnesium alloys are typically high because of their low density. The tensile strength of steels can vary from 50 to 300 ksi, while the strongest magnesium alloys have tensile strength in the 50 ksi range. Shear strengths are significantly less than tensile strengths, and

**Table 5.8** Symbols Used for Alloying Elements and Temper Designations

| Alloying elements | | Temper designations | |
|---|---|---|---|
| A | Aluminum | F | Fabricated |
| E | Rare Earths | O | Annealed |
| H | Thorium | H10, H11 | Strain-hardened |
| K | Zirconium | H23, H24, H26 | Strain-hardened and annealed |
| M | Manganese | T4 | Solution heat treated |
| Q | Silver | T5 | Artificially aged |
| S | Silicon | T6 | Solution heat treated and artificially aged |
| T | Tin | T8 | Solution treated, cold worked and artificially aged. |
| Z | Zinc | | |

**Table 5.9** Mechanical Properties of Some Magnesium Alloys

| Alloy | Type | Temper | Tensile strength (MPa) | Yield strength (MPa) | Shear strength (MPa) | Elongation (%) |
|---|---|---|---|---|---|---|
| AM100 | Cast | F | 150 | 83 | 125 | 2 |
| AZ63 | Cast | F | 197 | 94 | 125 | 4.5 |
| AZ81 | Cast | T4 | 275 | 83 | 125 | 15 |
| AZ92 | Cast | T6 | 275 | 150 | 150 | 3 |
| EZ33 | Cast | T5 | 160 | 110 | 145 | 3 |
| ZE63 | Cast | T6 | 289 | 173 | — | |
| AZ31 | Wrought | | 260 | 170 | 130 | 15 |
| AZ61 | Wrought | F | 295 | 180 | 145 | 12 |
| M1 | Wrought | | 250 | 160 | 110 | 7 |
| ZK60 | Wrought | T5 | 305 | 215 | 165 | 16 |

typically vary from 8 to 26 ksi. This property is an important consideration when designing joints or other applications involving shear loading.

A cast magnesium alloy's strength properties depend upon the casting method and cooling rate. Finer casting methods, such as investment casting, will produce higher strength properties. Generally, wrought alloys have better mechanical properties than cast alloys (Table 5.10).

**Table 5.10**   Common Temper Designations for Magnesium Alloys

| Designation | Description |
|---|---|
| F | As fabricated |
| O | Annealed, recrystallized (wrought products only) |
| H | Strain-hardened (wrought products only) |
| H1 | Strain-hardened only |
| H2 | Strain-hardened and partially annealed |
| H3 | Strain-hardened and stabilized |
| W | Solution heat treated; unstable temper |
| T | Heat treated to produce stable tempers other than F, O, or H |
| T2 | Annealed (cast products only) |
| T3 | Solution heat treated and cold worked |
| T4 | Solution heat treated |
| T5 | Artificially aged only |
| T6 | Solution heat treated and artificially aged |
| T7 | Solution heat treated and stabilized |
| T8 | Solution heat treated, cold worked, and artificially aged |
| T9 | Solution heat treated, artificially aged, and cold worked |
| T10 | Artificially aged and cold worked |

### Heat Treatability

Magnesium alloys are easily heat treatable and do not require any special conditions. Almost all cast alloys and many of the wrought alloys benefit from heat treating. Heat treatment can be used either to improve a mechanical property or to prepare the alloy for further processing.

### Fatigue Properties

The fatigue properties of magnesium alloys, like those of other ferrous metals, cover a wide, scattered band. The *S–N* curves indicate the endurance limits can range from 10 to 100 million cycles, depending on the applied load. Fatigue strength, as with tensile strength, is higher for wrought alloys than for cast alloys. Absence of stress risers is crucial to obtain good fatigue properties. Cast parts should be machined to improve fatigue properties. Further improvement in fatigue strength can be obtained by inducing residual compressive stress on the surface by cold working. This is done by using a shot-peening method to induce a plastic deformation at the surface, to produce which results in residual compressive stresses.

## Corrosion

In some environments magnesium parts can be severely damaged unless galvanic couples are avoided by proper design or surface protection. Unalloyed magnesium is not extensively used for structural purposes because of the possibility of corrosion. Chlorides, sulfates, and foreign materials that hold moisture on the surface can promote corrosion and pitting of some alloys, unless the metal is protected by properly applied coatings. The surface film that ordinarily forms on magnesium alloys exposed to the atmosphere gives limited protection from further attack.

Severe corrosion may occur in neutral solutions of heavy metal salts including those from copper, iron, and nickel. Such corrosion occurs when the heavy metal, the heavy metal basic salts, or both, plate out to form active cathodes on the anodic magnesium surface. Chloride solutions are corrosive because chlorides, even in small amounts, usually break down the protective film on magnesium. Fluorides form insoluble magnesium fluoride and consequently are not appreciably corrosive. Magnesium is rapidly attacked by all mineral acids except hydrofluoric acid and chromic acid.

## Machinability

The machining characteristics of magnesium alloys are excellent. Usually, the maximum speeds of machine tools can be used with heavy cuts and high feed rates. Power requirements for magnesium alloys are about one-sixth of those for mild steel. An excellent surface finish can be produced, and, in most cases, grinding is not essential. Standard machine operations can be performed to tolerances of a few ten-thousandths of an inch. There is no tendency of the metal to tear or drag.

## Weldability

Magnesium alloys typically melt between 600 and 640°C and only require two-thirds the energy required to melt aluminum. They are weldable by any process used for aluminum, including gas tungsten arc welding (GTAW) and laser beam welding (LBW). As with other materials, the weldability of magnesium alloys is influenced by both the process and the material composition and initial microstructure. Chemical fluxes are often used to improve GTAW welding. Suitable fluxes could increase the penetration of the weld pool over 100%.

## Fabrication Processes

Magnesium alloy sheets can be worked in much the same manner as other sheet metals, with the exception that the metal be worked while hot. The structure of magnesium is such that the alloys work-harden rapidly at room temperatures. The work is usually done at temperatures ranging from 450 to 650°F, which is a disadvantage because of the increased complexity of the process. Often this additional processing is offset by simplifying the design and related processing.

Sheets of magnesium can be sheared in much the same way as those of other metals, except that a rough flaky fracture is produced on sheets that is thicker than about 0.064 inches when sheared at room temperature. A better edge will result on a sheet over 0.064 inches thick if it is sheared hot. An annealed sheet can be heated to 600°F, but hard-rolled sheet should not be heated above 275°F. A method known as the Guerin process is the most widely used method for forming and shallow drawing, employing a rubber pad as the female die, which bends the work to the shape of the male die.

Magnesium alloys possess good casting characteristics. Their properties compare favorably with those of cast aluminum, with the added advantage that the dies used for magnesium alloys often last twice as long. Casting methods for magnesium alloys vary from sand casting and permanent mold to investment casting and other special methods such as metal injection molding. They can be cast by all the methods available but are most often used in die casting methods because of their rapid solidification characteristics. Magnesium alloys solidify rapidly and do not stick to the dies, allowing for higher production rates and longer tool life.

In forging, hydraulic presses are typically involved, although, under certain conditions, forging can be accomplished in mechanical presses or with drop hammers. Owing to the low densities of magnesium alloys, which result in forgings with very low weight, they are sometimes alternatives to aluminum when low weight is essential, eventhough material and processing costs are generally higher. Their modulus of elasticity, 45 GPa ($6.5 \times 10^6$ psi), is the lowest of that of all forging alloys. Both the density and modulus of elasticity are approximately two-thirds that of aluminum alloys, so that the specific stiffness is nearly equal.

Commonly forged magnesium grades:

- AZ31B — 3% Al, 1% Zn
- AZ80A — 8% Al, <1% Z
- ZK60A — 6% Al, <1% Th
- HM31A — 3% Th, 1% Al

The first three alloys are most commonly used at room or slightly elevated temperatures; HM31A is one of several designed for use at elevated temperatures.

## Raw and Stock Finishes

Magnesium is currently being commercially produced in five basic groups of alloying systems based on their major alloy elements. These five groups are the following: manganese, aluminum, zinc, zirconium, and rare earth. Materials based on these five basic groups are prepared for use in fabrication processes including sand casting, die casting, permanent mold casting, forging and making rolled/extruded products such as sheet, plate, bar, tube, and wire.

Casting products allow the forming of parts to near net shape, with minimal material removal being required depending on the specific process employed. The wrought alloys are available for part fabrication by conventional machining methods, forging, and extrusion. Forging grades allow for efficient use of material but require costly tooling and equipment, making them a wise choice for high volume production.

Machining magnesium and its alloys is often considered easier than machining other metals, since high tool speeds and feeds are allowed. Grinding is generally not required because machining produces good finishes, with surface roughness measurements of 0.075–0.125 $\mu$m (3–5 $\mu$in.) possible when using diamond-tip tools. Surface finishes of 0.25–0.75 $\mu$m (10–30 $\mu$in.) are common with standard tooling. It should be noted that when grinding or polishing magnesium, special care must be taken to avoid ignition of fine particles. Use of quality collection systems based on water-washed dust collection methods are best.

In many cases, magnesium can be used without any surface treatment or with a simple conversion coating. Often the surface is altered as needed by sanding, grinding, buffing, or vibratory finishing to achieve the desired appearance. When a coating is required, it will depend on the application as to which process is appropriate. The most commonly used are chromate conversion methods that provide a satisfactory paint base in mildly corrosive environments. Phosphate and anodic methods are also available for more aggressive environments.

Most cast or wrought magnesium alloys can be electroplated using a process similar to the plating of aluminum. Both material forms require a zinc immersion coating prior to plating. Zinc and nickel are the only metals that can be plated directly on magnesium.

## Processing Considerations

### Thermal Treatments

Magnesium alloys are heat treated either to improve mechanical properties or as a means of preparing it for further fabricating process. The specific heat treatment employed depends on the class of alloy (cast or wrought), its composition and the properties desired for the application. Heat treatments commonly used include annealing, stress-relieving and solution treatment, or aging.

Solution heat treatment improves strength and results in maximum toughness and shock resistance. Precipitation heat treatment can be done after solution treatment to provide maximum hardness and yield strength but will result in lower toughness. In the case of cast alloys, precipitation heat treatment can be used without prior solution treatment. However, the most important heat treatment for cast alloys is annealing and is performed on almost all cast parts. A detailed listing of temper designations for magnesium alloys is provided (Table 5. 10).

Wrought magnesium alloys in different conditions of strain-hardening or temper can be annealed by heating them to 290–455°C (550–850°F) for one or more hours depending upon the composition of the alloy. To free the cast components from residual stresses they are kept at 260–330°C for 1–2 h. This will provide stress-relief without significantly affecting mechanical properties.

### Hardening

Magnesium alloys are not selected for their hardness properties and consequently not used in applications where high hardness is a requirement. Their hardness characteristics are very similar to aluminum alloys and they are commonly solution treated and then aged (precipitation-hardening) in order to be hardened. The aging cycle depends on the composition of the alloy and the heat treatment given earlier. Aging temperatures range from 300 to 600°F with times from 2 to 48 h. Higher aging times are associated with lower temperatures and vice versa.

### Surface Treatments

Surface treatments are applied to magnesium parts primarily to improve their appearance and corrosion-resistance, similar to those applied to aluminum. The need for surface protection and the treatments selected vary widely with service conditions, appearance requirements, alloy composition, and fabricated forms. Chemical treatments, anodized coatings, paints, and metallic coatings are all used on magnesium and its alloys. Before any surface coating or treatment the surface is either cleaned mechanically or by using a chemical, as in acid pickling. Some common chemical treatments used are chrome, dichromate, chrome manganese, and ferric nitrate treatments. In addition to these, many others are used, which are developed for specific alloy system and application. Nickel and chromium plating and paints are also common, in addition to these treatments. Processes which can transform the surface into a magnesium oxide ($MgAl_2O_4$) are also becoming available. These involve the plasma discharge in a liquid electrolyte to form a surface similar to that provided by anodizing but with improved properties, such as denser surface, greater wear, and environmental resistance. Surface sealants such as PTFE can also be used.

## Commonly Used Magnesium

AZ80A and ZK60A (Heat Treatment T5) are used widely for forgings. Their applications include different kinds of automotive parts and some structural parts for small electronic equipment.

M1A and K1A are popular for their damping capacity and are used in electronic housings for aircraft and missiles.

AZ91 is used for die castings. It has excellent castability and high strength combined with moderate ductility and is usually the first choice for parts made by die casting.

AM50 and AM60 are commonly used in safety parts such as automotive panel supports, steering wheel armatures, and seat parts.

AS41B is used in transmission housings due to its good creep characteristics, strength, and damping.

## Tungsten

Tungsten is a reasonably hard metal (31 Rc) that is four times harder than stainless steel, silvery white in appearance and popularly known for its use in filaments for vacuum tubes and electric lights. It is a metal of unique properties, having the second highest melting point (3422°C) among all elements (carbon is first) and the highest melting point among all metals. It exhibits excellent mechanical properties at high temperatures. It has the lowest expansion coefficient of all metals and a density of 19.25 g/cm$^3$, making it one of the heaviest metals. In its purest form tungsten is relatively soft and pliable, which allows it to be easily processed. Tungsten obtains its hard and brittle characteristics from the addition of small amounts of carbon and oxygen. It is more corrosion-resistant than stainless steel.

Tungsten features the lowest vapor pressure of all metals, very high moduli of compression and elasticity (60 × 10$^6$ psi), very high thermal creep resistance, as well as high thermal and electrical conductivities. Because of these unique properties, tungsten, tungsten alloys, and some tungsten compounds cannot be substituted in many important applications in different fields of modern technology.

Tungsten is obtained commercially by reducing tungsten oxide with hydrogen or carbon. Tungsten and its alloys can be pressed and sintered into bars and subsequently fabricated into wrought bar, sheet, or wire. Many tungsten products are intricate and require machining or molding and sintering to near net shape and cannot be fabricated from standard mill products. Tungsten is commonly consumed in four forms: tungsten carbide, alloying additions, pure tungsten, and tungsten based chemicals.

Tungsten carbide accounts for about 65% of tungsten consumption, it is combined with cobalt as a binder to form cemented carbides, which are used in cutting and wear applications. Metallic tungsten and tungsten alloy mill products account for about 16% of its consumption. The most important alloys are the tungsten steels, which consume up to 18% of produced tungsten. Electrical contacts for switch gears are made from Cu and Ag–W alloys. Tungsten alloys are often found in applications where a high density material is needed, such as in counter weights, fly wheels, and high penetration armaments, as well as for cutting and wear applications. Additional applications include radiation shields and x-ray targets.

Wire formed tungsten is used extensively for lighting, electronic devices, and thermocouples. Tungsten chemicals are used for organic dyes, pigment phosphors, catalysts, cathode ray tubes, and x-ray screens. The high melting point of tungsten makes it an obvious choice for applications involving very high temperatures. Tungsten is frequently used as an additive to tantalum (max. 10% W) and niobium (max. 15% W) to improve their high-temperature strength and creep resistance.

## Properties and Designations

Three tungsten alloys are produced commercially: tungsten–thoria, tungsten–molybdenum, and tungsten–rhenium. The W–ThO$_2$ alloy contains a dispersed second phase of 1 to 2% thoria. The thoria dispersion enhances thermionic electron emission, which in turn improves the starting characteristics of GTAW electrodes. It also increases the efficiency of electron discharge tubes and imparts creep strength to wires at temperatures above one-half the melting point of tungsten.

In tungsten–molybdenum alloys, molybdenum helps refine the grain size of arc-cast tungsten. These alloys also are less dense than unalloyed tungsten, which is useful for aerospace applications. Molybdenum also lowers the melting point of tungsten, making production easier.

Rhenium-addition makes tungsten–rhenium alloys more ductile. These alloys are stronger than unalloyed tungsten. Rhenium is found to lower the DBTT of tungsten when it is added as an alloying element; it also serves to increase the recrystallization temperature. The addition of rhenium also provides increased resistance to thermal shock and thermal fatigue which makes the alloys suitable for making high-temperature thermocouples and rotating anodes for x-rays.

Cobalt–chromium–tungsten alloys, frequently called stellites, are a group of wear-resistant alloys that are found in applications where a tough wear-resistant material is required for bearings, high-speed cutting tools, valve seats, and pistons.

Other alloys containing tungsten are those known as super alloys, which include nickel, cobalt, or iron-based alloys with varying amounts of tungsten, molybdenum, tantalum and, more recently, also rhenium. Rhenium is commonly found in applications where extreme mechanical strength combined with excellent erosion-resistance at high temperatures is required, these applications include turbine blades for aircraft and space vehicle structures.

A form of tungsten known as tungsten heavy metal (heavimet, densalloy) is a sintered pseudoalloy of tungsten with an iron–copper or iron–nickel binder for applications where a high density material that is not brittle is required (e.g., for counterweights in aircraft, for darts, or for weights in golf club heads). It is also used in armament applications, such as high-velocity penetrators in anti-tank weapons. The densities of available grades range from 17 up to 19 $g/cm^3$, which is about 2.25 times that of steel and almost 7 times that of aluminum.

*Numbering*

Tungsten heavy metal alloys are defined by standard ASTM B459, which specifies their minimum mechanical

**Table 5.11**   Classes Defined on Basis of Composition

| Class | Tungsten content (%) | Density (g/cm³) | Hardness (HRC) |
|---|---|---|---|
| 1 | 89–91 | 16.85–17.25 | 32 max |
| 2 | 91–94 | 17.15–17.85 | 33 max |
| 3 | 94–96 | 17.75–18.35 | 34 max |
| 4 | 96–98 | 18.25–18.85 | 35 max |

**Table 5.12**   Types Defined on Basis of Tensile Properties

| Type | Tensile strength (MPa) | 0.2% yield strength (MPa) | Elongation (%) |
|---|---|---|---|
| I | 900 | 725 | 1.5 |
| II | 650 | 520 | 2 |
| III | 415 | 340 | 1 |

properties. These properties are usually specified at the time of purchase. These specifications of high density tungsten-based alloys usually divide them into four classes based upon their composition and three types based on their tensile properties (Tables 5. 11 and 5. 12).

### Strength Properties

Tungsten alloys are categorized into three types based on strength, with the first being the strongest. Strength properties are obtained primarily from solid-solution and dispersion-strengthening methods which also have an effect on the resulting alloy. Class 1 alloys are basically tungsten–nickel–copper or tungsten–nickel–iron alloys, the tungsten nickel copper alloys of this class typically contain 90% W, 6–7% Ni, and 3–4% Cu. Minor additions of other metals such as Mo or Co can be added to modify certain properties. Class 1 tungsten–nickel–iron alloy contains 90% W, 5–7.5% Ni, and 3–5.5% Fe. Classes 2, 3, and 4 are usually tungsten–nickel–iron alloys (Table 5.13).

### Mechanical Properties

Unalloyed tungstens are heavier than those that are alloyed, while tungsten heavy alloys have the highest density

**Table 5.13** Mechanical Properties of Tungsten Alloys

| Alloy | Class | Density $(g/c\ m^3)$ | Tensile strength (MPa) | Yield strength (MPa) | Elongation (%) | Hardness ($R_C$) | Modulus of elasticity (Gpa) |
|---|---|---|---|---|---|---|---|
| Tungsten–nickel–copper | Class 1 | 17 | 785 | 605 | 4 | 27 | 275 |
| Tungsten–nickel–iron | Class 2 | 17 | 895 | 615 | 16 | 27 | 275 |
| | Class 3 | 18 | 925 | 655 | 6 | 29 | 310 |
| | Class 4 | 18.5 | 795 | 690 | 3 | 32 | 345 |

among nonradioactive, nonprecious metals, and have a very high melting point. Heavy alloys are the only elastically isotropic metals known.

### Heat Treatability

Tungsten alloys can be heat treated as an initial treatment for mechanical working and also to change the microstructure. Heat treatment of tungsten alloys is not as prevalent as that of steels, primarily because of the cost involved. The details of the heat treatments are given under the heading of Thermal Treatments.

### Fatigue Properties

These are generally not a matter of concern for this material, as is not used in applications that require good fatigue properties.

### Corrosion

It has excellent corrosion-resistance and is attacked only slightly by most mineral acids. At room temperature it is resistant to most of chemicals but it can be easily dissolved with a solution of nitric and hydrofluoric acids. Elevated temperatures reduces tungsten's corrosion-resistance, making it more susceptible to attack. At about 250°C it reacts rapidly with phosphoric acid and chlorine. It begins to oxidize readily at 500°C and at 1000°C it reacts with many gases including water vapor, iodine, bromine, and carbon monoxide.

### Machinability

Tungsten alloys can be turned, milled, and drilled. As expected, due to high density, strength, and hardness, machining tungsten is difficult. For thick-walled sections slight preheating of the work piece up to 400°F is recommended, otherwise a highly chlorinated oil such as trichloroethylene is used. Carbide cutting tools have proven good for machining tungsten alloys. The final shaping of tungsten parts is done by grinding using silicon carbide wheels. To avoid surface cracking an oil emulsion type coolant is recommended during grinding.

## Weldability

Joining tungsten to itself or to other metals should be undertaken with caution and an understanding of the limitations involved. Tungsten can be welded to itself. However, the resulting weldment is always recrystallized and hence brittle. Even the use of a tungsten–rhenium filler rod does not eliminate the brittleness of the metal adjacent to the heat affected zone. On the other hand, if a leak-tight high-temperature joint is required, welding may be the only solution.

## Fabrication Processes

Tungsten mill products, sheet, bar, and wire are all produced via powder metallurgy. These products are available in either commercially doped or undoped versions. These dopings or additives improve the recrystallization and creep properties of these products especially important in case of filament use.

Powder metallurgy is employed because melt-processing tungsten, is not cost-effective. The first step in the process of PM is to produce a powder of the metal or alloy. The powder can either be produced by mechanical methods such as milling or by atomization. However, in the case of tungsten, oxide is the most widely used reduction material. This method is widely used for the production of iron, copper, and molybdenum powders from their oxides. Such powders are also known as sponge powders because of the presence of pores within the individual particles. Most commercial processes of oxide reduction employ stationary type reduction in belt furnaces. The reducing gas composition required for successful reduction depends on the particular metal oxide and the reduction temperature. The reduced oxide is actually a porous cake that must be ground to powder, which is sometimes annealed to further remove oxygen and/or carbon and to render it soft.

After the powder formation, it is screened and classified into different sizes. The desired powder size is mixed with binders and stabilizers and then compacted into the desired final shape. This compact is then placed into a furnace and sintered to form the completed part. Tungsten is consolidated to full density by two principal P/M processes: solid-state sintering and mechanical working (wrought P/M tungsten) and liquid-phase sintering of powders. Heat treatment after the sintering is not a requirement and only needed if some microstructural change is desired.

## Raw and Stock Finishes

Tungsten alloys are available in the form of rod, bar, tube, wire, plate, and sheet. Plate and sheet are only available for selected grades while other forms are available for all grades.

## Processing Considerations

### Thermal Treatments

The thermal treatments applied to tungsten are performed primarily for two reasons: softening for mechanical working and for microstructural change. Tungsten can be fabricated into many simple shapes. It can be folded, bent, formed, punched, and riveted. Because tungsten is a strong, hard, crack-sensitive metal that is usually brittle at room temperature, it requires special handling and skill beyond that necessary for most metals and alloys. The most important rule to remember when working with tungsten is that it must be formed or cut at temperatures well above its transition temperature. Failure to do so will generally result in cracked or laminated parts. Care should be taken to ensure that the metal remains at this temperature throughout the entire forming process. The use of cold tooling which rapidly chills the metal can be as harmful as not preheating the material.

The heat treatments used to obtain a change in microstructure depend on the initial microstructure. Heat treatments are common for increasing density by giving the alloy enough time at higher temperature so that the metal can diffuse into the spaces between the powder particles, increasing the density. There are many variations possible for this type of processing but it depends heavily upon the alloy, the present microstructure and the result desired as to the final processing approach that is selected. Because tungsten alloys are high-temperature materials, heat treatment must be very carefully chosen and executed to produce a quality product and avoid excessive processing cost.

### Hardening

Tungsten and its alloys are generally not extremely hard, with values of 30 Rc not uncommon. Hardness is generally not why tungstens are chosen when used in their unalloyed condition but it is used extensively to improve a materials hardness or wear. The hardening details are already given under the surface treatment heading.

### Surface Treatments

Tungsten oxidizes in ambient conditions and must be protected at elevated temperatures. Systems for protecting tungsten from atmospheric exposure at temperatures from 1650 to 2205°C have been developed which include roll cladding with tantalum–hafnium alloys. Roll cladding is a type of cladding in which the material is in the form of a roll and used to "coat" the parent metal. The process is used in cases where the area to be protected is large. The cladding material can cover a larger cross-section of parent material and is usually applied using an automated process where the roll is automatically fed into the machine and joined with the parent metal.

Functional or structural parts can be made completely from cemented carbide, stellite, or a superalloy. Alternatively, cemented carbide parts can be soldered onto selected surface areas. Hard-facing by welding is frequently used to protect selected areas with a hard layer in the millimeter thickness range. For this purpose, sintered or tube-type welding electrodes are available to produce wear-resistant surface layers. Cast tungsten carbide or tungsten monocarbide are the hard constituents, while most frequently iron, or an iron-based alloy, acts as a binder.

Thermal spraying is applied to make thinner wear-resistant surfaces. Powders for hard-facing are, for example, made of pure tungsten or more frequently tungsten monocarbide and/or cast carbide combined with cobalt, nickel, or a Ni–Cr–B–Si alloy. A large variety of grades are available to meet almost every requirement. Chemical vapor deposition using tungsten halogenides or alkoxides is a less common method to provide hard, acid-resistant tungsten coatings on bearings, dies, rolls, and gauges.

## Commonly Used Tungsten

Unalloyed tungsten has mechanical properties that are strongly dependent on deformation history, purity, and testing orientation (the processing of tungsten sheet, rods, and wire produces a unidirectional structure).

Unalloyed tungsten —this material is most commonly used for tungsten electrodes.

Tungsten carbides—this form of tungsten is used for machine tool inserts or similar metal working tools.

## SELECTION CONSIDERATIONS

Nonferrous metals have typically been found in less demanding industrial and consumer applications because of their lack of inherent corrosion-protection and infinite fatigue life. In general, nonferrous materials require a secondary manufacturing operation to impart corrosion-protection to the surface, which adds cost and processing time to the fabrication of the part. In contrast, ferrous metals include material families that provide outstanding corrosion-resistance without additional processing (i.e., stainless steels) as well as superior material properties including strength and fatigue resistance (Table 5.14).

**Table 5.14**  Comparison of Material Properties for Selection

| Material | Density (lbs/in.$^3$) | Modulus ($\times 10^6$ lbs/in.$^2$) | Yield strength H (lbs/in.$^2$) | Corrosion-resistance |
|---|---|---|---|---|
| Aluminum | 0.098 | 10.2 | 4–78 | Good |
| Copper alloys | 0.323 | 16.7 | 10–195 | Best |
| Titanium | 0.170 | 16.0 | 40–160 | Good |
| Magnesium | 0.065 | 6.5 | 10–24 | Poor |
| Tungsten | 0.695 | 60.0 | 45–100 | Good |
| Carbon steels | 0.283 | 29.0 | 29–300 | Good |

The selection of nonferrous metals for demanding industrial applications continues to expand as improved material coating processes are made available, allowing the use of these process-friendly materials. The primary benefit of nonferrous materials is their machining and forming properties, which offer reduced manufacturing times over ferrous materials of similar specific strength. Generally, this is made possible by the lower modulus of nonferrous metals (with exception of tungsten) which makes them easier to machine. They are less tolerant of surface imperfections because of the lack of infinite fatigue life, which means a small imperfection or scratch can result an undesirable latent fatigue failure.

The most commonly used nonferrous metals are aluminum and titanium, which offer low densities and can both be readily alloyed to produce materials of high specific strength and moduli. Copper alloys are widely used for bearing and corrosion-resistant applications that do not require exceptional strength properties. Copper has a similar density to that of ferrous metals but a lower modulus and higher material costs for the same specific stiffness. Magnesium, an alloy known for its low density but difficult processing, is finding more use in automotive applications as it offers a method of reducing vehicle weight. Improvements in processing and coatings have addressed longstanding issues involving manufacturing and corrosion-protection. Tungsten has the highest modulus of commonly available materials with a Youngs modulus of 60 Mpsi. Although it is the hardest and heaviest pure material with a hardness of 68 Rc and density of 0.697 lbs/in.$^3$, it has poor fabrication abilities.

Noticeably absent from this chapter are nickel-based alloys, which have not been included because they are a class of materials that are very unique and are not widely used because of their high cost and limited availability. When they are selected, it is typically for use in aerospace and energy producing applications where their excellent corrosion and heat-resistance are required. Nickel is frequently used as an alloy in stainless and alloy steels and copper alloys as well as a plating material.

Mechanical failure most commonly involves stress, fatigue, impact, or notch-based modes that are more sensitive in nonferrous metals. Although their tensile strength can be similar to many ferrous metals, the related strain can be up to three times as great due to difference in modulus ($E_{al} = 10$ Mpsi for aluminum versus $E_{carbon\ steel} = 30$ Mpsi for carbon steel). This additional strain adds to the surface sensitivity of the machine finish or other similar surface imperfections.

Fatigue failure is an important consideration because nonferrous metals do not exhibit a fatigue limit as observed in ferrous metals. Observation of acceptable loading and cyclic limits are paramount to the successful application of these materials to design environments prone to vibratory or cyclic operation. Aluminum is a material that is commonly selected for critical environments that are prone to fatigue failure, including aircraft and automotive applications. A careful consideration of the material's shortcomings is critical to the success of these applications. These include proper machining and handling to control surface roughness or damage, which can be the source of crack development, requiring the design to specify the allowable surface conditions and handling practices.

Surface wear is a common failure mode for engineered systems involving surface contact. The failure is due to three types of wear: sliding, impact, and rolling. Sliding wear is self-descriptive, as the wear occurs due to relative motion between two materials. The effects of this type of wear can be reduced by the use of lubrication or a material that offers similar properties. Selected nonferrous metals are chosen specifically for this purpose because they act as a lubricant between the surfaces. The most common material selected for this property are based copper alloys because they are easily alloyed with lubricating materials such as lead or zinc while providing a material of reasonable modulus (17 Mpsi).

Impact wear ranges from simple erosion to large magnitude surface contact. Erosion during the mass transport of solids or slurries often not only requires the use of materials that resist this form of failure, but also the selection of surface coatings. Copper and aluminum alloys do not perform well in these wear environments. The use of surface coatings can often reduce the material cost, but with the added expense of a secondary process, requiring a thorough cost analysis to determine the most effective solution. At the other end of this failure mode is that involving large loads. This is most often very difficult to design for, as the magnitude and frequency of the load can be difficult to quantify. Heavy equipment, such as a rock crusher or snow plow blade, are good examples of this problem.

Rolling wear failure is most often observed in ball bearings or other similar roller element systems. Providing adequate lubrication and protection from contaminants is paramount to preventing this form of failure. Shielded bearings and lubrication wipers are common application-specific solutions that are employed to extend the life of the roller elements.

Corrosion-failure modes for nonferrous metals include galvanic, intergranular, high temperature oxidation, and dealloying. Galvanic corrosion happens because of the presence of an electrochemical potential between two dissimilar metals when they are in a conductive solution. This form of corrosion is not typically a matter of concern within the same family of materials but exists when two dissimilar materials are in contact. For example, copper does not exhibit galvanic corrosion when in contact with other copper alloys but does behave as a cathode (electropositive) when in contact with aluminum or steel.

Intergranular corrosion is a failure mode that initiates between the grain boundaries, often extending to a depth of several grains. Unlike mechanical surface roughing, intergranular attack specifically produces stress concentrations between the grains which results in significantly earlier failure. The nonferrous metals most associated with this form of failure are Muntz metal (Copper/Zinc alloy, 60Cu–40Zn), admiralty metal (Copper/Zinc alloy, 70Cu–30Zn with 1% Sn), aluminum brasses, and silicon bronzes.

High-temperature corrosion generally manifests itself in the form of surface oxidation or "scale." Nonferrous metals are generally not used in excessively high-temperatures (with the exception of nickel based alloys) because of the inherently low melting point of the common alloys, such as aluminum and copper. When not exceeding their maximum temperature ratings, nonferrous metals do oxidize but generally exhibit good performance. In a static environment this may be of limited consequence, because the buildup of surface oxidation will likely retard additional surface degradation, resulting in little change in material performance. In environments involving wear or erosion, surface oxidation results in accelerated wear, as the surface oxide is easily removed continuously exposing a fresh surface for further oxidation. This cycle continues until failure occurs as a result of weakened structural capacity or contamination-induced failure, as would occur in a bearing.

Dealloying is a form of corrosion in which an alloy is degraded by the preferential removal of the more active alloying element(s). This form of corrosion of a solid-solution alloy is also known as selective leaching or parting and results in a weak, porous structure from which the more active metal has been removed. When this occurs to copper–zinc alloys it is called dezincification, in which the zinc is removed, leaving behind the copper in a porous form that does not possess the same structural integrity as the original material. This process will extend into the material and consume it if the source of corrosion is not removed and the material properly neutralized.

Stress corrosion cracking (SCC) is the spontaneous cracking of a metal under stress. It is most commonly an intergranular failure that occurs while the material is under load or stressed. The failure mode is dependent on two factors: the magnitude of the stress and the strength of the corrosive environment. Ammonia compounds are most often associated with copper alloys and their susceptibility to SCC. Aluminum alloys are affected by a variety of corrosive media ranging from sodium chloride to water. The presence of copper as an alloy in aluminum significantly retards the intergranular corrosion that results in SCC. Aluminum alloys with greater than 1% Cu are less resistant to general corrosion and include the 7XXX series. The 3XXX and 5XXX series have the highest general corrosion-resistance, with the 6XXX series following close behind and often the most used because of its general purpose properties including strength and weldability. The 5XXX series will outperform all of the others in marine exposure applications.

Titanium's susceptibility to SCC is dependent on its alloy content. Unalloyed titanium typically used in the chemical processing industry does not exhibit SCC unless the oxygen content is greater than 0.3%. Those titanium alloys typically used in aerospace applications are subject to SCC when a preexisting crack or discontinuity is present. Two additional factors influence the extent of SCC: alloy composition and heat treatment. Aluminum in levels greater than 6% and additions of tin, manganese, and cobalt increase sensitivity to SCC. Similarly, magnesium alloys with a composition that includes greater than 1.5% Al are also susceptible to SSC with wrought alloys being of greater concern than cast forms. Welding also increases the sensitivity to SCC.

Material aging for nonferrous metals has several meanings. These include the artificial aging, also called precipitation-hardening, of the material to control the microstructure and the long term aging affects of the material to exposure. All the nonferrous metals discussed here (excluding tungsten) have the ability to be artificially aged or strain aged with uniquely different results. These changes occur at ambient or elevated temperatures after processing (hot working, heat treatment or cold working) and result in a phase change but not a change in chemical composition. The aging of aluminum and copper alloys results in an increase in strength, while the aging of titanium and magnesium does not.

Conversely, the weathering form of aging is the result of prolonged exposure to an the operational environment and is highly dependent on the surface finish given to the material. Knowing this effect is critical in understanding the anticipated service life of the material. All nonferrous metals perform better when the surface is sealed, with aluminum and magnesium requiring a finish in all applications. Copper, titanium, and tungsten do not require a surface coating but do benefit from them. Failure because of this form of aging includes the modes discussed above, such as SCC and corrosion-failure modes.

### Performance Objectives

As a general rule, nonferrous metals have a lower yield point, lower modulus of elasticity and greater susceptibility to surface wear than ferrous metals. These properties translate into reduced manufacturing costs as machining and forming processes are less demanding, resulting in reduced tool wear and machining times.

The material properties of nonferrous metals vary widely, which makes it difficult to compare them directly with one another. To facilitate the decision process it is often necessary to use material indices which provide a method to compare and understand the material performance relative to one another and independent of the specific design.

For example, specific stiffness, $E/\rho$, and specific strength, $\sigma_y/\rho$, are two commonly used indices which quantify the mechanical performance relative to the density to provide an insight into the weight efficiency of the material. Carbon steel is included as a comparative reference point to the listed nonferrous metals. Details of these and other indices and their method of application are well documented in the given references [1] and [2] (Table 5.15).

There are some applications in which the selection is straightforward, as in the case of nonferrous bearing materials. Copper alloys are routinely selected for these applications and there are many to chose from. Similarly, noncorrosive spring material is most commonly beryllium–copper as it provides a reasonable spring constant in light to medium loading applications.

Secondary performance objectives include properties such as fatigue life, impact strength, notch-sensitivity, dealloying characteristics, etc.

Service life is also a significant consideration for nonferrous metals, as with other materials. Use of surface coatings is

**Table 5.15**  Comparison of Material Indices

| Material | $E/\rho$ ($\times 10^6$) | $\sigma_y/\rho$ |
|---|---|---|
| Aluminum | 104.6 | 41–800 |
| Copper alloys | 51.7 | 31–604 |
| Titanium | 94.1 | 235–941 |
| Magnesium | 100.0 | 153–369 |
| Tungsten | 86.3 | 65–144 |
| Carbon steel | 102.5 | 102–1060 |

paramount in extending the service life and these should be carefully selected based on the intended operating environment and life expectancy. Consulting with a knowledgeable coating process engineer is necessary to achieve the best result as they have a great deal of experience to facilitate the selection process.

## SELECTION METHODOLOGIES

Selecting a nonferrous metal can be a challenge because the designations are expectedly different than ferrous designations and generally they require some form of post-treatment to protect against oxidation. For example, when selecting an aluminum, the post-treatment most often involves either anodizing or chemical film treatment depending on the demands of the application (anodizing if environmental or wear protection is required and chemical film treatment if environmental protection in combination with electrical conductivity is required). Copper is less sensitive to surface oxidation but still may require treatment depending on the severity of the environmental exposure. Use in ambient conditions in an office or laboratory generally does not result in surface degradation or color change (to green), while exposure in an industrial application most certainly would. Tungsten is the most inert of the nonferrous metals, generally not requiring post-treatment, while titanium and magnesium require considerable attention to this detail. Although magnesium's use continues to expand, processing and corrosion-protect must be adequately addressed to produce a successful design.

The nonferrous metal selection process is different from that of ferrous metals in many respects because the performance objectives are not the same. Selection of ferrous metals is commonly driven by strength requirements while selection of nonferrous metals typically involves consideration for one or a combination of weight, wear and/or corrosion-protection. Prioritizing these performance objectives must be established first as it allows the selection process to then focus on the secondary performance objectives to drive the selection of candidate materials (Figure 5.11).

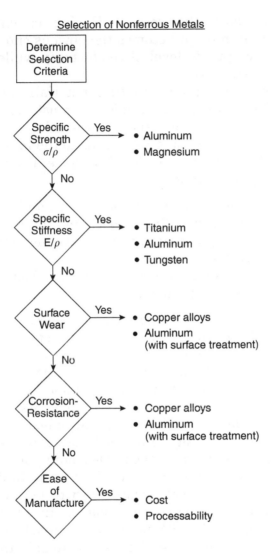

**Figure 5.11** Non-ferrous metals selection process flow chart.

Consider the selection of a material for use in long-distance, high-voltage power lines. At first pass, it would seem that conductivity is the most important parameter which would lead one to consider a copper alloy. But, upon closer examination, the ability of the material to support its own weight over great spans is the overriding requirement with conductivity being the second. Having established this requirement, aluminum becomes the obvious choice, because it offers an acceptable level of conductivity while providing the necessary strength.

When comparing and selecting materials, differences in failure modes between the candidates must also be considered. When nonferrous metals are selected or considered in substitution for ferrous metals it leads to the important need to consider these failure modes. Five general failure modes are of primary importance:

1. Mechanical: stress, fatigue, impact, and notch.
2. Surface wear: sliding, impact, and rolling.
3. Corrosion: galvanic, intergranular, high temperature, and dealloying.
4. Cracking: stress corrosion cracking and environmental.
5. Aging: embrittlement.

After having considered the performance objectives in the context of the above failure modes, the selection process can begin. Once several candidates are selected, detailed review of the available material databases must be performed to gain an understanding if the material is suitable for the predicted loads and environmental exposure requirements. Determining the values of the appropriate material indices is useful not only in providing an insight into the behavior of the candidate materials, but also to rank them. Depending on the application, detailed analysis may be required using finite element analysis techniques as well as laboratory test data to validate the selection and predicted performance. In addition to the above, working closely with material manufactures and suppliers is often the most effective method to gain the required knowledge to make an informed decision for the application.

Nonferrous material selection, as with other metallic materials, is typically not a straightforward process, requiring research and compromise and often risk due to the variable nature of the use environment. Use of engineering models and related testing is often a very good indicator of the materials performance in the final form if care has been given to accurately simulate the system level operation. To further gain confidence in the selection and performance characteristics, field-testing is a must at it allows the end user to interact with the design which often results in operational situations not considered during development that may lead toward early or unanticipated failure.

# 6

## Engineering Plastics

Plastic materials used in the design and manufacture of consumer, transportation, and aerospace products are made from synthetic polymers formulated to provide specific properties. These properties include:

- High specific strength (strength-to-weight ratio)
- Retention of mechanical properties at elevated temperatures
- Low creep
- Electrical properties (conductive or resistive)
- Chemical-resistance
- Thermal (conductive or resistive)
- Environmental-resistance

When a formulated plastic has one or more of these properties it is referred to as an *engineering plastic* because it provides the necessary chemical, electrical, and mechanical properties required of a material in an engineered design (Figure 6.1). Selecting the appropriate combination of these properties in the course of product development is the responsibility of the design engineer.

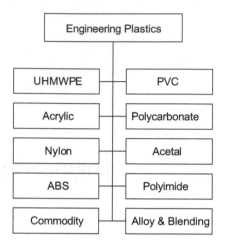

**Figure 6.1**   Common plastics used in engineering applications.

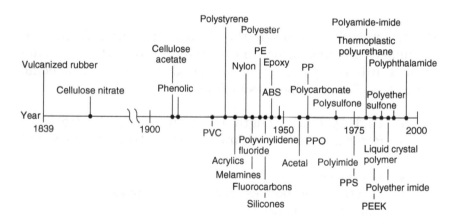

**Figure 6.2**   Chronological development of important engineering polymers.

The development of plastics has occurred relatively recently when compared to metallic materials and is generally considered to have begun with the vulcanization of rubber by Charles Goodyear in 1839 (Figure 6.2). The ability to modify natural rubber to improve its physical and chemical properties led to the belief that synthetically altered or fabricated polymers could be successfully produced. This was followed by the development of truly synthetic plastics called phenolic resins in 1909 by Leo H. Baekland. Plastics have grown from this simple beginning to form a class of materials that provide a wide range of properties that satisfy many design requirements.

Plastics generally have strength values that are a fraction of those achieved by metals with greater values of thermal expansion and lower operating temperatures. These less than favorable properties are offset by greater strength-to-weight ratios (specific strength) and chemical resistance properties not available from metallic materials. Composites of plastics that include organic materials such as glass or graphite fibers can produce materials with strengths comparable to or greater than similar sections made from metals, but at higher fabrication costs.

The basic chemical, electrical, and mechanical properties of plastic materials vary widely across the range of materials available, making them extremely versatile as engineering materials. Chemical resistance to acids, chlorinated solvents, and ultraviolet light (UV) can be achieved in plastics with many formulations. Electrically and thermally insulative, or conductive, properties can be obtained in varying degrees, as necessary for the application, by the addition of insulative or conductive filler materials. These varying resistivities are significant when working with semiconductors and related products, as the ability to control the static discharge characteristics is an important design consideration. The Triboelectric Series shows which materials will create the greatest amount of static electricity when rubbed together (Figure 6.3). When two of these materials are rubbed together under ordinary circumstances, the material listed on top becomes positively charged and the material listed later becomes negatively charged.

Mechanical damping and coefficient of friction characteristics can be varied over a greater range than with other materials by selecting the appropriate composition. All of these are possible while retaining significant engineering properties.

The ability to vary the properties of plastics allows them to be used in a wide range of design applications. They are

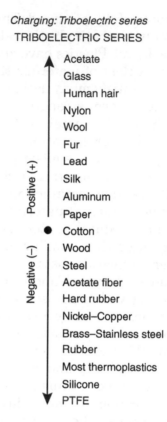

Charging: Triboelectric series

TRIBOELECTRIC SERIES

**Figure 6.3**    The triboelectric series.

*Source*:    Reprinted from Machine Design Magazine, Copyright Penton Media Inc. 2001.

often chosen for one or two specific properties, such as chemical inertness or ease of manufacture, rather than for their general properties as provided by their classification. Most often the applications require some particular mechanical property in conjunction with another physical property to meet the design requirement. For example, while a lens needs to be structurally rigid to allow it to be retained within a mount, it must also be optically clear. An exception to this generality would be that of a dielectric used in a capacitor, which must only provide the appropriate dielectric constant to give the device the necessary capacitance. Although it is evident that a secondary property is not needed in the device itself but may be beneficial to facilitate its manufacture. Unfilled or virgin polymer materials are known to be "neat," indicating that nothing has been added to alter the basic formulation.

Plastics are known to burn and produce toxic fumes when exposed to an open flame, making them dangerous when used in confined areas. By observing the material's behavior when a flame is applied to the surface, it is possible to identify the family it belongs to, which can be helpful when there is an immediate need to know what material a component is made of (Table 6.1). The Underwriters Laboratory (UL) classifies plastic materials in terms of their flammability by means of burning tests. These tests can be conducted in a vertical or horizontal orientation. The vertical tests measure the burn time and observe if the burning drops ignite a gauze placed under the sample. The horizontal testing also provides a method of measuring burn rate. The classifications are:

UL94V-0: Self-extinguishing within 10 sec, no flame-drips that ignite.

UL94V-1: Self-extinguishing within 30 sec, no flame-drips that ignite.

UL94V-2: Self-extinguishing within 30 sec, flame-drips that ignite.

UL94-HB: Reduced burning rate when burnt horizontally.

## PROPERTIES

Plastics are classified as either thermoplastic or thermoset materials. Thermoplastic materials are those that can be formed by the application of heat and pressure and subsequently reformed by applying heat and pressure once again; thermoset materials are formed as a result of a chemical reaction known as polymerization and cannot be reformed by the application of heat and pressure. These differences play a significant role in the material properties and performance of the plastic and, subsequently, its cost.

# Table 6.1 Flammability Characteristics of Plastic Materials

| Materials thermoplastics | No flame — Odor | Burns but extinguishes on removal of flame source — Odor | Color of flame | Drips | Continues to burn after removal of flame source — Odor | Color of flame | Drips | Speed of burning | Remarks |
|---|---|---|---|---|---|---|---|---|---|
| ABS | — | Acrid | Yellow, blue edges | No | Acrid | Yellow, blue edges | Yes | Slow | Black smoke with soot in air |
| Acetals | — | — | — | — | Formaldehyde | Blue, no smoke | Yes | Slow | — |
| Acrylics | — | — | — | — | Fruity | Blue, yellow tip | No (cast) Yes (molded) | Slow | Flame may spurt if rubber modified |
| Cellulosics acetate | — | Vinegar | Yellow with sparks | No | Vinegar | Yellow | Yes | Slow | Flame may spark |
| Fluorocarbons FEP | Faint odor of burnt hair | — | — | — | — | — | — | — | Deforms; no combustion, but drips |
| PTFE | Faint odor of burnt hair | — | — | — | — | — | — | — | Deforms; does not drip |
| CTFE | Faint odor of acetic acid | — | — | — | — | — | — | — | Deforms; no combustion, but drips |
| PVF | Acidic | — | — | — | — | — | — | — | Deforms |
| Nylons type 6 and 6/6 | — | Burnt wool | Blue, yellow tip | Yes | — | — | — | — | 6/6 More rigid |
| Polycarbonates | — | Faint, sweet aromatic ester | Orange | Yes | — | — | — | — | Black smoke with soot in air |
| Polyethylenes | — | — | — | — | Paraffin | Blue, yellow tip | Yes | Slow | Floats in water |
| Polyimides | —[a] | — | — | — | — | — | — | — | Chars; material rigid |
| Polypropylenes | — | Acrid | Yellow | No | Sweet | Blue, yellow tip | Yes | Slow | Floats in water; more difficult to scratch than poly-ethylene |
| Polystyrenes | — | — | — | — | Illuminating gas | Yellow | Yes | Rapid | Dense black smoke with soot in air |
| Polysulfones | — | —[a] | Orange | Yes | — | — | — | — | Black smoke |
| Polyurethanes | — | — | — | — | —[a] | Yellow | No | Slow | Black smoke |
| Vinyls flexible-rigid | — | Hydrochloric acid | Yellow with green spurts | No | — | — | — | — | Chars; melts |
| Polyblends ABS/carbonate | — | — | — | — | —[a] | Yellow, blue edges | No | — | Black smoke with soot in air |
| ABS-PVC | — | Acrid | Yellow, blue edges | No | — | — | — | — | Black smoke with soot in air |
| PVC/acrylic | — | Fruity | Blue, yellow tip | No | — | — | — | — | — |
| Melamines | Formaldehyde and fish | — | — | — | — | — | — | — | — |
| Phenolics | Formaldehyde and phenol | Phenol and wood or paper | Yellow | No | — | — | — | — | May crack |
| Polyesters | — | Hydrochloric acid | Yellow | No | —[a] | Yellow, blue edges | No | Slow | Cracks and breaks |
| Silicones | —[a] | — | — | — | — | — | — | — | Deforms |
| Ureas | Formaldehyde | — | — | — | — | — | — | — | — |

[a] Non-descript

*Source*: Port Plastics, Incorporated, 1997 Catalog, p. 295. Used with permission.

The material structure of a plastic is either crystalline or amorphous and in some forms a combination of both. A plastic with a crystalline structure has a repeating formation, which is similar to that found in metals, whereas an amorphous plastic does not. The materials with these properties extend the range of performance capabilities (Figure 6.4).

| Engineering Grade | Material Structure | Material | Uses |
|---|---|---|---|
| Commodity | Amorphous | ABS<br>PS<br>PVC | High volume applications involving low stress. |
| | Crystalline | HDPE<br>LDPE<br>PP | High volume applications requiring good chemical resistance. |
| Moderate | Amorphous | Acrylic<br>PC<br>PPO | General purpose for applications involving moderate loading. Bondable with minimal preparation. |
| | Crystalline | Acetal<br>PET<br>Nylon<br>UHMWPE | General purpose bearing materials with good chemical resistance. Dimensional stability varies. |
| Advanced | Amorphous | PEI<br>PPSU<br>PSU | High strength and hot water resistant materials with good formability. |
| | Crystalline | PEEK<br>PPS<br>PTFE | Superior chemical resistance with good wear resistance. Temperature and dimensional resistance varies. |
| Extreme | Imidized | PBI<br>PI<br>PAI | Superior temperature and chemical resistance. Typical uses involve highly demanding environments. |

| Abbreviation.... Material Name | | Abbreviation......Material Name | |
|---|---|---|---|
| PBI | Polybenzimidazole | UHMWPE | Ultra-High Molecular Weight Polyethylene |
| PI | Polyimide | | |
| PAI | Polyamide-imide | PC | Polycarbonate |
| PEEK | Polyetheretherketone | PPO | Polyphenylene Oxide |
| PPS | Polyphenylene Sulfide | PMMA | Polymethylmethacrylate (Acrylic) |
| PTFE | Polytetrafluoroethylene | | |
| PEI | Polyetherimide | PP | Polypropylene |
| PPSU | Polyphenylsulfone | HDPE | High Density Polyethylene |
| PSU | Polysulfone | LDPE | Low Density Polyethylene |
| PET | Polyethylene Terephthalate | ABS | Acrylonitrile Butadiene Styrene |
| PBT | Polybutylene Terephthalate | | |
| PA | Polyamide (Nylon) | PS | Polystyrene |
| POM | Polyoxymethylene (Acetal) | PVC | Polyvinyl Chloride |

**Figure 6.4** Amorphous and crystalline materials offer a variety of material properties for use in engineering applications.

Plastics offer material properties that cover a wide range of performance capabilities (Table 6.2). There are many variants of these materials with significantly altered material properties, which can lead to the presence of several appropriate candidates during the selection process. Careful consideration of the properties of the base polymer and its interactions in the application can narrow the list of available materials to those that will perform best.

## STRUCTURE

Plastics, like metals, possess chemical structures of various forms, which are the basis for their properties, performance, and manufacturing methods. The arrangement of the carbon-based polymer chains defines the linear and branched arrangement and resulting molecular weight which affects the extent a plastic can crystallize. Plastics are defined either by the arrangement of the atoms (linear or branched) or by the chemical reaction which formed them (cross-linked or net-worked). These forms are then the basis for the resulting structure of the solid material (amorphous or crystalline).

### Molecular Weight (MW)

Thermoplastic materials consist of discrete molecules that can be of varying lengths and are commonly known as polymer chains. The average length of these chains is determined by the polymerization process and can be expressed in terms of molecular weight (MW). As a general rule, the engineering properties of the polymer increase as the MW increases, because longer chains create a tangled structure, imparting greater tensile and compressive strength.

Thermoset materials are not considered to have a MW in the sense that it is defined for thermoplastic materials, because they do not have discrete molecules, but rather, consist of a large web structure. The material is then one large, interconnected molecule, which is not characterized by its MW.

**Table 6.2** Material Properties of Selected Plastics

| Material | Tensile strength @ yield (psi) | Tensile strength @ break (psi) | Impact strength (ft.lbs./in.) | Coefficient of friction | Moisture absorption (%) | Operating temperature long term (°F) | Deflection temperature (°F) |
|---|---|---|---|---|---|---|---|
| UHMW | 3,250 | 5,000 | 18 | 0.17 | 0 | 105 | 200 |
| PVC | 5,070 | 6,670 | 5 | n/a | 0 | 176 | 192 |
| Acrylics (cast) | 8,700 | 10,500 | 0.4 | n/a | 0.3 | 234 | 230 |
| Polycarbonate | 8,990 | 9,282 | 14 | 0.38 | 0.5 | 250 | 275 |
| Nylon 6 | 8,900 | 10,400 | 5 | 0.2 | 3 | 212 | 300 |
| Acetal | 8,412 | 8,847 | 1.4 | 0.25 | 0.5 | 320 | 302 |
| ABS | 6,190 | 6,526 | 5 | n/a | 0.3 | 140 | 210 |
| Polyimide | 17,400 | 13,300 | 0 | 0.3 | 1.3 | 525 | 625 |

## Linear and Branched Polymers

These polymers have a structure that consists of either linear or branched molecular chains that can be polymerized into materials of varying MW. Linear polymers form a structure of intertwined chains, which is dependent on the lack of order of the chains for its material properties. Branched polymers are linear polymers that have split into two chains somewhere along their length, during the polymerization process. Branching provides for greater strength and stiffening, because the molecules are intertwined to a greater degree. The bonding mechanism holding the polymer chains together includes Van der Waals forces and hydrogen bonding. These polymers forms have operating temperatures below 250°F (121°C).

## Cross-Linked or Networked Polymers

Cross-linked polymers are those in which the polymer chains are connected by a chemical bond, generally improving the material properties at the expense of processing flexibility. The strength of these bonds prevents remelting after the cross-linking has occurred because the bonds are too strong, but it does allow for improved high-temperature properties. In fact, heating of cross-linked polymers results in a chemical breakdown of the material, which is irreversible and destroys the material. Operating temperatures of nearly 500°F are obtainable with some cross-linked polymers.

## Crystalline

Crystalline and amorphous structures are the result of how linear, branched, and cross-linked polymer chains arrange themselves. The extent to which they exist in either of these states is dependent on many factors, including the amount of branching and cross-linking. Crystalline polymers are the result of linear or branched polymers folding over themselves in a regular manner, forming a repeating structure that is considered to be a crystal. These polymers are not typically entirely crystalline, but are semi-crystalline, as portions of the structure may be unable to fold up, remaining as amorphous regions. The extent to which a polymer is semi-crystalline is dependent on the polymerization process, post-processing thermal treatments, and the use of additives. Those with the most flexible chains generally form materials with the highest degree of crystallinity.

## Amorphous

Thermoset polymers exhibit an amorphous structure due to the fixed nature of their molecular form, which prevents folding of the polymer chains into a repeating structure. Because of this, thermosets are always amorphous. Thermoplastics can exhibit both amorphous and crystalline structures depending on the material properties and processing. Those that have polymers chains that are too rigid, or have excessive branching, are less likely to fold into a repeating structure that is necessary to constitute a crystalline polymer and remain amorphous. When selecting a material it is important to consider the differences between crystalline and amorphous materials (Table 6.3).

## Transition Temperatures

Amorphous polymers do not have a melting temperature but instead have a glass transition temperature, $T_g$. Below this temperature, the polymer remains a rigid solid and above it the material exhibits a gradual softening. Thermoplastics will continue to soften and melt as the temperature is increased, while thermoset plastics will begin to degrade, and possibly, burn when exposed to continued heating.

**Table 6.3** Crystalline and Amorphous Material Property Comparison for Plastics

| Crystalline | Amorphous |
| --- | --- |
| Higher tensile strength | Lower tensile strength |
| Distinct melting point | No distinct melting temperature |
| Opaque | Transparent |
| High shrinkage rate | Low shrinkage rate |
| Solvent-resistant | Solvent-sensitive |
| Fatigue/wear-resistant | Limited fatigue/wear-resistance |
| Greater toughness | Reduced toughness |
| Lower creep resistance | Higher creep resistance |
| Anisotropic | Isotropic |

Crystalline polymers have a distinct melting point, $T_m$, at which the polymer begins to lose its repeating order and flow. Processing of these thermoplastic polymers requires heating above this temperature.

## Porosity

Centerline porosity of large extruded plastic stock shapes is not uncommon with pore sizes ranging from less than 30 to above 150 $\mu$m (0.001–0.006 inch). There are two common reasons for porosity, both of which involve the thermal effects of processing. First, the thermal degradation of the core or center can occur due to the out-gassing of decomposition products, which can produce measurable porosity at the core. Second, the cooling of the large mass results in internal tensile stresses due to material shrinkage that can cause material cavitation, which is observed as porosity.

A simple dye penetration test method can be used to identify the extent of the porosity. This involves sectioning the material across its centerline and applying a dye to the surface. After allowing it to dry, wiping the surface with a suitable solvent to remove the surface dye will revile that which is trapped in the surface pores. A porosity of <30 $\mu$m will not retain the dye, and may still appear white, while a porosity >30 $\mu$m will retain the dye in amounts dependent on the pore size. The dye can be a simple red permanent marker or a more sensitive ultraviolet type.

## THERMOPLASTICS

Engineering thermoplastics are those materials made from linear or branched polymers and have a molecular weight. They are easily identified because they are capable of being reversibly softened by the application of heat. This characteristic allows them to be thermal formed, injection molded, and thermal welded, while reamining machinable from stock shapes. These materials are held together by means of secondary bonds such as Van der Waals forces and obtain their mechanical properties from the length of the polymer chain. The length of the chains are directly proportional to the molecular weight, meaning the longer chains produce a material with a higher molecular weight (MW). This higher MW provides improved engineering properties such as higher tensile and impact strength.

Thermoplastic polymers have either a melting temperature or a glass transition temperature, depending on their structure. Their structures range from fully amorphous to highly crystalline, with the former exhibiting a glass transition temperature and the latter exhibiting a melting temperature. Those materials that have a more uniform MW distribution will also have a structure that is more crystalline, because the polymer chains can fold upon themselves to form the necessary repeating structures. Although this may lead to greater crystallinity, the structure will never become entirely crystalline, as there will always be amorphous regions owing to the random nature of polymer chemistry, which will not produce repeating chains of exactly equal length.

Manufacturing with thermoplastics is typically more cost-effective, because the process time is shorter and the tooling costs are less. Thermoplastic materials do not require a cure time, as do thermoset plastics, but need only be cooled to a temperature that imparts sufficient structural strength prior to removal from the mold or fixture. They can also be recycled by applying heat and reforming, when working with panels and other shapes, or by regrinding and blending with virgin material, when working with injection molded products. It is worth noting that use of reground material can be limited by the product specification and also by processing techniques. Too much regrind can cause dimensional and cosmetic problems, because the material may change color or burn slightly when reprocessed and alter the thermal expansion characteristics.

Examples of thermoplastic materials include poly vinyl chloride (PVC), polystyrene (PS), polyamides (PA), and ultra-high molecular weight polyethylene (UHMWPE).

## THERMOSETS

Engineering thermosets are materials made from cross-linked polymers; their molecular weights are not defined, as they have a homogeneous structure. These materials cannot be reshaped, or reused, by heating because of the primary covalent bonds (electron sharing) that are formed during the polymerization or curing process, resulting in an irreversible chemical change in the material, in the form of cross-linking. If a thermoset is subjected to temperatures exceeding its $T_g$, it will burn or become charred rather than melt like a thermoplastic. Thermosets most often compete with metals when the polymer contains reinforcement materials such as glass or carbon fibers to increase their mechanical properties.

Thermoset plastics obtain their mechanical strength from their chemically formed cross-linked chain structure. They are stiffer as the chains are shorter and/or more numerous, producing a tangled web that resists movement. The addition of fillers and reinforcing materials, in some cases, increases the material properties by an order of magnitude. For example, a neat, or unmodified, unsaturated polyester resin has a tensile strength of 8.5 ksi while the tensile strength increases to 88.5 ksi when cured with a unidirectional woven fabric.

Processing of thermoset materials can range from room temperature forming or machining to injection molding, which requires heat and pressure to obtain full cure. Performing room temperature operations such as machining does not differ from what is done when using thermoplastics, while other operations, such as thermal, sonic, or solvent welding, cannot be performed on thermosets. Manufacturing complex and/or high-strength structures using thermosets requires complex fixturing and machines to apply heat and pressure to obtain the optimal desired properties once cured. Because of the added complexity of both equipment and cure time, thermosets can be a costly material selection that the design requirements must justify. Examples of thermoset materials include epoxy and unsaturated polyester.

## CLASSIFICATIONS

This chapter covers eight of the most common plastics found in engineering applications, as well as several commodity plastics. There is also a section on compounding, which involves the alloying and blending of plastics, as well as the use of fillers and reinforcements to improve desired material properties. It is useful to note that finished molded products will often include information, in the form of a symbol, that can assist in recycling (Figure 6.5).

PET–Polyethylene terephthalate is a transparent plastic with good gas barrier properties. It is also relatively cheap and strong. This makes it especially suitable for producing soft drink bottles.

HDPE–High density polyethylene is another plastic commonly used for drink bottles. It is translucent and strong with a good resistance against chemicals. Pigments are used to add color to HDPE bottles.

PVC–Polyvinyl chloride is frequently used for drainage pipes because of its superior chemical resistance. It is also a good insulator and is used for cable jacketing.

LDPE–Low density polyethylene is a tough and highly flexible plastic that is used for producing plastic bags, cable jacketing, and bottles. Other applications include products that can be heat sealed.

PP–Polypropylene is a strong and light plastic. It has a relatively high melting point and is used in products exposed to high temperatures. These include yogurt and ice cream containers that are filled while hot.

PS–Polystyrene is rather stiff and rigid. In this form, it is used in disposable cutlery and CD jewel cases. Packing foam is also produced by blowing air bubbles into PS while it is still setting.

Others–This designation is used for plastic products that do not fall into any of the other categories. Further sorting of these items is required at recycling centers. An example is melamine, which is often used in plastic plates and cups.

**Figure 6.5**   Material identification symbols for plastics.

*Source*: EarthOdyssey, LLC. Used with permission.

## Ultra-High Molecular Weight Polyethylene (UHMWPE)

Polyethylenes are members of the polyolefin family of polymers of which UHMWPE has the most desirable material properties. Although UHMWPE is often not considered an engineering thermoplastic because it lacks substantial structural strength, it does have other properties that make it suitable for engineering applications. There are other variants including ultra-low or very low density (ULDPE or VLDPE), low-density (LDPE), high-density (HDPE), and high molecular weight (HMWPE) polyethylenes. Although these are important materials they are not typically found in demanding applications because of their limited engineering properties.

### Properties and Designations

Generally, polyethylene is very lightweight, tasteless, odorless, and nontoxic, making it suitable for many food and beverage processing applications, when FDA and USDA compliant formulations are required. UHMWPE has properties that differentiate it from other polyethylene forms, which makes it an important engineering material (Table 6.4).

### Coefficient of Friction

Ultra-high molecular weight polyethylene has excellent friction properties, which led to its use as a solid lubricant and liner. The coefficient of friction at room temperature is 0.2–0.25 for static and 0.15–0.2 for dynamic friction. If the application requires a reduced coefficient of friction, silicone oils and molybdenum disulfide are used as additives. Low friction combined with its superior abrasion resistance is an ideal combination for liners in trucks and trailers for hauling sharp or abrasive cargo. It is used as a solid lubricant to protect metal surfaces in rail cars.

### Wear-Resistance

The wear-resistance of UHMWPE is superior to that of most other thermoplastics and arises from its molecular weight, which is described as "ultra-high" because it is in the range 3 to 6 × $10^6$, the highest for a polyethylene. Wear-resistance is

**Table 6.4** Engineering Properties of UHMWPE

| Property | Application considerations |
|---|---|
| Low coefficient of friction | This is one of the few polymers that exhibits a self-lubricating characteristic, resulting in a low coefficient of friction against metals and other plastics. It also results in a low stick-slip or break away behavior, with the difference between static and sliding coefficients being small |
| Wear-resistance | Because of its low coefficient of friction and semicrystalline structure, it has good wear-resistance against many aterials |
| Impact-resistance | UHMWPEs have a high impact-resistance or energy absorption capability which allows it to bend but not break when subjected to Izod impact notch testing Under tensile loading, it will elongate several hundred percent before failure |
| Chemical-resistance | Similar to all polyethylenes, UHMWPE exhibits excellent chemical-resistance to corrosive materials. Exceptions include fluorine and strong oxidizing acids. It is also low moisture absorbing, making it a stable material in many environments |
| Resistance to stress-cracking | Although UHMWPE has good resistance to stress-cracking, it is at the expense of the material's stability. When this polymer is used without additives or fillers, it cold flows |
| Dielectric properties | It has very good dielectric properties, making it useful as an insulator |
| $T_m$ | Limited branching and high molecular weight allows for a high melting point |

commonly provided in an index measured by the manufacturers test method. For example, an abrasion index of 10 indicates that the abrasion-resistance is 10 times that of carbon steel. Peroxide cross-linking UHMWPE further enhances its abrasion-resistance and reduces its deformation under load but affects impact strength and stiffness. UHMWPE is the primary material used in hip and knee joint replacements, with an estimated life of 15 years in the human body. Its performance does degrade to the point where it must be replaced, because wear particles inhibit the functioning of the joint. As mentioned previously, this wear-resistance is the primary reason it is selected for applications like liners, farm machinery, bearings, gears, mining industry, and joint replacements.

### Toughness

The mechanical toughness properties of the polyethylene family are excellent and UHMPWE is no exception. It has over 300% elongation at its break-point and about 20 to 27% at yield. The ultimate tensile strength is in the range of 5,800 to 6,500 psi (40–45 MPa). But its most outstanding mechanical property is impact toughness, which is probably the highest of that of any plastic. UHMWPE has excellent impact toughness even at cryogenic temperatures. Because of its very low temperature impact toughness, it is very commonly applied in parts dealing with liquid nitrogen or freezing temperatures such as edge protection on ice rinks.

### Stress Rupture

Like its other superior mechanical properties the stress rupture of UHMWPE is also excellent. It has an excellent resistance to environmental stress-cracking and to cyclic fatigue. No material is everlasting, but UHMPWE comes close in terms of cracking under stress. In the polyethylene family the yield strength and stiffness are a function of crystallinity which is about 45% for UHMWPE. The resulting properties are slightly worse than the linear polymers with lower molecular weight, but the increase in abrasion-resistance, impact strength, and fatigue strength compensates for the loss in yield strength.

### Machinability

UHMWPE can be machined easily with the use of sharp tools and common cooling fluids. For parts having thick sections or fine detail, machining is the most common fabrication method. This is mainly due to the difficulty in injection molding due to the long chains of the polymer. Operations such as drilling, milling, turning, sawing, planning, and screw-cutting are routine. For long production runs, carbide tools are recommended. Sheet metal punch methods can also be applied but features must be compensated for "spring-back," or the inherent deformation of the material that occurs during the process.

### Electrical

The electrical properties of UHMWPE are similar to those of most thermoplastics. It is a good insulator with an electrical resistivity in the order of $10^{16}$ $\Omega$ cm. It has a dielectric constant of about 2.3 with a dielectric strength of 28 kV/mm. The dissipation factor at lower frequencies is about 0.0002. It is also used in making conductive sheets; where sheets of UHMWPE are used for their superior mechanical properties, while the surface is coated with a conductive substance.

### Chemical-Resistance

UHMWPE has excellent chemical resistance to most of the solvents and environments. It has also excellent resistance to radiation and is regarded as a radiation-resistant polymer. Excessive doses will eventually cause embrittlement but the threshold is high. Owing to its superior resistance to different environments UHMWPE is the material of choice for many food and processing industries. It has been approved by FDA, USDA, and national bureau of standards sanctions for pure water and food handling. It is very stable and consists of extremely durable polymers, hardly attacked by non-oxidizing acids, alkalis, and many aqueous solutions. Nitric acid may oxidize the polymer leading to a deterioration of its mechanical properties. High dosages of UV radiation should be avoided and, in the case of applications where exposure to radiation is common, additives such as carbon black and certain pigments that absorb UV should be avoided.

### Flammability

The flammability characteristics of UHMWPE are not among the best. It is considered a self-extinguishing plastic, but is given a HB rating by UL94 testing which is the least fire retardant rating in UL94. When exposed to a flame it shows slow horizontal burning but it stops on its own. However, in applications where strict flame retardance is required, UHMWPE should not be used.

### Weatherability

UHMWPE has good weatherability with a high threshold for UV absorption, but will become brittle as it ages. Its resistance to moisture and temperature variations is excellent. Special grades of UHMWPE are available which are more UV-resistant by avoiding UV-absorbing additives.

### Water Absorption

Ultra-high molecular weight polyethylene typically has 0.01% water absorption. It is often selected for its chemical inertness and low water absorption characteristics for use in the food industry as well as in biomaterial applications.

### Thermal

Ultra-high molecular weight polyethylene has an approximate melting point of 270°F (130°C), with a maximum service temperature range in air of 140–180°F (62–82°C). Its minimum service temperature is −40°F (−40°C) with brittleness becoming significant below −103°F (−73°C), making it a good candidate for cryogenic applications. Its coefficient of linear expansion is 69–111 μin/in°F (125–200 μm/m°C). Processes such as sheet stamping, compression molding, and forging are carried out in the range of 290–310°F (140–155°C).

### Optical

UHMWPE is not commonly used in optical applications because it lacks sufficient transparency and rigidity. The crystallinity of UHMWPE is usually in the range of 40 to 45% and becomes clear when heated. It is mostly available in translucent grades. Its light transmission is about 80%, which is significantly below the 90% or greater required for most optical applications.

## Fabrication Processes

UHMWPE are fabricated from long, linear polymer chains, which allows for close packing and greater crystallinity. These high molecular weights are achieved by polymerizing ethylene gas under controlled temperature and pressure, with a metal catalyst. It is a costly process, which yields materials suitable for use in industrial applications, such as conveyor wear strips, product handling, feed screws and starwheels, gears, bushings, and other components requiring moderate strength and wear-resistance.

## Raw, Stock, and Mold Finishes

Most forms of UHMWPE are either cast or extruded into sheet, round, and rectangular shapes that can be machined using conventional or CNC equipment. Specifically, they include:

- Heavy gauge sheet ($1/_8$–6 in.) stress-relieved
- Rod ($1/_4$–4 in. diameter)
- Tubing (ID $5/_8$–3 in., OD $1 1/_8$–$3 1/_2$ in.)

The standard, or natural, unmodified color is white. Other colors are available as custom products but may have slightly reduced properties as a result of the fillers used for coloring.

## Additional Considerations

UHMWPE can be altered to improve selected material characteristics, such as stiffness and lubricity. The methods used to achieve them include chemical cross-linking, heat treatment, and including additives.

### *Stiffening Mechanisms*

Polyethylenes can be cross-linked either by chemical reaction or by high-energy irradiation, both of which have the same effects on the material's properties but at a loss of processing ease and increased cost. Improvements in material properties include greater stiffness at elevated temperatures, improved fatigue and abrasion-resistance, reduced friction and thermal expansion coefficients, and improved UV-stabilization without the use of a filler such as carbon black. Its electrical properties remain excellent, with improved dielectric strength at elevated temperatures. The natural color for this form may deviate from the virgin white depending on processing.

*Thermal Treatments*

Stress-relieving is used to improve material stabilization to allow machining of dimensionally stable parts.

*Additives*

Many additives can be included in UHMWPE to obtain improved properties but all result in increased costs. Additives such as oil or $MSO_2$ are used to reduce friction futher while maintaining the excellent wear properties of UHMWPE. Use of carbon black improves UV-stabilization. Addition of glass beads provides better abration-resistance and improved shock strength.

## Commonly Used UHMWPE

- Virgin UHMW
- Recycled UHMW
- Oil-filled UHMW
- Anti-static UHMW

## Polyvinyl Chloride (PVC)

Polyvinyl chloride is not commonly considered an engineering plastic because of its low load-deflection temperature compared to other higher performance plastics. PVC does have other properties that make it suitable for engineering applications, placing it in the engineering plastic category, including its most important, excellent chemical resistance. It also has a high strength-to-weight ratio and, in rigid forms, is very stiff and dimensionally stable with high compressive strength, making it an important material for structural applications requiring stiffness and strength at ambient conditions. Additionally, it is considerably less expensive compared to other engineering thermoplastics such as unmodified acetal, which has similar room temperature tensile strength and modulus properties.

Polyvinyl chloride is an amorphous thermoplastic that can be formulated into rigid, flexible, or expanded (foam) forms. The amorphous structure exhibits slight branching, which is responsible for the excellent stiffness and compression properties of PVC. Rigid forms are readily machinable with conventional or CNC tools to standard tolerances (within ±0.005 inch). It can also be thermal welded or solvent bonded, as well as assembled using conventional adhesives such as acrylic or epoxy but with variable results dependent on the amount and types of additives included in the PVC. Flexible PVC, also known as vinyl, is typically injection molded or cast and used for a wide range of applications, but not typically as an engineering material. Expandable forms are also available for non-loadbearing applications including prototype structures that may require dimensional accuracy without the added effort to machine a rigid material.

A variant of PVC is chlorinated PVC (CPVC), which has a higher heat distortion temperature and improved corrosion-resistance at 210°F (99°C) compared to 160°F (71°C) for standard PVC. This allows CPVC to be used in potable water piping systems in which regular PVC would not hold up when exposed to hot water. CPVC also exhibits greater fire-resistant properties than standard PVC.

## Properties and Designations

Vinyl polymers are among the oldest polymers and have the unique ability that they can be formulated with numerous compounding ingredients, producing a broad range of materials for use in numerous applications. The most notable property of PVC, which makes it an engineering polymer, is its excellent chemical-resistance. This property, in combination with being tasteless, nontoxic, and odorless, makes it suitable for applications such as in making factory automation components, packaging, building materials, tubing of various grades, and toys. It also has a high strength-to-weight ratio but is limited by its lack of fracture toughness.

The following properties are discussed in the context of rigid PVC as they apply to engineering design applications.

### Coefficient of Friction

This property is not commonly quoted in a manufacturer's data sheets, because PVC has poor lubricity in its unmodified form. Alloying with other elements can offer improved friction properties but is may be best to select another material such as an acetal or $MSO_2$ filled nylon to obtain the required properties cost-effectively.

*Wear-Resistance*

Polyvinyl chloride is not commonly considered a bearing material for use in applications involving sliding or rotating against metals or other plastics because of its poor lubricity and limited high-temperature properties. Alloying with other plastics or including lubricant fillers such as $MSO_2$ is not generally done because it reduces the glass transition temperature, which is important to maintain in a sliding friction environment.

*Toughness*

In its unmodified form, PVC has good impact strength and toughness. It can be formulated to be virtually unbreakable, with a notched Izod impact strength of greater than 0.5 J/mm (>10 ft lbf/in.) at $-40°F$ ($-40°C$). PVC without impact modifiers is sensitive to impact; typically 90% of the energy of impact is used in crack initiation and the rest in crack propagation. It is therefore very important to avoid sharp notches by giving sufficient radii.

The impact resistance or energy absorption capabilities can be improved by alloying with ABS or nitrile rubber but with some loss of tensile strength. Use of plasticizers is also effective in improving the toughness but at the expense of a reduced glass transition temperature.

Notch-sensitivity can be mitigated by avoiding the use of sharp corners or abrupt changes in features. The use of a generous radius to reduce stress concentrations is necessary to prevent early failure. As a general rule the radius should be 0.6 times the part thickness, but not less than 0.2 times in any case. Temperature also affects the toughness of PVC, high temperature results in higher Izod impact values.

*Stress Rupture*

All polymers have the potential to creep and eventually rupture when subjected to a constant strain condition. The data for this behavior is typically recorded until failure and plotted on a log-log scale at specific temperatures. As an example, consider PVC subjected to 5,800 psi (40 MPa) tensile load in room temperature conditions. Typically, in one year the stress in the sample will reduce to approximately 5,000 psi (35 MPa), a 12.5% reduction in stress. The curve can be extrapolated if desired but consideration should be made for nonlinear behavior that can occur over extended periods. It is common to design with a safety factor of 2, if the part design has notches, smaller radii, or weld lines and the safety factor is higher.

### Machinability

Polyvinyl chloride has outstanding dimensional control because of its high stiffness in compression, which is a result of the relatively low processing temperatures used when producing extrusions. Machining PVC is possible using traditional machine tools and water-based coolants, producing excellent surface finishes (<32 μin). PVC machining can be avoided by applying processes such as extrusion, injection molding, or blow molding where the parts produced in final form with superior finish. Lubricant additives help in processing and machining, internal lubricants help in reducing melt viscosity and prevent overheating, while external lubricants help in bringing it off the dies. High melt viscosity grades can be used to provide even higher dimensional control.

### Electrical

Unmodified PVC behaves as an insulative material, with a surface resistivity greater than $10^{12}$ Ω. It is produced in static dissipative forms to improve static control properties but at the expense of tensile strength and modulus. The fillers used to provide the electrostatic discharge (ESD) protection can vary from those that require moisture to improve their conductivity, such as nylon-based materials that absorb moisture to carbon black that inherently is conductive and easily dispersed uniformly within the material. Properties vary for different compounds. For example, the dielectric strength for rigid grades is up to 500 V/mil, whereas for flexible grades it is approximately 400 V/mil.

### Chemical Resistance

Polyvinyl chloride has excellent resistance to most chemicals, including sulfuric acid and chlorine bleach, but is less resistant to organic chemicals, such as acetone and carbon tetrachloride. Rigid PVC is commonly used in food and chemical processing plants due to its chemical-resistance as it does not exhibit cracking due to environmental stress and is resistant to most oxidizing inorganic acids and alkalis. PVC is much better than polyolefins (polyethylene and polypropylene) for resistance to mineral oils, petrol, and paraffin, but absorbs and is swollen by aromatic hydrocarbons, chlorinated solvents, esters, and ketones. Most manufacturers and their distributors can provide comprehensive rating lists of chemical compatibility for their product offerings.

### Flammability

Polyvinyl chloride is a self-extinguishing plastic because it will not sustain combustion after the ignition source has been removed. This desirable property is offset by the extreme toxicity of PVC when burned because of the chloride gas produced during combustion. PVC does not burn because of the high amount of chlorine in the compound and is occasionally used as a flame-retardant additive to ABS. Most of the grades have a UL94 V0 and 5V rating. Flexible grades are not as self-extinguishing as the rigid grades and they need additional flame-retardant additives to avoid burning. The production of chloride gas while burning presents a serious threat, and hence it is necessary to avoid burning. No flame-retardant can make any plastic fire-proof, however, it can reduce the susceptibility to fire by increasing the ignition temperature and reducing the rate of flame spread. The common additives for reducing flammability are alumina trihydrate, antimony trioxide, and borates.

### Weatherability

Polyvinyl chloride compounds that are properly designed and processed have outstanding weatherability, including good color and impact retention, good tensile and flexural strength retention, and no loss in modulus. For example, rigid vinyl exterior window profiles and house siding installations have accumulated more than 30 years of weathering history with good color and physical property retention. Good resistance to UV and outdoor exposure can be obtained from unmodified PVC.

### Water Absorption

When compared to other plastics, PVC has lower than the average absorption which contributes to the high degree of dimensional stability. PVC does absorb water when exposed for a substantial time period. The amount of water absorbed is different for different compositions and in general, the rigid grades absorbs less water, about 0.04–0.4% during 24 h, while the flexible grades can absorb 0.15–0.75% during 24 h.

### Thermal

As previously discussed, the low thermal properties of PVC have limited its use as an engineering plastic. It is intrinsically unstable because of molecular defects in some of the polymer chains, and when subjected to heat they initiate a self-accelerating dehydrochlorination reaction. Stabilizers neutralize the HCl produced and introduce nucleophilic substation reactions that prevent further degradation. The common heat stabilizers used are lead salts, lead soaps, organo-sulfur tin, Ba–Zn liquid, Ca–Zn liquid, diketones, and epoxy esters. Rigid PVCs are normally used upto 60°C but can be used up to 100°C by external reinforcement with glass reinforced polyester or epoxy resin. Chlorinated PVC (CPVC) can be used up to 120°C due to its higher chlorine content. A heat deflection temperature of 82°C (179°F) is low compared to many of the other thermoplastics which often are above 121°C (250°F).

### Optical

Rigid PVC is not a material typically considered for use in optical equipment because it is, at best, a translucent or opaque material. PVC, depending on the composition and process, is available in opaque to transparent formulations. Most of the grades available are translucent or opaque with less than 5% haze. Flexible grades are available in transparent formulations for use in tubing or other molded forms. It can be made available in all colors.

## Fabrication Processes

The primary production method involves reacting natural gas with sodium chloride acid in the presence of an appropriate catalyst. It cannot be processed neat because there are additives required for processing that must be included to make the resulting product useful. PVC is commonly compounded with plasticizers to impart toughness and stabilizers to improve UV-resistance.

## Raw, Stock, and Mold Finishes

Polyvinyl chloride forms can be processed by injection molding, extrusion, and foam casting. Extrusions are available in sheet, bars, and rounds, while foam or expanded products are available only in sheet form. The most common color available is gray, because it retains the highest material properties. Adding fillers to change color will reduce the already lower than average material properties, but is done for less demanding or cosmetic applications.

Extruded forms are commonly available in the following sizes, sheet (1/16–4 in. thick), rod (1/4–10 in. diameter), tube (1/4–6 in. diameter), bar (1/4–6 in., also available in hollow bars), and custom shapes. The overall size or length can vary depending on the cross-section of the form and due to the fillers included in the formulation. For example, ESD materials are generally not available in large extruded shapes and do not have good mechanical or thermal properties.

## Additional Processing Considerations

It is important to remember when considering the use of a modified PVC that the material properties are going to be reduced from their already low values.

### Stiffening Mechanisms

Polyvinyl chloride is inherently stiff and brittle, so it is typically used in its unmodified form and not commonly combined with fillers such as glass or carbon fibers. When toughness is required, plasticizers are added to effectively temper the polymer properties sufficiently to improve the energy absorption properties. The quantity of plasticizer determines the flexibility of the formulation, which can range from rigid to extremely flexible.

### Thermal Treatments

The two widely used processes for making PVC products are injection molding and extrusion. The melt temperature for injection molding is 196–204°C (385–400°F). In the case of extrusion, the melt temperature is a little lower: 170–200°C (340–390°F). Because of these relatively high processing temperatures for PVC, vinyl degradation takes place at these temperatures for unstabilized PVC. Heat stabilizers are added to make processing possible without degradation. Because of its lower tolerance for temperature, thermal treatments are usually not applied to PVC products. In the case of a process that causes severe strains to be left in the final product, a stress-relieving heat treatment could be provided, usually in a range less than 212°F (100°C). Addition of heat stabilizers is crucial for avoiding degradation if heat treatment is necessary.

### Additives

Plasticizers are the primary additives used with PVC but others, including UV stabilizers, are necessary to improve properties such as weatherablity.

## Commonly Used PVC

- PVC, Type 1 (standard)
- PVC, Type 2 (impact resistant)
- PVC foam grade
- PVC, alloyed with ABS or polypropylene

## Acrylics

The clarity of acrylics have long been their most notable property and the primary reason for their selection in optical applications. High clarity in combination with UV-resistance, modest impact strength, and abrasion-resistance make them useful in consumer applications such as windows, skylights, and protective lenses. The need for these properties justifies the high cost of acrylics over its closest competitor, polycarbonates. Acrylics generally cost more than polycarbonates because their polymerization process is more time consuming.

Acrylics are available in thermoplastic and thermoset forms, with the former being the most common. Thermoplastics allow for thermal forming, blow molding, casting, and adhesive bonding, while thermosets are processed more like unsaturated polyesters, cured at room temperature with the help of a catalyst. Typical applications of these thermosets are resins for use in place of polyesters and electronic encapsulation and potting compounds. Thermoplastic acrylics will be the focus of the remainder of this chapter.

The trade names synonymous with acrylics are Plexiglass (GE) and Lucite (DuPont) which were both pioneers in the development of commercial grades of acrylic based on the polymethyl-methacrylate (PMMA) monomer. These are available in numerous forms, including general purpose, high impact, bullet-resistant, abrasion-resistant, UV-filtering, non-glare, and textured forms. If it is important to make an on-sight determination of whether the material in question is an acrylic, exposing it to a ketone solvent such as MEK will fog the surface. Acrylics are not only rigid thermoplastics, they are also the basis of cyanoacylate adhesives.

## Properties and Designations

Acrylics provide a unique blend of optical clarity and toughness along with common fabrication techniques to make them very desirable candidates for meeting requirements that may previously have been satisfied by glass products. For items requiring optical clarity and lightweight material, acrylics are nearly 50% lighter than glass.

The following properties are discussed in the context of acrylic (PMMA) as it applies to engineering design applications.

### Coefficient of Friction

This property is generally not considered relevant when designing with acrylics because the behavior of this material does not lend itself to wear or friction applications. The optical properties of acrylics are the primary reason for their selection.

### Wear-Resistance

Acrylics are also generally not chosen for their wear-resistance, consequently, ratings for this property are generally not included in manufacturers specifications. There are abrasion-resistant grades which include a surface coating to resist scratching, but are not intended for applications requiring significant wear-resistance.

### Toughness

Acrylics in their unmodified form are not very tough and when compared to polycarbonates, may appear to be brittle. The impact strength of acrylics is approximately one-fifth that of polycarbonates and acetals, making them a poor choice if similar optical properties are required. Use of generous radii is necessary to avoid stress concentrations that could initiate cracks.

Their tensile strength properties are generally higher than polycarbonates and their elongation at yield lower, as would be expected with a material of limited toughness. Impact modified grades are available, which significantly improve toughness but at the expense of tensile strength and optical clarity.

### Stress Rupture

They possess good creep-resistance under moderate stress levels and have high initial tensile strength, but over the long-term, stress rupture is common. Limiting the stress to 10% of the tensile strength is preferred.

### Machinability

Machining acrylics is possible using traditional machine tools and water-based coolants, producing excellent surface finishes (<32 μin.) that allow for excellent post-processing flame treatments. Because of the limited toughness of acrylic, surface roughness should be minimized to reduce the possibility of introducing stress concentrations.

### Electrical

Acrylics have a very high electrical resistivity and dielectric strength, which makes them useful as insulators. Formulations to reduce these properties to improve static dissipation for ESD considerations are not generally available because they would effect the optical properties.

### Chemical-Resistance

Acrylic plastics are generally resistant to inorganic solvents such as water and ammonium hydroxide which are often used to clean them. Acrylics are not tolerant of chlorinated or ketone-based solvents, such as acetone, methyl ethyl ketone, or trichloroethylene, as they attack the material, ruining the surface structure.

### Flammability

Acrylics are very flammable and will continue to burn after the ignition source is removed, unlike polycarbonates, which do not continue to burn after the ignition source is removed.

### Weatherability

Acrylics have excellent resistance to outdoor exposure, including a high tolerance for UV radiation, without significant loss of clarity or yellowing. Its UV-resistance is the highest compared with that of all transparent plastics.

### Water Absorption

Water absorption is slow for acrylics and can take several weeks to reach equilibrium. Its absorption is typically 0.3% by weight, which is similar to acetals or ABS and significantly less that of nylons.

### Thermal

The thermal properties of acrylic are not exceptional, with a maximum service temperature of 180°F (82°C), similar to acetal and ABS. There are heat-resistant grades that raise the maximum temperature by approximately 20%, but at the expense of impact-resistance. The rate of thermal expansion exceeds that of glass 8 to 10 times but is similar to many plastics and repeatable to the point that it is considered a dimensionally stable material, allowing it to be a useful optical material.

### Optical

The exceptional clarity of acrylics in combination with their low density, and a toughness that exceeds optical glass makes them ideal candidates for many engineering applications. Colorless grades have total white light transmittance of 92%, which is as good as the best plate glass. The remaining 8% is due to reflective losses. Applications including focusing lenses, windows, and aircraft canopies are made using the appropriate grade of acrylic.

## Fabrication Processes

The typical commercial polymerization method for acrylics is based on a reaction involving acetone and hydrogen cyanide. Different grades of acrylics are produced using additives and modifiers during the polymerization process. The entire process is time consuming, which is a significant factor in the higher cost of this plastic.

There are three basic methods of forming usable acrylic: cell casting, extrusion, and continuous casting. Casting between two glass plates is the original method of production and is used for small volumes, specialized grades, and thicker sheets. The method includes pouring a partially polymerized "syrup" into a mold and curing at an elevated temperature. Extrusion is used to produce standard or uncommon shapes and sheets up to ¼ in. thick. Continuous casting combines these two techniques into an economical process, capturing the cost efficiency of extrusion and most of the physical property benefits of cell casting. Extruded and continuous casting processes have superior thickness tolerances.

## Raw, Stock, and Mold Finishes

Acrylics are available in many forms, including raw pellets used to create extruded forms. Much of the industrial grade product is produced by continuous casting or extrusion into common forms such as:

- Heavy gauge sheet (⅟₁₆–12 in.) stress-relieved
- Rod (⅟₁₆–24 in. diameter)
- Tubing (ID ¼ in. minimum, OD 27 in. maximum)
- Custom extruded and cell cast shapes
- Injection molding

Low volume production and custom shapes are more economically produced by cell casting because the tooling and processes are designed around making a single part rather than many. High-volume products can be cost effectively produced from acrylics by injection molding with the desired acrylic resin.

Acrylics are most often produced as clear materials and used in applications requiring high clarity and strength. Custom grades of various colors are also produced, including white, the second most common form.

## Additional Processing Considerations

Acrylics are most often chosen for their clarity, but they also have surface properties that allow them to be coated and embossed. When decorative coatings are required they are achieved by painting, hot stamping, and vacuum metalizing.

Special gamma radiation resistant grades designed for use in medical devices are available but with tensile strength reduced by 10%. Generally, these grades maintain exceptional clarity after 5 Mrad of gamma radiation, with no adverse effect on physical properties.

### Stiffening Mechanisms

The most common stiffening material added to acrylic to increase its physical strength properties is glass fiber. When 10 to 40% glass fiber (by weight) is added, the tensile strength can increase by as much as 100% from unmodified material, with corresponding increases in impact strength and deflection temperature but ranging from 10 to 20%.

### Thermal Treatments

Extruded or molded acrylics that have been machined using conventional machine tools may benefit from annealing to stress-relieve the material while retaining good dimensional stability throughout the life of the part.

### Additives

Including additives such as UV-stabilizers and colorants is not uncommon, but this does affect tensile strength and toughness properties. Addition of UV-stabilizers results in a slight increase in tensile strength and a corresponding loss of toughness. Colorants will also effect the material properties, but it is more difficult to generalize their effect. The manufacturer or distributor should be consulted and the product data sheet reviewed for specifics of the material.

## Commonly Used Acrylics

- Plexiglass (GE)
- Lucite (DuPont)
- Cast acrylic
- Optical grade acrylic
- Impact grade acrylic
- Acrylic/Polycarbonate alloy

## Polycarbonate

Polycarbonate is an amorphous and linear thermoplastic with limited crystallinity, transparent in its unmodified form and is generally lighter than acrylic. It is thermoformable, weldable by heat and solvents and available in injection moldable resins. Polycarbonate has an impact strength that is 10 to 30 times that of acrylic and a white light transmittability of greater than 90%, depending on the grade.

The mechanical strength properties of polycarbonate are similar to those of nylon and acetal with the additional optical properties that make it a useful material for environments requiring high clarity and excellent toughness as well as lightness of weight. The most widely recognized commercial polycarbonate's trade name is Lexan®.

## Properties and Designations

The properties of polycarbonates are very similar to acrylics in many respects but with a few significant differences. These include greater toughness, self-extinguishing, reduced optical clarity, and reduced cost. Their densities are similar at 1.19–1.2 g/cc (0.043–0.0434 lb/in.$^3$) with slight variations based on processing methods.

The following properties are discussed in the context of polycarbonates as they apply to engineering design applications.

### Coefficient of Friction

As with acrylics, polycarbonates are not generally used in bearing applications. Its thermal properties under load are greater than PVC but substantially less than preferred bearing materials, such as acetals or filled nylons. Alloys with PTFE are available but are generally not a substitute for a good bearing plastic.

### Wear-Resistance

Lack of property data for wear-resistance follows from polycarbonates' limited use as bearing material. In unique designs which require this property it is best to consult the material manufacturer or distributor and review the specific property data sheet.

### Toughness

Polycarbonates have impact strengths up to 16 times greater than acrylics, making them more attractive for applications requiring tough materials and clarity only slightly less than that provided by acrylics. Notch-sensitivity can be a problem if not properly addressed in the detailed phase of the design. Use of generous radii and smooth surface finishes can significantly reduce this concern.

### Stress Rupture

Polycarbonates offer very good creep-resistance compared with other engineering plastics. Care is required to avoid stress raisers or similar defects to improve material life in constant load environments.

### Machinability

Polycarbonate machines similar to acrylics using conventional machine tools to produce surface finishes less than 32 $\mu$in.

### Electrical

Unmodified polycarbonate is an insulating material with an electrical resistivity greater than $10^{15}$ $\Omega$. Additives such as carbon black and stainless steel fiber are used to improve conductivity to produce anti-static and conductive forms.

### Chemical-Resistance

The chemical-resistance of polycarbonate is generally good with general limitations involving chlorinated solvents and acetone. Some chlorinated solvents dissolve polycarbonate while acetone induces cracking. Exposure to petroleum-based products is not recommended.

### Flammability

Polycarbonate is a self-extinguishing plastic as it will not burn once the ignition source is removed, unlike acrylic which can sustain combustion. For this reason, it is more desirable than acrylic for applications that may involve potential for combustion.

### Weatherability

After extended outdoor exposure and UV conditions it is not uncommon for polycarbonate to take on a translucent yellow color. Although the color may be a detracting byproduct, property retention after yellowing is generally good.

### Water Absorption

The hydroscopic nature of polycarbonate is not a significant factor for parts in their final form because it is generally less than 0.2% (volume), making it a dimensionally stable plastic. This property is a concern for raw stock pellets used for injection molding because of the adverse effect of moisture on the molding process. These materials must be thoroughly dried prior to molding.

### Thermal

The thermal properties of polycarbonate are generally greater than acrylic but not as good as some advanced thermoplastics such as PEEK or polyamide imide. The maximum service temperature in air and heat deflection temperature is greater than acrylic by 20 to 40%, making it preferrable over acrylic for applications requiring non-optical grade clarity.

*Optical*

White light transmittance is near 90% for unmodified polycarbonates, less than that for acrylics but greater than other clear grades of plastics such as polyurathane. Polycarbonate has poor UV performance in unmodified grades but is available in grades with UV-stabilizer additives.

## Fabrication Processes

Commercial forms of polycarbonate are produced from a chemical reaction between phenol and acetone under acidic conditions. The MW properties are controlled as necessary to product the desired end product. For example, a MW of 25,000–35,000 is used for injection molding grades, and higher MWs for castings and materials requiring uniform electrical properties.

## Raw, Stock, and Mold Finishes

Polycarbonate is a very versatile material and available in a large number of grades and shapes ranging from thin film to large rotational molded parts. The quality ranges from general purpose grades for use in signs or display cases to aircraft grades with low smoke and abrasion-resistance. Common sizes are available as follows:

- Sheet (1/32–4 in.)
- Film (0.001–0.029 in.)
- Rod (1/4–6 in. diameter)
- Injection molding resins
- Thermoform sheet
- Rotational molding resins
- Blow molding resins
- Abrasion-resistance coatings

## Additional Processing Considerations

Numerous additions can be made to polycarbonate to alter its unmodified performance with many having only small impact on physical properties.

### Stiffening Mechanisms

Many options exist to improve the tensile strength of polycarbonate, including the addition of glass, carbon or aramid (Kevlar) fibers, and glass beads. The effect of glass beads is to reduce the notch-sensitivity with a small reduction in tensile strength, but also results in improved machinability. Machining fiber-filled grades can be problematic because exposing the fiber matrix may increase notch-sensitivity and result in poor finishes.

### Thermal Treatments

Annealing polycarbonate to eliminate residual stresses is routinely performed when dimensional stability is of prime importance. It adds to the processing costs and therefore must be justified to meet the design requirements.

### Additives

Polycarbonate can be processed to include many additives to produce a desired combination of properties. These additions include UV-stabilizers to minimize the yellowing and brittleness that can result from long-term exposure to UV radiation; ignition-resistant fillers, such as silicone, that also improve impact strength by greater than 50%; conductive materials, such as stainless steel and carbon black, to produce anti-static and conductive properties; and lubricating materials, including PTFE, to reduce friction.

## Commonly Used Polycarbonates

Lexan®
Optical grade polycarbonate
Foam polycarbonate
Anti-static/conductive grade polycarbonate

## Nylon (Polyamides)

Nylon is a semicrystalline polyamide thermoplastic with poor dimensional stability in unmodified grades. The primary contributor to poor dimensional stability is its water absorption properties. They are the highest of all the engineering plastics and can result in dimensional changes of up to 2%. The high moisture absorption lowers tensile strength and stiffness while increasing toughness.

Nylon's excellent toughness, good abrasion-resistance, and low coefficient of friction make it a popular replacement ferrous and nonferrous materials. Its density is an order of magnitude less than that of steel or bronze, allowing it to be a lightweight alternative to traditional bearing materials. Nylon 6/6 has been available since the 1950s, making it one of the oldest engineering plastics and one of the most modified.

Unmodified nylon also has excellent resistance to fatigue and repeated impact, and even greater impact- and thermal-resistance when reinforced with glass, allowing it to be used in physically demanding engineering environments. Nylons exhibit a operating temperature range of 69–228°C (151–440°F). Industrial products, such as pumps, motor housing, gears, and under-hood automotive components benefit from the fatigue- and thermal-resistance while consumer products, such as toys and sports equipment, benefit from the resistance to repeated impacts.

## Properties and Designations

The designations used to identify different grades of nylon are unique in that they provide details of the chemical makeup of the specific formulation, by employing a suffix to refer to the number of carbon atoms in each of the reacting substances involved in the polymerization process. Use of a single monomer is noted by including the number of carbon atoms it contains following the word nylon. For example, nylon 6 was formed from a monomer with 6 carbon atoms per chain. When two monomers are used to form a copolymer, the carbon atoms in each are included in the designation, such as nylon 6/6, separated by a slash. The number of carbon atoms can range from 2 to 10 for the first monomer and from 2 to 18 for the second.

Knowing the designations can help to understand the basic material properties of the particular grade in question. Generally, the higher the number of carbon atoms, the lower the melting temperature. For example, unmodified nylon 6 has a maximum service temperature in air of 230°F while nylon 12 is 190°F. This difference is offset by the lower moisture absorption of nylon 12, making it a more stable material at the expense of thermal-resistance. Copolymers generally temper the properties of monomer nylon grades as evident in the reduced melting temperature and increased flexibility of two of the most popular forms, nylon 6/6 and nylon 6/10.

### Coefficient of Friction

Unmodified nylon has an inherently low coefficient of friction against most plastics and metals, making an ideal candidate for gears and other sliding surfaces. Values of 0.2–0.3 against steel are common for unmodified grades. Even lower coefficients can be obtained when additives such as $MSO_2$, oil, and PTFE are included to produce grades that also have reduced stick-slip behavior. Coefficients of 0.16–0.2 against steel are obtainable.

### Wear-Resistance

The pressure versus velocity rating of nylon and its various grades are typically very good because of nylon's lower friction coefficient and material hardness. Nylon is not recommended for high-speed applications as lower speeds and moderate loads are where nylons perform best.

### Toughness

The toughness of nylon can increase two to three times from ambient conditions due to its water absorption properties which allow it to have excellent toughness in moist conditions. In drier conditions it loses some of its toughness and can exhibit brittle behavior in very dry environments.

### Stress Rupture

Nylon offers excellent creep-resistance compared to other engineering plastics. Care is required to avoid stress raisers or similar defects, in order to improve material life in constant load environments.

### Machinability

Nylons can be easily machined using conventional machine tools because of their low friction properties and reasonably good hardness.

### Electrical

Nylons exhibit good insulative properties at low temperatures and humidity, similar to but not as good as PVC or polyethylene. These properties deteriorate rapidly as both temperature and humidity increase, preventing them from being used reliably for anti-static or conductive applications. Nylon has been used as a filler in materials designed for anti-static or conductive applications but with limited range of use because of the moisture effects.

### Chemical-Resistance

Nylons have exceptional resistance to hydrocarbon-based materials including fuels, oils, and lubricants. They are attacked by acids, some salts, and glycols and may swell or dissolve when exposed to alcohols. Nylons are considered biologically inert, which makes them suitable for many food and beverage-related packaging applications. Additionally, their outstanding gamma radiation resistance allows for inspection and sterilization of the package and its contents. Generally, chemical-resistance of homopolymers is greater than that of copolymers.

### Flammability

Nylon is a very flammable, similar to acrylic, with the exception that it will not continue to burn when the heat source is removed. This self-extinguishing property is desirable for most applications as it affords a certain level of safety.

### Weatherability

Nylon's mechanical properties make it a desirable material from a machine design standpoint but it exhibits poor weatherablility, including damage from high exposure to UV radiation.

### Water Absorption

Nylons are known for their susceptibility to absorb moisture to such a great extent that they should not be used in designs requiring significant dimension stability. Nylon 6/12 exhibits the lowest moisture absorption of all nylons but with reduced properties including lower toughness than nylon 6. Nylon 11 and nylon 12 have the best dimensional stability and lowest water absorption but also have reduced properties including lower deflection temperatures.

### Thermal

Nylons exhibit a wide range of thermal behavior depending on the additives and processing. Unmodified grades have deflection temperatures of 200°F (93°C), while those that have glass or carbon fiber additives can have deflection temperatures above 350°F (176°C).

### Optical

Nylon is not known for its optical clarity and is not considered for applications requiring it. There are nylon grades available that are transparent with good toughness.

## Fabrication Processes

Nylons are formed by a controlled reaction of a formulated salt solution under high temperature and pressure to produce a plastic with long carbon chains with recurring chemical groups, similar to polyesters. Their MW ranges from 2,000 to 10,000 and is controlled by altering the ratio of the mixture. Nylons are very versatile materials that are formulated into many different materials ranging from liquid coatings to adhesives and solids. The properties and use of nylon as they apply only to a solid plastic for use in engineering applications are discussed in this section.

## Raw, Stock, and Mold Finishes

Nylon is available in many unique grades and forms because of its ability to be formulated with various additives and fillers. It is extruded and cast to net shape for use as formed or machined as necessary to obtain the required shape. Nylon is also available in resins for injection molding and in final form as a film for use as a packaging wrap material such as Dartek by DuPont. Common sizes are available as follows:

- Sheet, extruded and cast (1/32–4 in.)
- Cast disks (28–80 in. diameter and >2 in. thickness)
- Rectangular bar (up to 8 × 8 in.)

- Film (0.001–0.029 in.)
- Extruded rods (¼–6 in. diameter)
- Cast rods (2–38 in. diameter)
- Tube (¼ ID × 5.0 OD to 10.0 ID × 18.0 OD)
- Injection molding resins
- Blow molding resins
- Film for food packaging

## Additional Processing Considerations

Because nylon has been in use for engineering applications for longer than most plastics, many alterations to the unmodified resin have been experimented with, resulting in a large number of variations.

### Fillers and Stiffening Mechanisms

The strength performance of nylon improves remarkably with the addition of glass or carbon fibers, with tensile strength increasing 100% from a typical value of 8 to 16 ksi. The heat deflection temperature also increases from 190°F for unmodified nylon to above 350°F for nylon with 30% fibers.

The addition of fibers also increases creep-resistance while reducing notch-sensitivity and improving wear-resistance by more than 20 times. Other fillers include spherical beads to improve stiffness (modulus) and strength. Cross-linking is also possible for increasing stiffness but is not commonly preferred because of the added processing and expense.

### Thermal Treatments

Heat treatment of parts fabricated from nylon is not common. However, as with acrylics, acetal and PTFE-filled nylon can benefit from annealing and normalizing heat treatments to relieve stresses when needed. Conditioning temperatures in the range of 180 to 280°F are used for short periods time.

### Additives

Several common additives are used to improve the properties of nylon, including $MSO_2$, oil, and PTFE. $MSO_2$ is included for improved wear and reduced friction while increasing tensile strength. It also results in a lower coefficient of expansion than straight nylon. Oil impregnated grades reduce friction by 25% and last five to seven times longer than plain cast nylon. PTFE grades increase wear-resistance and lubricity while still benefiting from nylon's strength and stiffness properties.

Commonly Used Nylons

Nylon 6 and nylon 6 with $MSO_2$.

## Acetal

Acetal is widely regarded as the most universal engineering plastic because of its properties which combine high stiffness, excellent elevated temperature, solvent- and fatigue-resistance, and low coefficient of friction on most materials. It also has a low stick-slip and creep behavior. Acetal has high crystallinity (75 to 80%) which allows for a dense structure that results in tough, resilient material over a large range of temperature and humidities. The temperature range for continuous use is from 32 to 212°F (0 to 100°C) and for short-term thermal loading it can be used in the range of −76 to 200°F (−60 to 140°C). Thermal stability and continuous use temperatures are, however, estimates and must always be considered in the context of the relevant application.

Polyoxymethylene (POM), as acetal is sometimes called, often replaces metals, such as steel, zinc, and brass in moderate load applications because acetal offers improved wear and weather-resistance and reduced weight. Nylon typically has better impact and wear-resistance, while polycarbonate has better creep resistance. Although acetal may not have the highest performance in these categories, it performs within 20% of these maximum values, when combined with its other properties such as tensile strength, low coefficient of friction and stick-slip, it becomes an attractive plastic for many applications. Radiation degrades the material properties of acetals, making them not suitable for use in medical devices requiring ionizing sterilization methods or other applications involving exposure per ASTM D 638 test method (3.5-Mrad gamma radiation).

Properties and Designations

Acetal is polymerized into two grades, a homopolymer and a copolymer. The most notable grades of acetal available are homopolymer, copolymer, and PTFE-filled copolymers. Generally, the homopolmer has greater short-term mechanical properties while the copolymer's properties are reduced but has better long-term stability. Injection molding grades exhibit high shrinkage compared to other engineering plastics.

The most common homopolymer is marketed under the trade name Delrin while the copolymer grade names are Celcon or Aceton. Cost of the homopolymer is less than the copolymer. Many grades are FDA or NSF approved materials making them useful in food processing or potable water systems.

The following properties are discussed in the context of unmodified acetal as they apply to engineering design applications.

### Coefficient of Friction

Unmodified grades of acetal exhibit a low coefficient of friction against metals (0.15) and most plastics but not against itself (0.35). This makes it a good bearing material and a strong candidate for use in place or oil impregnated brass or bronze. Friction properties can be further reduced to 0.05 by including additives such as PTFE or other proprietary lubricants.

### Wear-Resistance

Acetal has only moderate resistance to wear in unmodified grades but is significantly improved with the addition of a lubricant such as PTFE. Modified grades exhibit one of the best PV values for engineering plastics, exceeded only by high priced polyamide-imide bearing materials. Acetal also exhibits a resistance to cold flow, unlike other low friction plastics such as UHMWPE.

### Toughness

Unmodified grades exhibit a poor notch-sensitivity when compared to nylons or polycarbonates, making it seem out of line when considering acetals other mechanical properties. This property can be improved with the addition of elastomers and by design, limiting transition radii to the largest possible values or at least greater than ⅛ in. Acetal exhibits high hardness and rigidity due to its structure which is the reason for its lower notch-sensitivity. The degree of crystallinity in copolymers is less than homopolymers resulting in 10 to 12% higher hardness and rigidity in homopolymers. Its toughness property is valid down to −76°F (−60°C), below that the embrittlement range begins. The copolymers have the advantage of a better toughness, thermal, and chemical stability while homopolymers have higher hardness and rigidity.

### Stress Rupture

The stress rupture properties of acetal are excellent, owing to its superior mechanical properties. However, it should be kept in mind that stability of a structural component for a given purpose depends not only on the mechanical properties of the raw material but to a large extent on its structural form and shape. A poorly designed component would fail under load in spite of its higher mechanical properties.

### Machinability

Acetal has excellent machinability, often requiring little deburing when machined with conventional tooling. Care must be taken to minimize tool marks that may act as stress concentrations, increasing the notch-sensitivity of the finished part. Close tolerance parts may require stress-relieving to improve stability due to the stress imparted during processing.

### Electrical

Unmodified grades do not have high volume resistivities but are considered nonconductive because they are greater than $10^{12}$ $\Omega$-cm. It is available in static dissipative and conductive grades. The mechanical properties of the static dissipative grades area are reduced by nearly 50% over unmodified grades while the conductive grades retain nearly their full property values.

### Chemical-Resistance

Acetal has good chemical-resistance against fuels, hydrocarbon-based oils and most organic solvents such as acetone or MEK, showing little or change after prolonged exposure. Copolymers exhibit a superior resistance to strong bases over homopolymers, while neither should be used in acidic solutions of less than 4.0 pH.

### Flammability

Acetal is flammable and continues to burn after removal of the flame source. It is classified as a slow burn material, which burns slower as the section width increases. Glass-filled grades burn at slower rates than unfilled grades.

### Weatherability

Unmodified acetal has very poor UV-resistance and will become brittle with a chalk-like surface coating when over-exposed. Acetal grades with carbon black added show the greatest resistance against UV exposure, with little or no loss of material properties.

Exposure to super-heated steam is not recommended because the permeability of acetal is higher for hydrogen, $CO_2$, and water. Conversely, it has very low permeability for propane, butane, aliphatic, aromatic, and hydrogenated hydrocarbons, alcohols, and esters which promotes its high usage in cigarette lighters, aerosol containers, and automotive parts in contact with fuels. The mechanical properties of acetal are not affected by prolonged exposure or immersion in boiling water.

### Water Absorption

Acetals exhibit low water absorption, although slightly greater than polycarbonate it is not nearly as absorbant as nylon. The homopolymer (Delrin) absorbs only about 50% of the quantity absorbed by the copolymer (Celron), resulting in a greater dimensional stability for moist environments.

### Thermal

Thermal properties of acetal are good in the unmodified grades and excellent in those with glass or carbon fibers added. Unmodified grades have a maximum air temperature exposure of 180°F (82°C), while glass-filled grades are good up to nearly 350°F (176°C) with heat deflection temperatures above 300°F (150°C).

### Optical

Acetal is not available in clear structural shapes or molding resins. It is natural color is a translucent white. The light transmission of 2 mm thick injection molded plates is 50%.

## Fabrication Processes

Simply stated, acetal is formed when methanol is catalytically oxidized in the vapor phase over a metal oxide catalyst to produce a formaldehyde solution. This solution is then polymerized in any of a number of ways as determined by the manufacture.

## Raw, Stock, and Mold Finishes

Acetal is available in numerous grades that include fillers and additives to improve its performance. Because acetal is moldable, there are resin grades as well as extruded shapes that can be machined to their final form. The common sizes available are as follows:

- Sheet, extruded and cast (⅜₂–4 in.)
- Rectangular bar (up to 8 × 8 in.)
- Extruded rods (¹⁄₁₆–12 in. diameter)
- Tube (¼ ID × 0.5 OD to 2.0 ID × 2.5 OD)
- Injection molding resins
- Rotomolding resins
- Film (0.001–0.029 in.)

## Additional Processing Considerations

Acetal is typically formulated using a variety of fillers and additives to enhance key physical properties or processing requirements. Much like nylon, acetal has a unique ability to accept a wide range filler and additives, making it a versatile engineering plastic.

### Stiffening Mechanisms

The most common filler is glass fiber, which is used for its capacity for stiffening and improved high-temperature properties, while it decreases the susceptibility for creep and reduces the coefficient of thermal expansion. Mineral or glass bead fillers are included to improve dimensional stability and reduce shrinkage during injection molding. More recently, mineral fillers, such as titanium oxide and other proprietary materials, are used to improve the effects of laser marking on the plastic. For example, black Delrin can be compounded to produce a white mark when exposed to a ND : YAG laser.

### Thermal Treatments

Acetal can be annealed to increase dimensional stability after molding or machining. Heating to 200°F (93°C) for 10 min (depending on the part thickness) can reduce internal stresses that are present in molded parts and highly machined forms. Thermal treatments for injection molded parts are not necessary, only highly machined parts require post-treatment.

### Additives

PTFE, is the most common additive to acetal because it reduces the coefficient of friction from 0.15 on steel for an unmodified grade to 0.05 for a 25% PTFE-filled grade. The PTFE is typically in fiber form as used in Delrin AF, providing improved wear-resistance and low friction. $MSO_2$ is generally not added because of problems with acidic impurities that degrade the acetal. Oil impregnated grades as well as proprietary lubricants are available to reduce friction properties.

Carbon black is the primary additive to improve UV protection, and is also used as a colorant. Other colorants can be used for decorative purposes and supplied as a special order. Silicone is added to improve temperature properties by 20 to 30%, resulting in a heat deflection temp of 230°F (110°C) rather than the 180°F (82°C) available in unmodified grades.

## Commonly Used Acetals

- Delrin® Acetal, homopolymer, unfilled, extruded;
- Delrin® AF Blend, acetal, homopolymer, PTFE-filled, extruded.

## Acrylonitrile Butadiene Styrene

Acrylonitrile butadiene styrene (ABS) is an amorphous thermoplastic terpolymer that is characterized by ductile yielding down to temperatures as low as −40°F (−40°C), while subjected to high strain rate loading. Each polymer component contributes to the material's usability: acrylonitrile provides heat and chemical-resistance and surface hardness properties; butadiene provides toughness and impact-resistance properties at all temperatures, but most importantly at low temperatures where many plastics become brittle; styrene provides rigidity, strength and processability. ABS costs less than most engineering plastics and bonds well using a variety of methods including solvent, thermal or with epoxies or cyanoacrylates.

ABS has moderate tensile properties, excellent impact characteristics, and good toughness, allowing it to be used in a wide range of applications. It can be notch-sensitive but much less than most engineering plastics. Using generous radii between transitions and machining to a smooth surface without tooling marks can eliminate this concern.

ABS is available in both extrusion and injection molding grades. It has limited chemical-resistance when exposed to petroleum-based machine oils and concentrated oxidizing acids such as nitric, sulfuric, perchloric, and phosphoric. Injection molded grades offer better dimensional stability than extruded grades and high quality surface finishes because it has a lower melt temperature. Special electroplating grades are available that offer a lower coefficient of thermal expansion to reduce the differential thermal-induced stress between the plate material and substrate. Once molded, the surface must be prepared using an acid etch process prior to plating to improve adhesion.

### Properties and Designations

ABS offers a good balance of engineering properties. Impact strength is its most notable property because it offers one of the best low temperature performance of all the engineering plastics. All plastics have a glass transition temperature below room temperature (if not, it is considered an elastomer) at which point they become brittle and fracture easily when rapidly loaded. ABS offers a reduced impact strength at temperatures as low as −40°F but does not become brittle.

### Coefficient of Friction

ABS should not used in applications requiring low friction because its terpolymer formulation with butadiene rubber does not lend itself to a slick surface. There are also more appropriate materials of similar cost such as acetal that would be better candidates for applications requiring reduced friction.

### Wear-Resistance

Wear-resistance of ABS is good but not its best property. Because ABS is used mostly for structural applications for which surface loading is generally low or nonexistent, wear is not a significant design parameter.

### Toughness

ABS is a cost-effective plastic that has high impact strength for a very wide temperature range, especially at low temperatures. Impact performance can be varied by altering the rubber content of the terpolymer. At room temperature, polycarbonate exhibits better impact properties but it is considerably more expensive and lacks the low temperature properties. As a general rule the cost of polycarbonate is more per cubic inch than ABS.

### Stress Rupture

ABS offers excellent creep-resistance in both extruded and molded forms. This behavior along with its impact tolerance make it a versatile material.

### Machinability

ABS has excellent machinability, which allows it to be easily processed with conventional equipment using carbide-tipped tools. It also is a stable material with a thermal expansion coefficient similar to polycarbonate as well as low moisture absorption properties.

### Electrical

The electrical-resistance properties of ABS are good but it is generally not used as a primary insulator because there are plastics such as polycarbonate that have greater volume resistivity. ABS is often used in applications such as electrical machine covers or bodies that require dissipation of static electrical charges. There are static, dissipative, and conductive grades available that are formulated using carbon black or stainless steel fibers to achieve the desired electrical properties. The mechanical properties will be altered as a result of the additives and should be reviewed carefully to avoid unwanted results. For example, the addition of carbon black increases tensile strength but at the expense of reduced impact strength, the primary property for which ABS is often selected for initially.

### Chemical-Resistance

ABS has limited chemical-resistance to petroleum-based machine oils while resistant to mineral, vegetable, and animal oils. It also has good resistance to inorganic salts and acids such as hydrochloric and nitric acids. ABS is attacked by solvents such as ketones (MEK) and chlorinated hydrocarbons such as tricholorethylene, methylene chloride, chloroform, and carbon tetrachloride.

### Flammability

In unmodified grades, ABS is highly flammable and will burn after the flame is removed. There are grades available that include flame-retardants such as PVC or halogenated additives that can reduce the flammability to levels similar to a flame-retardant polycarbonate. The additives used to reduce the electrical resistivity will, typically, also reduce the flammability.

### Weatherability

ABS has moderate weatherability characteristics and is generally not considered for continuous use in outdoor applications. Continuous exposure to UV can cause some embrittlement and yellowing.

### Water Absorption

ABS is hydroscopic, with similar moisture absorption characteristics as polycarbonate or acetal but only half of that exhibited by nylon. They can absorb up to 1.8% weight water on storage in aqueous media due to residual emulsifier and the polarity of the nitrile side groups. Under normal conditions, dimensional or property changes are negligible. Thermal form grades require oven drying to improve their processability due to the effects of residual moisture. While the mechanical properties are not greatly effected by this moisture, its presence during processing can affect part appearance. A dehumidified air drier could be used to reach the suitable moisture levels before processing.

### Thermal

The thermal behavior of ABS is unique because of the range of its temperature performance. After many plastics become brittle, ABS maintains comparatively high impact strength at temperatures as low as −40°F (−40°C). Its heat deflection and continuous service temperatures are between 185°F (85°C) and 200°F (93°C) with heat-resistant grades offering a maximum temperature of nearly 240°F (116°C).

### Optical

Clear ABS grades substitute the styrene with methyl methacrylate (MMA) to obtain light transmission levels of nearly 88%. These can compete with polycarbonates as they have similar impact and tensile strengths but at a lower cost.

## Fabrication Processes

ABS is a graft terpolymer in which the butadiene chain attaches itself to the side of the styrene acrylonitrile (SAN) chain, producing a product with improved strength and toughness of the SAN polymer, while maintaining desired strength properties.

## Raw, Stock, and Mold Finishes

ABS is a widely used plastic that is available in numerous grades that include the necessary additives to improve particular physical properties as required. Injection molding grades are of lower MW to reduce their melt temperature and shrinkage. There are many specialty grades that provide improved performance for properties such as elevated temperature- and flame-resistance as well as grades that improve the ability to plate ABS.

- Sheet, extruded and cast (⅛₂–4 in.)
- Extruded rods (¼–6 in. diameter)
- Tube (¼ ID × 0.5 OD to 2.0 ID × 2.5 OD)
- Injection molding resins
- Injection molding resins for electroplating
- Rotomolding resins
- (0.001–0.029 in.)
- Foam grades
- Blow molding resins
- Compression molding sheets
- Clear ABS

## Additional Processing Considerations

Acrylonitrile butadiene styrene is often alloyed with other plastics to obtain the desired engineering properties because of its structure which allows the grafting of polymer chains to the SAN chain. In addition, it can be processed using some of the more common stiffening materials such as glass or carbon fiber, as well as common additives including PTFE and carbon black.

### Fillers and Stiffening Mechanisms

Glass and carbon fillers are commonly processed with ABS to improve stiffness and high-temperature performance. Grades ranging from 10 to 40% fiber are common which produce an increase in tensile strength of 100% and increase in operating temperature of greater than 20%. The added stiffness is obtained at the expense of reduced toughness and loss of low-temperature impact-resistance, excluding its use in these environments.

Alloying ABS with other plastics such as PVC and polycarbonate can also result in improved stiffness. These alloyed plastics extend the stiffness range while providing a secondary benefit such as improved dimensional stability. These grades of plastics reduce the cost-effectiveness of using ABS as they are more expensive but not as costly as a polycarbonate or nylon.

*Thermal Treatments*

Stress-relieving by thermal annealing is the most common method of improving the dimensional stability of ABS for those applications requiring close tolerance.

*Additives*

The most common additives to ABS are PTFE for reduced coefficient of friction and carbon black to improve static dissipative properties. PTFE reduces the coefficient of friction from >0.3 to 0.2, placing it in the range of an acetal. The addition of halogenated compounds reduces flammability.

## Commonly Used ABS

- ABS, extruded;
- ABS, molded (many grades available).

## Polyimide

Extreme wear applications requiring high-temperature performance are often best resolved using polyimide plastics such as Vespel, Upilex, Pyralin, Pyre-ML, and Kapton (DuPont, USA and Ube industries, Japan) to provide low friction surfaces and stable substrates up to 680°F. Brief excursions up to 900°F can be tolerated. Unmodified grades have tensile strengths exceeding 12,000 psi, low coefficients of thermal expansion and excellent electrical resistivity. All polyimide grades are expensive because of their complicated polymerization process.

Polyimides are available as both thermoplastic and thermoset plastics. Thermoplastic grades are crystalline or semi-crystalline ones that can be injection molded and easily machined using conventional machine tools. Thermoplastic molding requires processing temperatures that can be difficult for conventional equipment to attain without modification. Thermoset grades are available as B-staged, solvent-based varnishes, laminating resins, adhesives, and prepregnated laminates that require elevated temperature to cure. Laminates are often used for forming high-performance structures most often found in aerospace applications.

Thermoplastic polyimide parts can be fabricated by a method known as direct forming, which uses a process similar to that found in powder metallurgy. Direct formed shapes are produced under high temperature and pressure allowing the polyimide powder to coalesce and form a solid. This solid can be machined and further processed in a similar manner as brass. Polyimides are also available under the trade name Kapton (DuPont, USA) as thin films for PCB laminates and adhesive tapes.

## Properties and Designations

Polyimides offer exceptional mechanical properties at temperatures ranging from cryogenic to above 600°F. It is a material with excellent chemical-resistance, which has zero flammability and is a very good electrical insulator. The following properties are discussed in the context of thermoplastic polyimide as they apply to engineering design applications.

### Coefficient of Friction

Low coefficient of friction throughout a wide temperature range is the most notable property of polyimides. Values ranging from 0.20 for unmodified grades to 0.12 for PTFE-filled grades are common. PTFE-filled grades provide the lowest friction-resistance and are best for low-speed applications. Graphite-filled grades also provide low friction in higher speed applications, where they remain stable in the presence of considerable friction-generated heat. $MSO_2$ is also useful in reducing friction and is most commonly used in dry environments where graphite can become abrasive due to lack of moisture.

### Wear-Resistance

Unmodified grades of polyimide plastics have good wear-resistance that can be further improved by the addition of graphite. The addition of graphite allows the highest PV values of any modified polyimide.

### Toughness

Unmodified polyimide is not a high energy absorption plastic and consequently exhibits low impact-resistance as well as low toughness. Graphite and PTFE additives make little contribution to improving the toughness, while the addition of stiffening mechanism such as glass or carbon fiber significantly improve toughness by more than five times that observed with unmodified grades. Although this may appear significant, polycarbonate typically has an impact-resistance more than ten times that observed with unmodified polyimide.

### Stress Rupture

Nylons offer exceptional stress rupture resistance which makes them suitable for many structural applications, including molded housings and supporting parts.

### Machinability

Polyimides are easy to machine because of their inherent mechanical strength, stiffness, and dimensional stability. They can be machined using standard metalworking equipment and the processes and techniques applied to machining brass to produce parts with tolerances that were once thought to be too close for plastics. As with any machining process used with plastic materials, it is necessary to avoid overheating the material by taking large cuts or using high surface speeds.

### Electrical

Unmodified grades provide the best electrical insulation properties. PTFE-modified grades retain the excellent electrical-resistance properties while providing low frictional resistance at low speeds. Conductive or antistatic grades are not produced owing to the expensive processing required to manufacture polyimides, and as other materials are better suited for obtaining these properties.

### Chemical-Resistance

Polyimides are resistant to most organic solvents and bases but are attached by oxidizing acids. They are not recommended for long-term exposure to caustic solutions, including ammonia and hydrazine. They exhibit very low outgassing, making them desirable for space-based applications.

### Flammability

The flammability rating of all grades of polyimides is the lowest possible rating as defined in UL94. Polyimides do not burn but become charred when exposed to an open flame.

### Weatherability

Polyimides generally have poor UV-resistance but because most applications involve internal bearings or high-temperature environments, this is rarely a matter of concern. It does exhibit good resistance to ionizing radiation, allowing it to be a desirable candidate for space-based applications or those involving radiated product.

### Water Absorption

Unmodified grades and graphite/PTFE grades have absorption values of 0.25% which are similar to acetals but far below those observed with nylons. The addition of glass or carbon fiber reduces the absorption by half, further improving the dimensional stability of this material.

### Thermal

The thermal properties of polyimides are by far the most desirable. They can tolerate continuous operation ranging from cryogenic temperatures to 600°F (315°C) with excursions to 900°F while still providing low friction and wear properties as well as electrical-resistance. Unmodified grades provide the greatest thermal-resistance.

### Optical

Polyimides are not used in applications requiring optical properties because its chemistry does not lend itself to this physical property.

## Fabrication Processes

Thermoset polyimides are most often supplied in a low molecular weight, prepolymer state or B-staged. It can then be applied as a varnish or used in prepregnated composite materials.

Thermoplastic polyimides are also supplied in a completely polymerized form, requiring only heating above its glass transition temperature for molding.

Raw, Stock, and Mold Finishes

Thermoplastic polyimides are available in many forms for use in injection molding and machined parts. They are even available in ball-bearing diameters for use in low load, high-temperature applications. Film grades are also available for use as laminates in PCB fabrication and as insulation tapes in motors and other environments requiring high-temperature insulation.

- Rings ⅝–2.5 in. in width and ⅛–¼ in. in thickness
- Disks
- Plates
- Bars (2 × 4 × 38 in.)
- Balls (⅛–⅝ in. diameter)
- Extruded rods (¼ to 3¼ in. diameter)
- Tube (1.1 ID × 1.6 OD to 5.6 ID × 7.1 OD)
- Injection molding resins
- Film (0.0005–0.005 in.)

Additional Processing Considerations

Polyimides are best known for their high-temperature properties and are generally modified to extend some characteristics to a greater performance in the high-end of the temperature range. The cost of these modified grades are greater than the unmodified grades, making an already expensive plastic even more costly. Careful consideration of the operating environment and required part configuration can minimize the necessary material volume, consequently, controlling cost.

*Fillers and Stiffening Mechanisms*

The addition of glass or carbon fiber to polyimides results in a 100% increase in tensile strength and similar increase in impact strength. Unmodified grades offer tensile values in the range of 12 ksi and fiber-filled grades, twice that value. A 50% increase in compressive yield from nearly 20 to 30 ksi is also obtained which is an important improvement for bearing grade applications. Additions of 10 to 40% fiber are common. The increase impact strength goes against conventional thinking because generally, an increase in tensile strength results in a decrease in impact strength. This result indicates polyimides are inherently brittle and the added fiber offers a method to not only stabilize them, but also to improve them.

### Thermal Treatments

Polyimides generally do not require post-processing thermal treatment for stress-relieving. Polyimides require high-processing temperatures, meaning that their stress-relieving and softening temperatures are also high and post-treatment is not desirable.

### Additives

The most common additives to polyimides are graphite and PTFE. Graphite offers improved wear-resistance and high PV values for applications involving moderate speeds and high loads because of the increased material stability and hardness of graphite. When moderate amounts are added (15% by weight) improvements in lubrication and dimensional stability are obtained. Greater stability can be obtained with greater amounts of graphite but with a considerable loss of physical strength.

PTFE is added to reduce sliding friction at near zero velocities for moderate load applications. It does not offer the improved wear-resistance of graphite but does provide an extreme temperature material with low coefficient of friction. Grades combining PTFE and graphite are available and offer a greater balance of properties.

$MSO_2$ modified polyimides are formulated as a substitute for graphite in dry or vacuum environments requiring reduced friction because graphite can be extremely abrasive in dry environments. Additions of 15% by weight are also common.

## Commonly Used Polyimides

- Vespel® SP-1 Polyimide, machined;
- kapton® Polyimide film.

## COMMODITY PLASTICS

Engineering plastics have been defined as those providing properties relevant to an engineered design which includes strength, chemical, and thermal properties. Plastics that fall short of the standard used to define engineering plastics are considered a commodity plastic, typically found in low load, ambient environmental applications. They have limited ambient strength and weatherability but are very inexpensive compared to engineering plastics.

There are three commodity plastics that complement engineering plastics by providing similar properties but at a reduced level for a lower cost. Two are polyolefins (polyethylene and polypropylene) and the third is a styrenic (polystyrene). Careful review and testing of their properties in the context of their intended use may reveal their ability to be substituted for engineering plastics at a considerable cost savings.

## Polyethylene

Ultra-high molecular weight polyethylene has the best physical properties of all the polyethylene (PE) grades and was included above as an engineering plastic because of this. Lesser grades of PE are semi-crystalline in structure and classified by density, ranging from very low density (VLDPE) to high density (HDPE) with the general range of densities distributed as follows: VLDPE (0.905–0.913 $g/cm^3$); LDPE (0.916–0.923 $g/cm^3$); MDPE (0.926–0.961 $g/cm^3$); and HDPE (0.941–0.965 $g/cm^3$).

As a general rule, the lower the density, the lower the melting point and tensile strength. They are often used for clear films and wire coatings, linings in chemical, cutting boards, and tanks. Higher density grades are more crystalline and correspondingly stronger with tensile strengths of 4,400 psi (30 MPa).

The properties of HDPEs are also affected by their MW. Those grades with a MW < 200,000 are considered as general purpose materials with moderate properties. Supermarket bags are made from HDPE. When the MW is in the 200,000 to 500,000 range, the material exhibits superior strength, stiffness, chemical- and abrasion-resistance. High stress crack-resistance is also a desirable property of PE and often is a result of its tendency to cold flow under high loads.

- VLDPE — tensile = used mostly for sheet
- LDPE — tensile = 2000 psi (13.7 MPa)
- MDPE — tensile = 3300 psi (22.7 MPa)
- HDPE — tensile = 4400 psi (30.3 MPa)

Generally, PE is known for its low moisture absorption, low coefficient of friction, high energy absorption, and excellent chemical-resistance throughout a wide range of temperatures. UV-resistance is poor in unmodified grades but is significantly improved by using fillers such as carbon black. Its tensile strength is low compared to engineering plastics and proportional to material density and MW. Unmodified grades are flammable and will remain burning after the removal of the source. Variations are available throughout the density ranges that leverage one or several of these properties and, additionally, can be produced in FDA approved grades.

Because of its low density, all unmodified grades float in water. They are easily machined using conventional machine tools and processes and can be welded using thermal or ultrasonic techniques. PE is difficult to bond because of its low surface energy and requires special surface treatments such as flame or corona discharge to activate the surface for bonding. Bonding with a solvent adhesive is not possible.

Fillers are not commonly used in PE because of the variation on performance afforded to it by the various densities and MW. When they are, HDPE is most often selected. Glass fiber is added for strength, carbon black for improved UV protection and static dissipation, and wood fillers as extenders.

## Polypropylene

The other commodity plastic from the polyolefin family is polypropylene, a semi-translucent, milky-white plastic that floats when in its unmodified form. It offers a low density (0.90 g/cm$^3$) material with moderate properties, excellent fatigue properties at a low cost. Polypropylene is the plastic most often used to create molded hinges for injection molded products. They are created by including a thin section between the base and cover which is filled perpendicular to the hinge axis during molding, creating flow lines perpendicular to the axis. Immediately after molding, while the part is still warm, the hinge must be flexed properly to correctly orientate the molecular chains, creating a hinge that is extremely fatigue-resistant.

Homopolymer and copolymer grades are available. Homopolymer grades offer higher tensile and superior working temperatures when compared with either LDPE or HDPE but with reduced impact-resistance and low temperature brittleness. Copolymers offer improved impact-resistance by additional copolymerization with ethylene to produce a range of end use characteristics.

Polypropylene is similar to polyethylene in its machinability, weldability, and flammability but offers a harder surface finish which is more difficult to scratch than polyethylene. There are flame-resistant and no-smoke grades that offer reduced toxicity but are of a higher density and do not float in water. Its electrical characteristics are excellent when used as an insulator but is not a good high-frequency dielectric.

The physical properties of polypropylene can be modified by the addition of a wide range of fillers. These include glass and carbon fiber for improved tensile strength, glass beads for reduced water absorption and thermal conductivity, carbon black for improved UV protection and static dissipation, stainless steel and copper for greater conductivity, and wood and mineral fillers as extenders.

## Polystyrene

The use of polystyrene in engineering applications is most common when its low cost can be aligned with a particular required property. It is an amorphous thermoplastic with high hardness and rigidity, with limited heat-resistance and which can be cross-linked. Unmodified cross-linked grades can achieve a transparency of 87%, providing a lightweight (1.05 g/cc) material of use in low-grade optical environments that weighs 15% less than acrylic. It can be further improved by alloying with SAN to improve its weatherability, mechanical, and thermal properties.

Polystyrene has excellent machinability and resistant and dielectric properties. Its electrical-resistance of $10^{-19}\Omega$ is considerably higher than $10^{-15}\Omega$ observed with acrylics and urethanes. Its flammability is comparable to the polyolefins with some grades available that are self-extinguishing.

Fillers are also used with polystyrenes, obtaining similar results as described for the polyolefins. These include grades that are cross-linked with fiber reinforcement.

## COMPOUNDING

Altering a plastic from its unmodified forms is referred to as compounding. The purpose of compounding is as varied as the processes and materials involved. It is most often done to improve a specific material property while retaining the properties for which the material was originally selected. Compounding is also done to improve the processing or post-processing stability while, again, minimizing the effect on the base material.

### Alloying and Blending

These are two unique processes that have similar objectives, to improve some desired physical property. They are defined as follows:

*Alloying*: a plastic formed from two polymers that results in a plastic with a single glass transition temperature and synergistic properties derived from the individual polymers.

*Blending*: a plastic formed from two or more polymers that results in multiple glass transition temperatures and properties that are average for the individual polymers.

Alloys and blends become coherent plastics without chemical bonding, but as a result of mechanical mixing. Improvements in impact strength, weather-resistance, and flammability are the most common reasons to use an alloyed or blended plastic (Table 6.5).

**Table 6.5**  Commercial Plastic Alloys and their Notable Properties

| Alloy | Notable Properties |
| --- | --- |
| PVC/acrylic | Flame, impact, and chemical-resistance |
| PVC/ABS | Flame-resistance, impact-resistance, processability |
| Polycarbonate/ABS | High impact strength at low temperatures, flame resistance, balanced mechanical properties |
| Polyphenylene oxide/ polystyrene | Improved processability, lower cost |
| Nylon/elastomer | Impact resistance |
| Polybutylene terephthalate/ polycarbonate | Impact resistance with toughness, chemical resistance and high temperature dimensional stability |
| Polyphenylene sulfide/nylon | Lubricity |
| Acrylic/polybutylene rubber | Clarity with impact resistance and flexibility |

## Fillers and Reinforcements

Many plastics are modified by the addition of a filler or reinforcement material. A filler can be any material added for the express purpose of either altering the bulk material property or improving the surface property. Fillers such as wood flour, calcium carbonate, silica, or clay, added in small percentages of the overall volume, dimensionally stabilize the plastic and can improve machining properties. When added in quantities larger than 35%, they are considered extenders, added for the sole purpose of increasing the resin yield to reduce costs. Metallic fillers are added to improve the electrical or thermal conductivity of the plastic, often with the desirable side effect of improved strength. Fillers added to primarily improve surface properties include PTFE, $MSO_2$, and graphite which often reduce the material properties a small amount from their modified form.

Reinforcements are those materials added for the express purpose of improving the tensile strength of a plastic (Table 6.6). Most common are glass, carbon, or aramid (Kevlar) fibers added in percentages ranging from 10 to 40%. The greater the percentage fiber, the higher the tensile

**Table 6.6**  Common Fillers or Reinforcements and their Applications

| Fillers | Applications |
|---|---|
| Barium sulfate | Used as a filler and white pigment, increases specific gravity, frictional-resistance, chemical-resistance |
| Calcium carbonate | Filler, low cost |
| Carbon black | Filler, used as a pigment, antistatic agent, or to aid in cross-linking, conductive |
| Metal fillers | Filler, improves both electrical and thermal conductivity as well as magnetic properties |
| Microspheres, hollow | Filler, provides weight reduction, improved impact and stiffness properties |
| Microspheres, solid | Filler, improves stress distribution and flow properties |
| Organic fillers | Filler, also used as an entender, including rice, corncobs and peanut shells |
| Silica | Filler, improves heat and chemical resistance with high dielectric strength |

| Reinforcements | Applications |
|---|---|
| Boron fibers | High tensile strength and compressive load-bearing capacity, expensive |
| Carbon/graphite fibers | Reinforcement, high modulus and strength, low density, low coefficient of expansion, low coefficient of friction, conductive |
| Ceramic fibers | Reinforcement, very high temperature-resistance, expensive |
| Glass reinforcement (fiber, cloth, etc.) | Largest volume reinforcement, high strength, dimensional stability, heat-resistance, chemical-resistance |
| Metal filaments | Used to impart conductivity (thermal and electrical) or magnetic properties or to reduce friction, expensive |
| Polymeric fibers | Reinforcement, lightweight |
| Talc | Extenders/reinforcements/fillers, higher stiffness, tensile strength and resistance to creep |

strength, and the lower the toughness. The addition of fiber increases the brittleness of the composite material, leaving it less tolerant of impact loading.

Fibers come in four common forms: A-glass, E-glass, E-CR glass, and S-glass (Table 6.7). A-glass is a soda–lime–silica

**Table 6.7**  Properties of Different Filler Types

| | Specific gravity | Tensile strength | | Tensile modulus | | Elongation at break (%) | Shear modulus | |
|---|---|---|---|---|---|---|---|---|
| | | MPa | ksi | GPa | $10^6$ psi | | MPa | ksi |
| E-glass | 2.58 | 3450 | 500 | 72.5 | 10.5 | 4.8 | 30 | 4.3 |
| A-glass | 2.50 | 3310 | 480 | 69.0 | 10.0 | 4.8 | 29.1 | 4.2 |
| E-CR-glass | 2.62 | 3450 | 500 | 72.5 | 10.5 | 4.8 | n/a | n/a |
| S-glass | 2.48 | 4585 | 665 | 86.0 | 12.5 | 5.4 | 35 | 5080 |

fiber which offers good chemical-resistance. E-glass is a calcium alumino-borosilicate glass fiber that offers good electrical properties and is the most widely used glass fiber. E-CR glass fiber provides a combination of good electrical and chemical properties (E-type, corrosion-resistant). S-glass is a magnesium-alumino silicate fiber most often used in aerospace because its high-temperature properties exceed that of the other fiber types.

Two additional fiber types are gaining popularity, C-glass and R-glass. C-glass has greater chemical resistance than A-glass while R-glass offers higher strength and modulus.

## Additives

Additives are those chemicals or materials included in the polymerization process to improve the end product or support the production of it. The most commonly used are plasticizers, heat stabilizers (often PVC), antioxidants, UV stabilizers, flame-retardants, and colorants. The specifics of these additives vary greatly between each application that it is best to review them with resin manufacture on a case by case basis.

## SELECTION CONSIDERATIONS

The selection of an engineering plastic can be a daunting task because of the many material families and their overlapping properties. Compounding this selection process are the additional property variations that exist within a material family, made available by the addition of many fillers and additives that can significantly alter the base properties of the plastic. The advantages of plastics over metallic materials are well known and include superior corrosion-resistance, lower density, and often lower cost for a finished product with similar properties. Coatings or finishes applied to metals can produce similar performance properties but at an increased cost due to the additional processing. Additional advantages include ease of machining of raw stock forms and the ability to form to a net shape when molding, as well as color, finish, and texture options that are generally not available with other materials.

Unlike ferrous metals, plastics are not selected primarily for their strength or corrosion-resistance, but, similar to non-ferrous metals, for their wear-resistance and stiffness.

## Performance Objectives

Plastics exhibit a slightly viscoelastic behavior when below their $T_g$ and consequently do not follow Hooke's law as observed by metals.

$$\sigma = E\,c$$

The viscoelastic effects become more pronounced as the operating temperature nears the $T_g$, resulting in reduced strength and greater toughness (Figure 6.6). The opposite also follows, lower temperatures result in increase strength and a loss of toughness.

Amorphous polymers exhibit greater viscoelasticity than crystalline or cross-linked forms and are considered to follow Newton's law of constant viscosity. Two mathematical models are commonly used to describe the viscoelastic behavior of plastics:

- The Kelvin/Voigt model of creep: a dashpot in parallel with a spring. This model always returns to its original position much like an actual viscoelastic solid.
- The Maxwell model: a dashpot in series with a spring. The maxwell element produces an exponential stress-decay relationship with time.

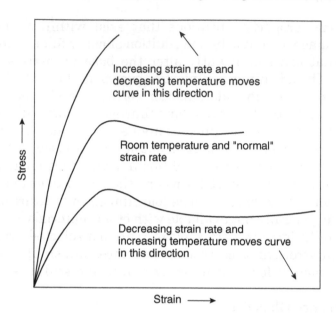

**Figure 6.6** Mechanical stress response to changes in thermal and strain rate conditions.

*Source*: ASM International, Online Metals Handbook, Desk Edition, Introduction to Engineering Plastics, General Characteristics: Metal verses Plastics, Figure 2.

The significance of these models becomes important when studying the creep behavior or long-term effects of material behavior in a specific application environment. A large base of literature is available to support this type of analysis.

Unmodified plastics offer similar anisotropic properties as those observed by metals with the added ability to be made isotropic (directionally varied) by adding fibers or controlling the molding process. Whether a plastic is amorphous or crystalline in structure is a function of the material family, with some plastics having the ability to be both depending on their processing.

Creep properties are also readily observable with engineering plastics and a function of their structure. Because plastics are inherently viscoelastic to some degree, their ability to sustain a load over a long period of time is proportional to the extent of viscoelasticity exhibited by the plastic. Materials such as nylon or acetal have good creep-resistance because of their highly crystalline structure, while UHMWPE is amorphous and readily cold flows or creeps under light loading (Figure 6.7). Extended exposure to creep loading conditions combined with adverse environmental conditions results in a phenomenon known as stress rupture or material failure at load levels below the tensile strength of the material. Plastics are much more susceptible to this form of failure than other material families.

Cyclic loading of plastics is generally not recommended because the fatigue effects can be highly variable. The viscoelastic nature of the material allows for permanent strain when cyclically loaded, regardless of the applied level of stress. Fatigue failure of plastics is not fully understood and can be highly variable depending on the base polymer properties and the effects of fillers and additives. Most often plastics fail in high-cycle fatigue due to the resulting loss of physical strength from hysteresis heating effects in the stressed region. Low-cycle fatigue failure can happen as a result of a

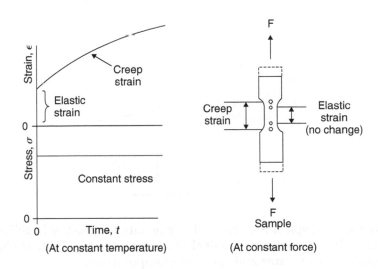

**Figure 6.7** Plastics subjected to constant stress and temperature will exhibit some creep if the loading is above the elastic limit of the material.

significant stress concentration from preexisting crack, sharp corner or poor surface finish. Limited fatigue data is available for material families and even less for the many variations. Careful analysis of the application requirements is necessary when selecting a plastic for a highly cyclic application and should be followed by representative field testing to assure it will be successful (Figure 6.8).

Selecting the appropriate material requires an understanding of how the materials of a given family relate to one another in the context of common design parameters. Strength, density, modulus, temperature, and cost are the most commonly used parameters for judging a plastic, for use in an engineering application. How these parameters for each material class relate to one another across the family of materials provides insight into their comparative performance. Plotting these on a log–log scale allows for this comparison. For example, engineering plastics are available in a range of densities extending from 0.8 to 2 Mg/m$^3$ for which their strengths range from 5 to 150 MPa

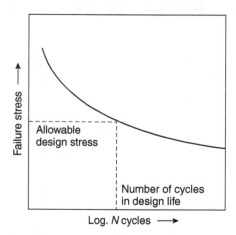

**Figure 6.8** Typical S–N curve. Fatigue data for plastics is difficult to generalize because mechanical properties are highly dependent on applied strain rates and operating temperatures.

*Source*: ASM International, Online Metals Handbook, Desk Edition, Introduction to Engineering Plastics, General Characteristics: Metal verses Plastics, Figure 3.

(Figure 6.9). If the design requires a plastic material that floats (density <1.0 Mg/m³) and strength is not important, then a LDPE will likely be the primary candidate. Consideration for designing to minimum weight can also be made by plotting guidelines of the appropriate strength to density ratio (see Selection Methodologies).

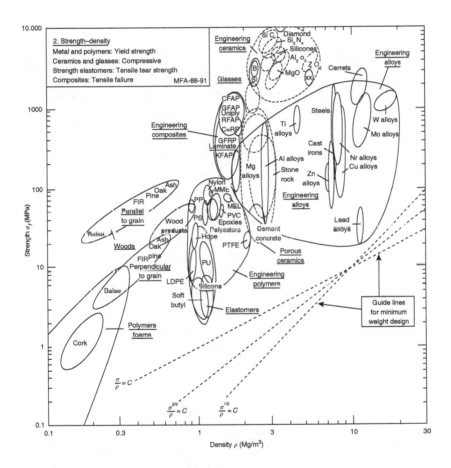

**Figure 6.9** Materials property chart of strength verses density for a wide range of materials. Note that engineering polymers are bounded by wood products on the low end and metallics on the high end.

*Source*: Reprinted with permission from Materials Selection in Mechanical Design, Michael F. Ashby, p. 39 Fig. 4.4, 1999, with permission from Elsevier.

A similar case can be made for a plot of material modulus versus strength. When stiffness is important to the design of a structure, a chart of this type will assist the design engineer to narrow the material choices down to a manageable number of plastics (Figure 6.10).

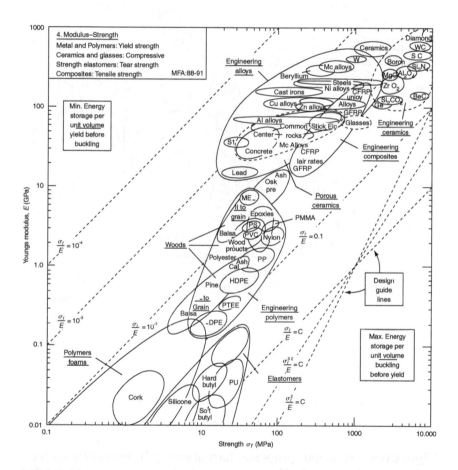

**Figure 6.10**  Materials property chart of material modulus verses density for a wide range of materials. Note the separation of the material families with respect to their moduli.

*Source*: Reprinted with permission from Materials Selection in Mechanical Design, Michael F. Ashby, p. 37, Fig. 4.3, 1999, with permission from Elsevier.

Engineering plastics exhibit the widest range of moduli of any material family, making the selection of an appropriate material even more difficult. The modulus of many unmodified plastics can be significantly improved by the addition of mineral fillers and glass or carbon fiber stiffening mechanisms, resulting in an increase as high as ten times the base value. Fillers provide a modest improvement in modulus for a relatively low cost while fiber reinforcement can have a significant effect on modulus, especially when the matrix exceeds 40%, while adding significant cost due to added materials and processing. These improved properties combined with plastics' inherent low density results in highly efficient materials ideal for high stiffness to weight ratio applications. The downside of the improved modulus is loss of toughness, making these materials prone to catastrophic failure with little or no warning.

Combining these past two charts into one that presents the material modulus and strength that includes density results in a chart with $E/\rho$ versus $\sigma/\rho$, known as specific stiffness versus specific strength (Figure 6.11). The value of this chart is that it plots candidate materials with consideration of their density, which is ideal for selecting materials for those designs requiring minimum weight while maximizing stiffness and strength. Aerospace and automotive designs are most likely to have these requirements. Additional considerations, must then be used to further narrow the choice of materials.

All materials are affected by temperature but none more than plastics and elastomers. Unlike metals and ceramics, the transition temperature for plastics and elastomers are much closer to ambient temperature conditions that common excursions, such as boiling water, can exceed the maximum operating temperature of many of these materials. The strength and creep behaviors of these materials are closely coupled to their operating temperature, which results in one of the most significant limitations of plastics when used as engineering materials. Their elevated temperature strength can be improved, often significantly, by the addition of filler and stiffening materials but these do increase their cost and reduce processability. Plotting strength versus temperature provides insight as to the performance of the material from purely a strength standpoint,

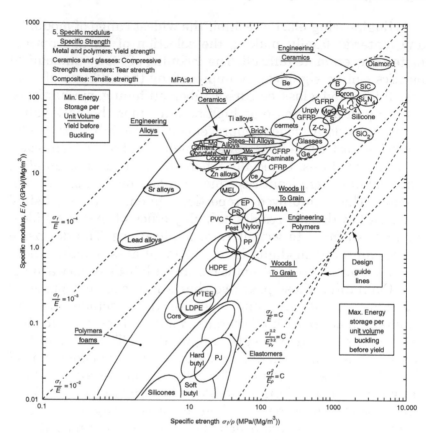

**Figure 6.11**   Materials property chart of specific modulus verses specific density for a wide range of materials. Note the inter-mixing of material families due to the efficiencies of specific material grades.

*Source*: Reprinted with permission from Materials Selection in Mechanical Design, Michael F. Ashby, p. 44, Fig. 4.7, 1999, with permission from Elsevier.

without consideration for creep effects (Figure 6.12). Additional charts can be generated with thermal conductivity, heat deflection temperature, or linear thermal expansion coefficients in place of strength when their properties are part of the selection criteria. Further study and testing must be performed to validate the material will provide the desired performance.

Material cost and availability are most often the primary drivers for the final selection of an engineering material.

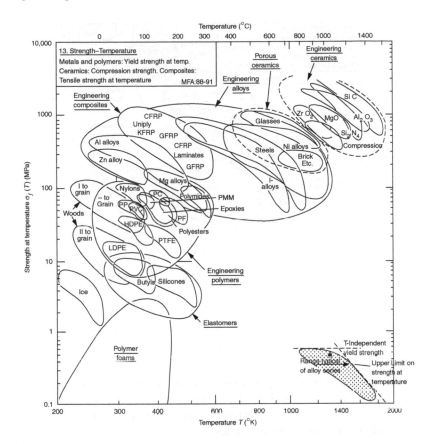

**Figure 6.12** Materials property chart of strength at temperature verses temperature for a wide range of materials. Note that engineering alloys dominate the greatest portion of the temperature range.

*Source*: Reprinted with permission from Materials Selection in Mechanical Design, Michael F. Ashby, p. 56, Fig. 4.15, 1999, with permission from Elsevier.

Plotting strength versus cost provides a chart that can be used to support this final decision which will likely lead to further discussion regarding the material selection involving performance compromises that could be made to further reduce costs (Figure 6.13).

During this process it is important not to lose sight of the requirements that drove the selection of the final candidates,

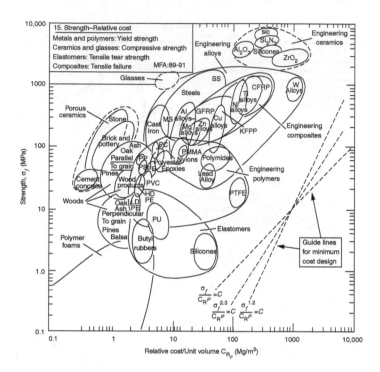

**Figure 6.13**    Materials property chart of strength verses unit cost for a wide range of materials. Note the overlap of the materials families indicating the need for much consideration for all potential candidates durin the selection process.

*Source*: Reprinted with permission from Materials Selection in Mechanical Design, Michael F. Ashby, p. 37, Fig. 4.3, 1999, with permission from Elsevier.

namely the preceding charts. Once a final candidate is selected, the design engineering must backtrack through his decision tree to validate the selection, providing confirmation that the final selection holds true to the initial selection criteria. This step is often the most critical and, unfortunately, the most overlooked.

The use of plastic materials in applications involving relative motion requires careful consideration of the operating environment and the predicted material response to that environment. Friction, wear, and erosion are material properties that are defined, based on a controlled interaction with another material or substance and are not unique for each material, but rather a property that is a result of the interaction. Standardized testing for determining these properties are used by manufacturers to normalize the presentation of material performance data to allow comparison between material types. The coefficient of friction is either stated as static or dynamic with dynamic friction generally the lower of the two. When considering this property one should take care to differentiate between the two. The classic definition of friction is given by

$$F = \mu N$$

where $F$ is the driving force; $\mu$ is the coefficient of friction; and $N$ is the force normal to surface.

This definition holds for materials of high hardness, such as metals and ceramics, and is independent of contact area. It is also typically derived at near room temperature conditions and low velocities with little consideration for thermal effects.

For plastic materials, because of their inherent low hardness and limited thermal properties when compared to metals or ceramics, this relationship is altered. The observed frictional properties are altered by pressure, velocity (frictional heating), surface finish, and material fillers as well as ambient temperature conditions. The frictional model becomes one related to contact area and material shear strength, namely

$$F = S A$$

where $F$ is the driving force; $S$ the shear strength of the weak member; and $A$ the true contact area with

$$A = N/P$$

where $N$ is the normal force and $P$ the penetration hardness. This can be reduced by consolidating the above relationships

$$\mu = S/P$$

The resulting equation more closely aligns the material property with the operating condition which is critical to the application of plastics because of their behavior at near ambient conditions (when compared to metals). Plastics do not wear well against themselves because frictional heating is difficult to dissipate when nonconducting surfaces are involved. Introducing a conducting surface such as metal, the friction behavior is substantially improved. In either case, if surface velocities and loads become extreme, heating will occur and result in a loss of hardness and shear strength, further reducing the coefficient of friction, leading to material failure. Although the above theory is adequate to emphasize the frictional behavior of plastics, rigorous application of this method is difficult to apply as the penetration hardness values are not readily available in the manufacturer's literature as is a scaled hardness such as Rockwell M or R.

Lubricant additives are commonly included in plastics, as observed above in the general discussion of each material type. These are included primarily for reducing the frictional interaction that produces undesired heating but they also offer improved dimensional stability. These materials include: paraffin, PE wax, PP wax, fatty acids, long-chain alcohols, stearone, stearamide, EBS, stearate esters, complex esters, and stearate salts.

Wear and friction are not mutually exclusive properties, and when considering wear of plastic materials, they can be related by the formulation of the base material rather than a result of a lubricating substance placed between two materials, as most often found with metals. Material wear is generally defined as the progressive removal of material from contacting surfaces involved in relative motion. Plastics exhibit five distinct wear processes:

- Abrasive — hard material against a soft plastic
- Adhesive — shearing of local welding or cohesive bonding
- Fatigue — cyclic loading that may cause pitting, flaking or spalling
- Chemical/corrosive — change in surface due to chemical reaction(s)
- Fretting — small amplitude oscillatory motion

**Table 6.8** Material Wear Characteristics as a Function of Operating Environment

| Wear process | Operating parameters |
|---|---|
| Abrasive | Sliding of extremely rough surface with smooth plastic, resulting in material removal |
| Adhesive | Sliding of smooth surface with smooth plastic, resulting in material welding |
| Fatigue | Cyclic loading of surfaces in contact while under tensile or compression loading, resulting in surface fatigue and related failure(s) |
| Chemical/corrosive | Chemical reaction at surface that produces a change in the surface, resulting in swelling or corrosive byproducts |
| Fretting | Contacting surfaces involved in oscillatory motion, resulting in heat build-up and surface failure |

Consideration for these forms of wear must be evaluated when using plastics in dynamic design environments that involve sliding of surfaces (Table 6.8). Differentiating between the operating parameters that produce these forms of wear is important and will help define the inherent wear mode of the system.

The materials involved in the wear process are also of significant concern. When they are plastic-on-plastic, unmodified grades generally perform poorly. Lubricated grades used against one another perform best, while reinforced grades should always be used against lubricated grades. Reinforced grades against each other should be avoided because the reinforcing materials work in an abrasive manner, leading to rapid failure. Plastic-on-metal applications are where plastics excel and allow for use a wide range of grades depending on the surface speeds and loading. Use of the pressure × velocity equation is helpful. In both these material interfaces, wear at elevated temperatures only accelerates the process because of the loss of material properties. Crystalline polymers generally perform better than amorphous ones because of their structure. The most notable plastics that perform best in wear are UHMWPE and $MOS_2$ filled nylons. Where UHMWPE lacks significant structural strength to be useful in high loads or elevated temperatures the nylons can.

Erosion of plastics is generally not a concern for typical machine design applications but is a form of wear that must be considered when using plastics in material transport applications. The process of erosion is one of material removal due to the impingement of high velocity particles on a surface. These can be in the form of liquid slurries or dry powder traveling in an open chute or enclosed ducting or piping. The particle size, shape, roughness, and angle of impact all have a profound effect on the erosion rate and resulting material performance. UHMWPE is generally the most resistant plastic for this type of wear because of its low frictional properties and compliant surface properties which allows the material to give, but not erode, when subjected to high-velocity particles.

The behavior of plastics when exposed to various environments is quite different from that observed with metals and more like that observed with adhesives. Metals are known to corrode in the form of surface oxidation (rust) and pitting when exposed to water and various liquid chemicals. Sealing these surfaces by additional processing to prevent the material from damage can be done using many methods, including painting, black oxidized coatings for steels and chemical film, and anodizing for aluminum. Plastics offer several advantages over metals with one of them being, they do not require post processing.

Environmental exposure can be divided into two categories: ambient and chemical. Both conditions can alter the properties of an engineering plastic to such a degree as to lend it useless for its intended application. The time and degree to which this occurs is as variable as the materials themselves. What is important is to understand the duration of exposure the material is to see during its lifecycle. Many materials can handle brief periods of excessive exposure to adverse environments, for example, brief temperature extremes are a common form of deviation that many designs require and can be managed by proper material selection. Of course, the selection can only be successful if a truthful assessment of the extremes has been made with consideration of a factor of safety to provide the boundary temperatures in which the device must operate.

Electrical properties of engineering plastics have become increasingly important in the design of electronic devices and related enclosures. Traditionally, plastics are considered insulators due to their inherent insulative properties, both electrically and thermally. The plastics industry has matured in parallel with the electronics industry, leading to increased demands as design requirements have become more sophisticated. Plastics not only are applied as insulators, but also as conductors (Figure 6.14).

The ability of a plastic to conduct is controlled by its volume resistivity, a bulk measure of the material resistance which can be varied by the types and quantities of additives (Table 6.9). There are three categories of electrical properties for plastics: insulative, static dissipative (or antistatic), and conductive. Most plastics are by their chemical composition insulative in nature both electrically and thermally. The nonconductive base polymer can be modified to have varying degrees of resistance to suit the desired application, ranging from $10^{15}$ to $10^{-3}$ $\Omega$.

An unavoidable byproduct of electrical conductivity is thermal conductivity because the physics of those materials that offer the lowest electrical resistance also applies to thermal resistance. Although these two properties generally are complementary, there are applications that require some level of static dissipation while remaining thermally insulated.

How the plastic will be processed into its usable form needs to be given equal consideration as that applied to its property selection. There is little value in selecting a material that may possess specific desired properties only to find it requires extensive nonrecurring costs to produce it in the necessary design configuration. Included in these considerations should be processing methods appropriate for the material, stock forms, and availability.

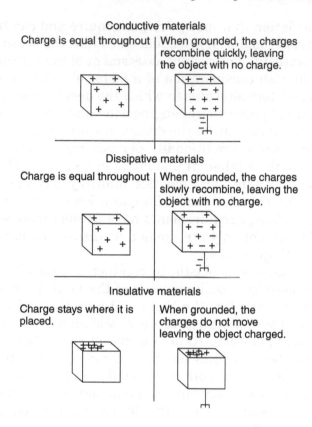

**Figure 6.14**    Electrical behavior of plastic materials.

*Source*:    Reprinted from Assembly Magazine, Copyright BNP Media Inc. 2002.

**Table 6.9**    Conductive Ratings for Plastics

| Insulative (high resistance) | Static dissipative (antistatic– high impedance) | Static dissipative (intermediate impedance) | Conductive plastics (low impedance) | Conductive Material (minimal impedance) |
|---|---|---|---|---|
| $>10^{15}\Omega$ | $10^{11}–10^{2}\Omega$ | $10^{6}–10^{2}\Omega$ | $10^{5}–10^{2}\Omega$ | $<10^{-2}\Omega$ |
| Unfilled plastic/ rubber | Filled with non-carbon anti-static material (SD-A) | Filled with carbon based material (CF-F fibers & CN-P powders) | Filled with stainless steel (CN-SS) | Bulk metallic materials |

Lifecycle environmental considerations must be included in the selection process as governments throughout the world continue to require safer work environments and end of use recovery methods for plastic products. Many materials are proprietary and therefore their classification is not easily determined or revealed by the manufacturer, making understanding of their composition and performance difficult to gauge. Working closely with a qualified supplier can facilitate the development and manufacturing process by providing the appropriate detailed information in a timely manner.

## Selection Methodologies

Plastic materials are among the most difficult to select of all the material families presented in this text, because of the numerous formulations available. Custom formulated materials are also possible when the application justifies the added development time and cost. Plastics are often selected as replacements for ferrous and nonferrous materials as a way of saving material cost and fabrication time. Careful study of their operational environment in terms of mechanical strength, wear, fatigue, chemical exposure, and electrical properties is necessary to make an informed decision of an appropriate replacement plastic. They are also selected to reduce or eliminate maintenance-related activities such as eliminating the need for lubrication in gear or sliding wear applications or eliminating the need for cosmetic coatings by use of colorized materials.

### Mechanical Properties

Accurate determination of the loading properties is essential in the selection of plastic materials because the factor of safety is typically not as high as that which is possible for metallic materials. The development of fiber reinforced plastics has afforded an improved tolerance to intermittent loading extremes but these must be carefully considered. Determining the desired loading values and impact requirements is the first step in the selection process (Figure 6.15) The fatigue property of a material is also an essential consideration for plastic parts because they behave much like nonferrous metals and exhibit no real fatigue limit as seen in metallic metals. Operating temperature as a function of material strength is also an important consideration when establishing the required mechanical properties (Figure 6.16).

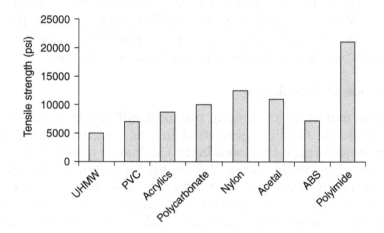

**Figure 6.15**   Tensile strength for selected plastics.

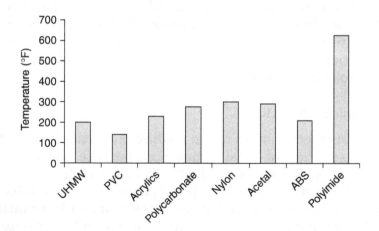

**Figure 6.16**   Operating temperature for selected plastics.

## Chemical-Resistance

The chemical-resistance characteristics of plastics are defined by their based chemistry and therefore must be considered during the initial selection process (Figure 6.17). Unlike ferrous and nonferrous metals, plastics are generally not post-processed for corrosion-resistance because of their limited adhesion properties, making it difficult to effectively protect them by plating or painting. Although it is possible to plate plastics with a decorative metallic coating, it is generally not sufficiently bonded to survive industrial applications. Resistance to specific chemical solvents and acceptable levels of exposure should be reviewed for all possibilities during the selection process with the final material candidate being selected based on the material property datasheet provided by the manufacturer. A general listing of material behavior can be reviewed as an initial guide. Consideration for moisture absorption should also be included in the selection process for chemical compatibility (Figure 6.18). These amount of absorption is dependent on the chemical make-up of the polymer, the relative humidity and the length

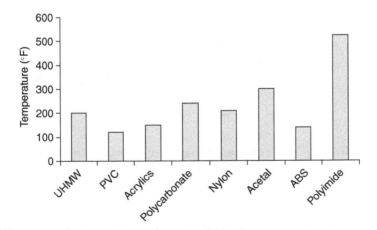

**Figure 6.17**  Long-term operating temperature for selected plastics.

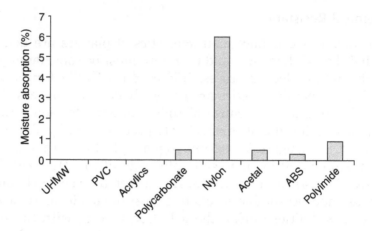

**Figure 6.18**    Moisture absorption for selected plastics.

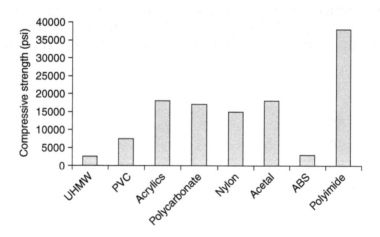

**Figure 6.19**    Compressive strength for selected plastics.

of exposure. Careful review of manufacture test data is essential to successfully evaluating this selection criteria.

Secondary properties including coefficient of friction, wear behavior, electrical properties, material density, flammability,

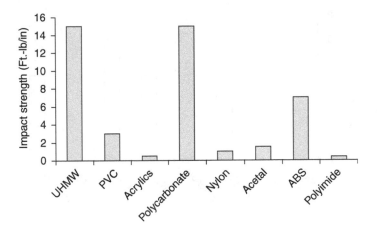

**Figure 6.20**   Impact strength for selected plastics.

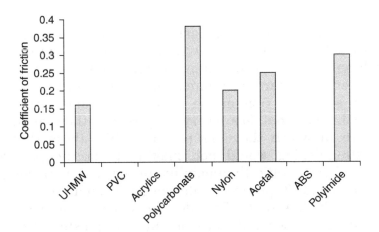

**Figure 6.21**   Coefficient of friction for selected plastics.

and transparency also factor into the selection process. With the exception of transparency, these properties generally offer a wider latitude in their material properties because they can be enhanced by additives to improve the base material properties.

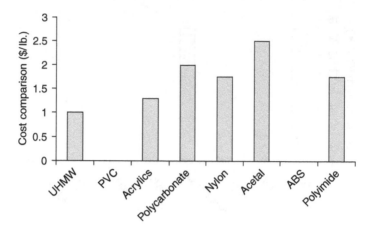

**Figure 6.22**    Cost comparison for selected plastics.


The flammability of a plastic is also an important selection consideration because the resulting combustion products are hazardous or even fatal. An important distinguishing characteristic regarding flammability is whether or not the plastic supports combustion when the flame source is removed. Those that do are very dangerous as they would feed any fire in which they are present, while those that are self-extinguishing would not, making them the preferred choice when specific flammability rating is a requirement. For example, PVC produces chlorine gas when it burns, which is not easily detectable. The odor threshold is approximately 0.3 to 0.5 parts per million (ppm) but distinguishing toxic air levels from ambient air levels may be difficult until irritative symptoms are present. Similar concerns regarding the combustion effects of other plastics must be considered in their selection. Various additives are known to reduce a material's susceptibility to burn, raising the minimum combustion temperature to acceptable levels.

At times it may be necessary to identify a material sample without the aid of sophisticated laboratory equipment.

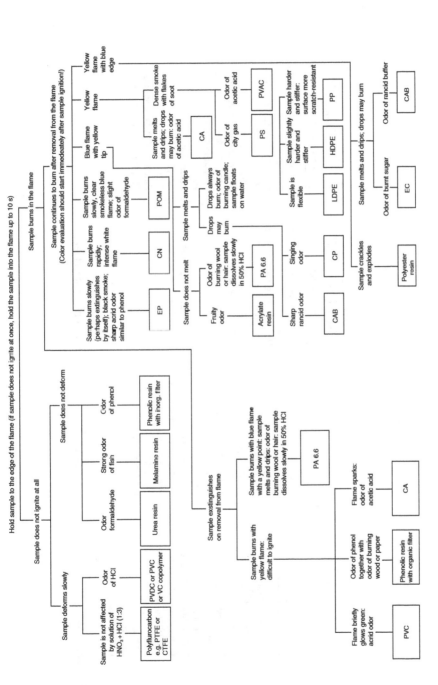

**Figure 6.23** Flammability and odor testing for plastic materials.

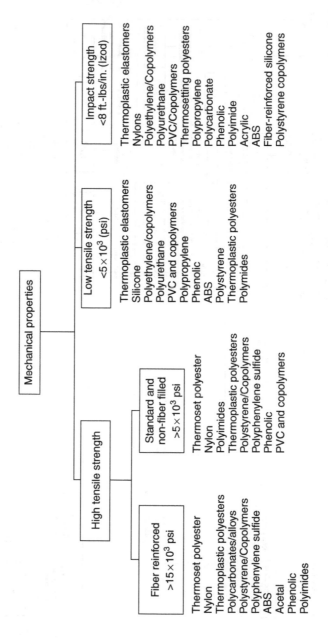

**Figure 6.24** Plastics selection diagram based on mechanical properties.

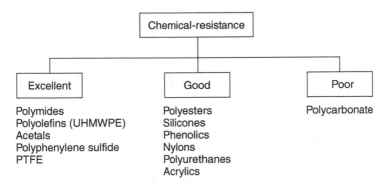

**Figure 6.25** Plastics selection diagram based on chemical-resistance properties.

Performing a simple flame and order test can quickly classify a plastic's material family which can then be used in aiding the selection by either confirming the required material or helping to avoid a under performing one (Figure 6.19).

Figures 6.15 through 6.22 provide a comparison of the material properties of those plastics presented in this chapter. Using these to gain a basic understanding of the differences in material properties can be helpful in selecting the appropriate material family for a specific design.

Figures 6.23 through 6.26 provide a graphic layout of how different plastics relate to one another for various property values and should also be used to further refine the selection. The materials listed also include those not discussed in this chapter, providing a greater range of possible candidates.

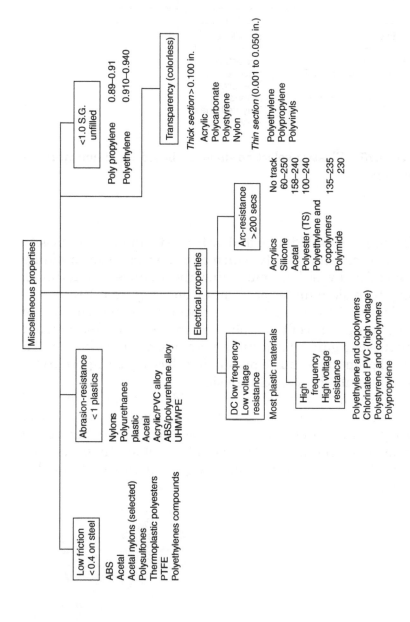

**Figure 6.26** Plastics selection diagram based on various secondary properties.

# 7

## Adhesives

### INTRODUCTION

Adhesives have existed for thousands of years but only in limited form prior to the 1900s. Most were formulated from organic materials such as animal or vegetable byproducts or inorganic minerals. These primitive adhesives included natural gums and waxes such as rubber and beeswax, which were selected for their good moisture-resistance. Traditional woodworking adhesives were made by boiling animal bones to produce granules or flakes that were dissolved in water prior to use. These had limited resistance to moisture and temperature, in contrast to modern woodworking adhesives made from synthetically produced polymers. Soluble sodium silicate is an inorganic adhesive used for bonding porous materials while providing high temperature and moisture-resistance properties. In general, these primitive adhesives produce bonds with limited strength and variable performance.

The development of modern adhesives has paralleled the plastics industry because of their common science, polymer chemistry. As plastics began to be produced in varied forms in the early 1900s, so did adhesives. Initially, adhesives were developed for domestic and industrial applications as a means to simplify assemblies of similar or dissimilar materials or to provide a water-resistant barrier. Phenol formaldehyde adhesives are an early example, which provided both bonding and sealing in the fabrication of plywood. The advancement of adhesive technologies also followed by the aerospace industry because of demanding applications requiring high strength-to-weight ratios. Figure 7.1 summarizes the development of modern adhesives.

The consumption of adhesives continues to increase as their performance properties are better understood and properly integrated into the design solution. Matching the application to the adhesive is essential to an adhesives successful performance (Figure 7.2). Often, the painful experience of a failure prevents design engineers from exploring this joining method during future design applications. This is unfortunate because adhesives can provide an extremely cost-effective design solution when appropriately selected and applied.

A typical adhesive bond (Figure 7.3) includes the adhesive, a primer applied to each mating surface (when necessary) and the adherends. The term "adherend" is defined as a material that is attached to another material by an adhesive. When not attached, the material is often referred to as a substrate. The joining of two or more substrates with an adhesive produces a bonded joint that has properties unique to the composite combination and the processes used when joining them. An additional influence known as a weak boundary layer has also been found to affect bond strength and performance. This region is found between the adhesive and adherend and is mainly the result of adherend surface conditions. The effect of this region on the bond varies greatly as it is dependent on the substrate's chemical composition and processing methods that may produce undesirable surface conditions. For example, metallic substrates often have metal oxides, such as rust, that were formed either during processing or storage. Plastics can have impurities at their surface related to processing or the bonding process that reduce the integrity of the bond.

Material fastening has been traditionally performed by mechanical methods such as screws/rivets or by brazing/welding,

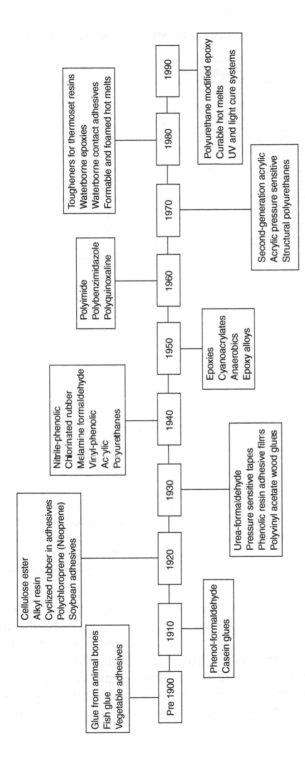

**Figure 7.1** Adhesive development timeline. The evolution of adhesive development closely followed the advances in chemical and process engineering during the 20th century. Although this figure presents the full spectrum, the adhesives presented in this book are representative of those commonly available.

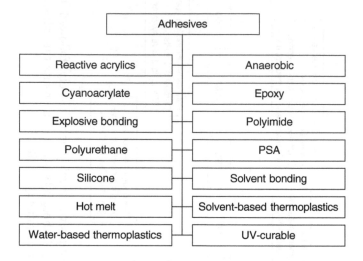

**Figure 7.2**   Common adhesives used in engineering applications.

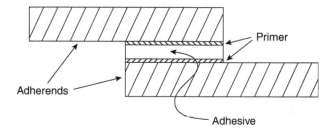

**Figure 7.3**   Typical adhesive bond. In its simplest form, an adhesive bond involves two elements, the adhesive and related adherents. Additional elements may be required to improve the bond, including primers or activators to assist the initial bonding process and accelerators (applied to the exposed adhesive) to help complete the cure.

both of which have undesirable limitations. Screws and rivets create unwanted stress concentrations which can result in premature failure due to overload or fatigue. Welding and brazing offer a similar solution but their application process is significantly different. Welding involves a great amount of heat which can result in material property changes and distortion. Brazing, although it requires less heat, produces a bond of lesser strength.

Adhesives offer an attractive alternative to other fastening methods for several reasons, including even distribution of applied loads, shock and vibration isolation, sealing, weight reduction and cost savings. Additionally, adhesives provide the ability to bond dissimilar materials without significant preparation or alteration and, in the case of metallics, bonding without fear of galvanic corrosion. Fastening materials without alterations to the substrate can also be a strong argument for their use. Often, a combination of these provides the motivation behind the selection of an adhesive over other fastening methods (Figure 7.4).

The disadvantages of adhesive bonding must not be overlooked. If the substrate is not properly prepared or if an adhesive is not properly applied to a substrate that is correctly prepared, no amount of adhesive or cure time can salvage the bond. Unlike a threaded fastener which can be retorqued if

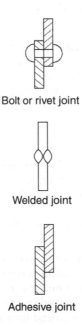

Bolt or rivet joint

Welded joint

Adhesive joint

**Figure 7.4** Joining methods. The substrate material often dictates the joining method that would be most effective. For example, bolts or rivets work well with metals but are not usually effective with plastics because of the high local stresses.

improperly assembled, or if it becomes loose, adhesives provide only one chance. The consequences can be costly if done incorrectly and this fact alone is often the only reason to discourage their selection. Table 7.1 provides a summary of the advantages and disadvantages of adhesive bonding.

When compared to other mechanical fastening methods, the load carrying ability of adhesives are typically less for the same working area. This limitation can often be surmounted by proper joint design and adhesive selection. It is also acceptable to create a hybrid joint which includes both a threaded fastener and an adhesive. This approach can eliminate the concerns associated with the sole use of an adhesive by offering a mechanical backup, while also providing a clamping method that is necessary during joining. Adhesives cannot be disassembled for the purpose of inspection or if an error was made

**Table 7.1**   Advantages and disadvantages of adhesive joining

Advantages
- Provides more uniform distribution of stress and larger stress-bearing area than conventional mechanical fasteners
- Joins thick or thin materials of any shape
- Joins any combination of similar or dissimilar materials
- Minimizes or prevents electrochemical corrosion between dissimilar metals
- Resists fatigue and cyclic loads
- Provides smooth contours
- Seals joint, insulates (heat and electricity), and damps vibration
- Frequently faster and less expensive than conventional fastening
- Heat required to set adhesive is usually too low to affect strength of metal parts
- Post assembly cleanup of parts is not difficult

Disadvantages
- Requires careful surface preparation of adherends
- Relatively long times are sometimes required for setting adhesive
- Limitation on upper service temperature is usually 175°C (350°F), but materials are available for limited use to 370°C (700°F)
- Heat and pressure may be required for assembly
- Jigs and fixtures may be required for assembly
- Rigid process control is usually necessary

*Source*: ASM Engineered Materials Handbook Desk Edition: Online Desk Edition, Adhesives, Fundamentals of Adhesive Technology, Limitations of Adhesives, Table 1.

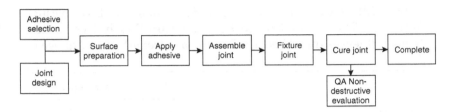

**Figure 7.5** Generic bonding process. The steps necessary to join two substrates is shown. The quality assurance process can range from simple visual inspection by the assembler to test coupons for pull tests, sectioning, aging or other analysis.

during assembly because it would destroy the bond (effectively the fastener) and likely damage the adherend(s).

The adhesive bonding process follows the progression shown in Figure 7.5. Although they may vary for any given adhesive, it is important to consider these critical processes and how they may affect the final design. The requirements for the substrate and how they are manufactured can directly impact the allowable surface preparation and assembly methods.

Although it is difficult to classify any adhesive property as a typical one because of the ability to formulate so many different compounds, they can be listed as approximate or typical values (Table 7.2). Included is a qualitative classification of the bond line and strength during curing or strength build.

## RHEOLOGY OF LIQUIDS

How an adhesive interacts with an adherend is an important consideration in their selection and application. Interpreting the manufacturer's specification as it applies to the application is the beginning of the selection process of which a portion involves rheology. Rheology is the study of flow characteristics of fluids and in the context of adhesives defines its behavior when interacting with the adherend and the bond geometry.

**Table 7.2** Lap Shear Strength for Common Engineering Adhesives. The Values Offered here can Vary Widely Due to the Many Influences that Affect the Lap-shear Strength of an Adhesive Joint

| Adhesive | Lap shear strength | |
| --- | --- | --- |
|  | psi | MPa |
| Silicone | 150–500 | 1.0–3.5 |
| Hot Melt | 200–700 | 1.4–4.8 |
| Pressure Sensitive Adhesive (PSA) | 100–1,500 | 0.69–10.3 |
| Polyurethane | 500–3,000 | 3.5–20.7 |
| Ultraviolet (UV)-Curable | 1,200–3,500 | 8.3–24.1 |
| Polyimide | <3,600 | <24.8 |
| Cyanoacrylate | 400–3,900 | 2.8–26.9 |
| Anaerobic | <4,000 | <27.6 |
| Reactive Acrylics | 750–4,000 | 5.2–27.6 |
| Epoxy | 250–6,000 | 1.4–41.4 |
| Solvent Bonding | 85–100% of the base material | |

An adhesive's ability to bond with an adherend is largely dependent on the adhesive's wetting characteristic. Wetting is the adhesive's ability to flow across and into surface of the adherend. The extent of the interaction is dependent on the characteristic of the adhesive as a liquid and on the surface energy of the material (more on surface energies latter).

The important aspect of rheology in the context of adhesives is viscosity which is defined as the resistance of a liquid to flow and represented by the relationship:

$$\eta = \frac{\tau}{\partial \varepsilon / \partial t}$$

where $\tau$ is the shear stress and $\partial \varepsilon / \partial t$ is the shear strain rate. If this relationship is true at all values, then the fluid is considered Newtonian and the relationship between shear stress and shear strain rate is linear (Figure 7.6). Viscosity is measured in units of poise or stokes. Water is a Newtonian fluid having a viscosity of one centistoke at room temperature 72°F (23°C). Viscosities of other fluids are presented in Table 7.3.

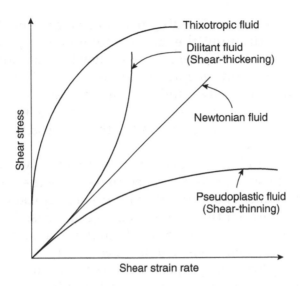

**Figure 7.6** Viscosity effects on shear rate.

*Source*: Alphonsus V. Pocius. *Adhesion and Adhesive Technology: An Introduction.* New York: Hanser, 1997, p. 23.

**Table 7.3** Room Temperature Viscosity Comparisons

| Fluid | Viscosity (cps) |
|---|---|
| Water | 1.0 |
| Blood | 10 |
| #10 Motor oil | 400 |
| #40 Motor oil | 4,500 |
| Honey | 10,000 |
| Hershey chocolate syrup | 25,000 |
| Peanut butter | >60,000 |

There are other non-Newtonian fluid behaviors that are important to describing adhesive properties prior to their cure state. The most notable is a thixotropic liquid, which is one that displays a yield stress. This behavior provides the adhesive an ability to remain in place until a small force is applied, offering a sag-resistance that allows an adhesive to remain in place when bonding to a vertical or angled surface. The act of positioning the mating adherend is sufficient to exceed the threshold shear stress, allowing the adhesive to flow into the joint geometry. Materials such as bearing grease, vaseline jelly, and toothpaste are examples of thixotropic materials. Adhesives formulated with this property are identified as such under their viscosity rating as being thixotropic.

Dilatant and pseudoplastic properties are two other less common adhesive properties used to describe the behavior of the uncured material. The dilatant or shear-thickening property is one in which the fluid exhibits an increased resistance to shear at higher shear rates. Examples of this are some pastes and colloidal suspensions. In contrast, the pseudoplastic or shear-thinning property is one which behaves in an opposite manner, exhibiting less resistance to shear at higher shear rates. Examples include ball point pens, nondrip-paints, and lubricants.

## ADHESION THEORY

The ability of an adhesive to provide the joining interface between two adherends (or substrates) is contingent on many factors that may or may not be related to the materials being joined. Unlike mechanical fasteners, which are readily predictable because of their inherent properties, adhesives depend on material properties including both mechanical and chemical behaviors as well as the material surface condition. The resulting bond strength will vary significantly depending on these conditions.

Bond strength is a good indicator of the success of the bonding process, but will vary depending on the substrate and adhesives used to form the joint. Joint failure also provides insight as to the strength of the bond and the level of adhesion achieved within it. Joints fail either by adhesive or cohesive methods (Figure 7.7). Adhesion failure occurs when the bond separates cleanly between the adhesive and the adherend with little visible retension of adhesive to the adherend.

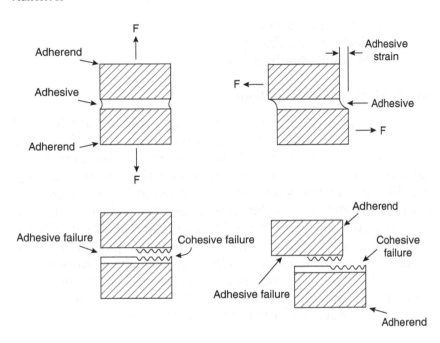

**Figure 7.7** Adhesive and cohesive failure. The reasons for joint failure can be numerous while the actual failure mechanism is either adhesive or cohesive failure within the joint.

Cohesive failure occurs when the adhesive fails and remains bonded to the adherends.

The bonding of an adhesive with the substrate is believed to be the result of a combination of interactions between the two materials. The mechanisms involved can be divided into two categories (primary and secondary): those which are most widely accepted as the primary bonding mechanisms and those which enhance the bonding process but are not significant contributors or occur infrequently.

Primary bonding mechanisms
- Mechanical Interlocking Theory
- Wettability Theory
- Weak Boundary Layer Theory
- Diffusion Theory

Secondary bonding mechanisms
- Electrostatic Theory
- Covalent Bonding Theory

## Mechanical Interlocking Theory

Surface roughness plays an important role in the level of adhesion obtained by physically interlocking the adhesive to the adherend. All surfaces have some level of irregularity or imperfections when viewed microscopically. These can vary significantly in range and form from micron-sized grooves or random scratches to submicron dendrites (minute hair-like features) as shown in Figure 7.8. The resulting effect of these features on adhesion is to provide a tortuous path which prevents the separation of the adhesive from the adherend. An additional benefit is that it also increases the surface area available for bonding, adding strength to the joint as a result of the area-dependent theories discussed below.

| Surface topography of copper foil | | Mean peel load lb/in |
|---|---|---|
| $3\mu$ high angle pyramids | Diagrammatic representation | |
| Flat | ———————— | 3.75 |
| Flat + $0.3\mu$ dendrites |  ⊿⊿⊿⊿⊿⊿⊿⊿⊿⊿⊿ | 3.8 |
| Flat + $0.3\mu$ dendrites + oxide |  ⊿⊿⊿⊿⊿⊿⊿⊿⊿⊿⊿ | 4.4 |
| $3\mu$ pyramids (high angle) |  /\/\/\/\ | 5.9 |
| $2\mu$ low angle pyramids + 0.3l dendrites |  | 7.3 |
| $2\mu$ low angle pyramids + 0.2l dendrites + oxide |  | 8.8 |
| $3\mu$ high angle pyramids + 0.2l dendrites + oxide |  | 13.5 |

**Figure 7.8**   Mechanical interlocking is the most dominant mechanism of adhesion between two materials. Proper selection of the adhesive for the materials to be bonded is required to achieve the best performance.

*Source*: Transactions of the Institute of Metal Finishing Vol. 48, Arrowsmith, D.J. (1970), p.88. Used with permission.

## Wettability Theory

The ability of a substrate and adhesive to come into intimate contact is critical to the success of an adhesive bond. The term "wettability" is used to describe the interaction of an adhesive with a substrate and how readily the adhesive flows into the surface of the adherend (Figure 7.9). Incomplete flow into the surface grooves and crevices results in a nonhomogeneous boundary between the adhesive and adherend; complete or spontaneous flow into the surface results in a homogeneous or uniform bond between the adhesive and adherend. The advantage of the latter result is increased contact area and no stress concentrations at the bond–adherend interface. The wettability is a result of the interaction of the surface tension of the materials (Figure 7.10).

For good wetting:

Adhesive surface tension ≪ Substrate surface tension

This relationship results in the liquid adhesive flowing across and into the surface of the substrate, filling the microscopic surface imperfections.

For poor wetting:

Adhesive surface tension ≫ Substrate surface tension

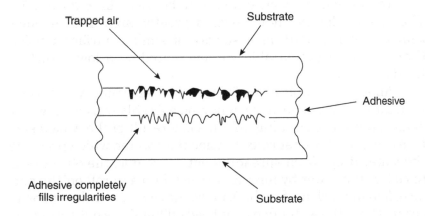

**Figure 7.9** Effects of wetting. Gas bubbles can form if the surface is too rough or if the adhesive surface tension or energy is too high.

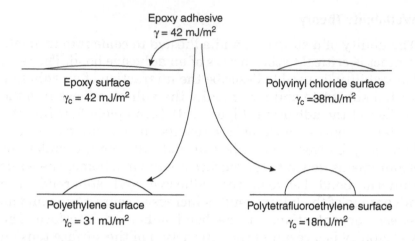

**Figure 7.10**    The surface energy influences the material's ability to fully contact the adhesive. The angle between the adhesive and surface is a function of the wetting, or surface tension, and increases as the surface energy decreases.

*Source*: Alphonsus V. Pocius. *Adhesion and Adhesive Technology: An Introduction.* New York: Hanser, 1997, p. 131.

The resulting behavior when this relationship occurs is the beading of the liquid adhesive which prevents it from flowing across the surface. An example that demonstrates the effect of surface tension on wettability can be seen in Figure 7.10. The epoxy adhesive flows over a greater area and becomes thinner when placed on a surface of similar surface tension than a drop of comparable volume placed on a lower surface tension substrate.

Another example of surface tension behavior involves common car wax. We have all seen how on a freshly waxed car water beads up prior to running off. This is a result of the waxed surface being of a lower surface tension than water, which prevents the water drops from spreading out. Of course, the objective is to collect the water by forcing it to bead up and roll off the surface, leaving a clean finish. And, we have all seen the opposite occur, where the water does not bead off but flows out across the surface, leaving undesired water marks. The surface energy for some common materials is provided in Table 7.4.

**Table 7.4**   Surface energy of some commonly bonded materials

| Material | Surface energy | |
|---|---|---|
| | oz/in | dyn/cm |
| Metals | | |
| Copper | 0.100 | 1100 |
| Stainless steel | 0.082 | 900 |
| Aluminum | 0.077 | 840 |
| Zinc | 0.069 | 750 |
| Tin | 0.048 | 525 |
| Lead | 0.042 | 460 |
| Glass/porcelain | 0.034 | 375 |
| Plastics (high surface energy) | | |
| Kapton | 0.0046 | 50 |
| Phenolic | 0.0043 | 47 |
| Nylon | 0.0042 | 46 |
| Polyester | 0.0039 | 43 |
| Polyurethane | 0.0039 | 43 |
| ABS | 0.0038 | 42 |
| Polycarbonate | 0.0038 | 42 |
| PVC | 0.0035 | 38 |
| Plastics (low surface energy) | | |
| Polystyrene | 0.0033 | 36 |
| Acetal | 0.0033 | 36 |
| Polyethylene | 0.0028 | 31 |
| Polypropylene | 0.0026 | 29 |
| PTFE | 0.0016 | 18 |

## Weak Boundary Layer Theory

The condition of the substrate surface significantly influences the level of adhesion that is achieved by the adhesive. When the surface is contaminated or contains elemental byproducts of the substrate, its inability to form strong adhesion is evident by reduced bond strength or premature failure. Often adhesives are evaluated under carefully controlled conditions that result in the desired performance levels. Once the bonding activity is transitioned to production, a reduced level of performance or early failure is observed due to poor or inadequate preparation. The presence of a weak boundary layer can often be attributed to this phenomenon.

The weak boundary layer theory addresses the observed reduction in cohesive strength at the bond line that can occur in both metallic and plastic materials to varying degrees. Some problems are caused by oils, while others are a direct result of poor surface or near-surface conditions. Boundary layer effects are material-dependent and must be removed to achieve the highest level of adhesion possible.

### Diffusion Theory

In most bonding applications, the adhesive and substrate are significantly different in terms of their material properties, resulting in stress concentration at the bond line at any temperature other than the one at which the bond was formed. If this mismatch could be eliminated, a bond of higher integrity could be achieved. The diffusion theory applies to bonding applications in which the adherends dissolve into each other, unencumbered by an adhesive material (Figure 7.11). The resulting bond interface is free of stress concentrations that are

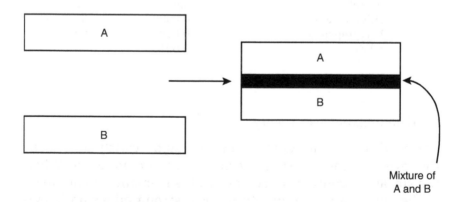

**Figure 7.11**   Diffusive bonding involves the dissolving of two materials into one another such that the bond line is a mixture of the two. Solvent bonding is the most common form of this method.

*Source*: Alphonsus V. Pocius. *Adhesion and Adhesive Technology: An Introduction.* New York: Hanser, 1997, p. 122.

commonly part of the bonding process. This theory has been observed in only a few thermoplastic materials and then only to a limited extent because of the necessary criteria that the materials have similar solubility. Diffusion requiring the assistance of either chemicals or heat is more widely applied in solvent bonding, thermoplastic welding, and ultrasonic welding.

## Electrostatic Theory

The adhesion obtained in a bond is believed to include an electrostatic component. Although little is understood about how this may actually be described, several examples confirm that some form of electrostatics is involved in the formation and breaking of a bond. Consider the following examples:

> Electrostatic charging of a balloon. This familiar example involves rubbing a balloon against your clothing to create a highly charged surface (positive or negative) which causes the balloon to adhere to non-conductive surfaces.
>
> AM radio amplification. In this experiment, tape is applied to an AM transistor radio and then rapidly stripped off, producing an amplified sound of the debonding process.

These are compelling examples that support the theory, but, unfortunately, they do not explain it. Although electrostatic adhesion is an unexplained phenomena in terms of mechanical adhesion, it is an acceptable theory for biological cell adhesion.

## Covalent Bonding Theory

The idea of an adhesive forming a chemical bond with the adherends provides an explanation for the observed strength in selected applications. Strong bonds have also been observed in the absence of any chemical surface bonding, making this theory an elusive one, universally applicable to adhesion. Covalent bonding involves the sharing of electrons between atoms which produces a very strong attraction between the materials involved. Although it would be desirable to have these strong bonds involved in the bonding process, it is not necessary to obtain strong adhesive bonds.

## FORMS

Adhesives are available in many forms that may require mixing or some other mechanism to initiate the curing process. There are three generic forms: one-part, two-part, and tapes.

The first two involve adhesives with a wide range of viscosities and chemical compositions and are typically available in squeeze tubes or a bulk form. These bulk forms are of varying size from several milliliters to large liter containers, depending on shelf-life and dispensing requirements. The third, tape, is also available in more than one form.

One-part adhesives are the most convenient to dispense because they typically are packaged in a squeeze tube or syringe, not requiring precise measurement for mixing. The common criticism with this type is the limited shelf-life because the curing mechanisms are either already at work but just at a reduced rates (as with an epoxy) or are readily available in the surroundings such as the moisture required for RTV. Preapplied adhesives are a common form of a one-part adhesive used to retain fasteners (Figure 7.12).

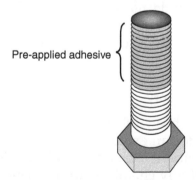

**Figure 7.12**  Preapplied adhesives for threadlocking applications simplifies the use of a one-part adhesive. The form involves microballoons of adhesive which are broken when the fastener is installed, releasing the appropriate amount of adhesive for the application.

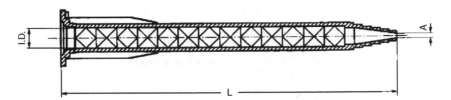

**Figure 7.13** A mixing nozzle is commonly used when dispensing two-part adhesives to assure proper mixing. An important requirement of this form of application is to discard a minimum of one nozzle length of product to eliminate any adhesive that is not completely mixed.

Two-part adhesives require controlled mixing of precise quantities if a quality bond is to be produced. The proportions are metered by the either volume or weight. The most commonly used method of volumetric metering is achieved by the use of a static mixing nozzle which progressively mixes the two components as they travel through it (Figure 7.13). These can be hand or pneumatically pumped as appropriate for the application. Alternatively, the components can be measured by weight and then combined. This method is more cumbersome and prone to error because it involves additional equipment and greater operator skill.

There are also two-part adhesives that require an activator rather than direct mixing to initiate the room temperature curing process. The activator is applied to the surface opposite that of the adhesive and allowed to "flash-off" (dry) prior to assembly. This form is simpler to use than other two-part systems because it does not require accurate mixing to be effective, but works best on larger surface areas which allow even flow and accurate placement of the low viscosity activator.

Adhesive tapes do not always meet the common definition of a "tape," which implies an adhesive bearing backing material used to stick to a surface with the exposed side untreated (Figure 7.14). In this arrangement, the backing material, also known as a carrier, is the loadbearing material.

Adhesive tapes are constructed in two generic configurations: with a carrier material and without, as shown in Figure 7.15. The adhesives and carriers are the loadbearing

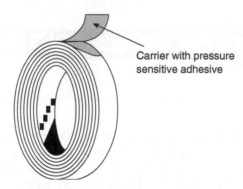

Carrier with pressure
sensitive adhesive

**Figure 7.14** The most common form of adhesive tape construction consists of a pressure-sensitive adhesive on a carrier material and dispensed from a wound roll.

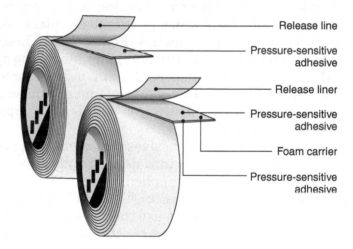

Release line

Pressure-sensitive
adhesive

Release liner

Pressure-sensitive
adhesive

Foam carrier

Pressure-sensitive
adhesive

**Figure 7.15** Adhesive tapes are available with foam carriers as well as without any carrier. This second form includes a release liner that is coated with adhesive, which is applied directly to the surface, and the release liner, discarded.

elements and are varied to meet the needs of the application. The adhesive is a pressure-sensitive adhesive (PSA) form, while the carrier can be made from a range of materials including paper, polyester film, metallic foils, fabric, or synthetic material. A release liner is used to separate the rolled adhesive layers and is generally the least important component, because it is removed and discarded during the application of the tape.

## STORAGE

The success of an adhesive bond is dependent on many factors. The storage environment can play a significant role in the ability of the adhesive to perform as intended. Adhesive manufacturers typically recommend a storage life or shelf life, which defines the duration for which the materials will perform as intended if applied properly. After that period, the materials can become increasingly viscous or separate irreversibly, rendering them useless.

Control of the storage temperature is also an important factor in the shelf life of an adhesive. At a minimum, storage conditions should be cool and dry, with room temperatures of 60–80°F (15–27°C). This condition works well for multiple-component adhesives, because they are stable and require mixing before the curing reaction is initiated. Single component adhesives require storage at reduced temperatures because they are less stable and are curing slowly at room temperature. Reducing their temperature by refrigeration (35–40°F; 1.7–4.52°C) or freezing (0–10°F; −18 to −12°C) significantly slows the reaction, prolonging shelf life. If the adhesive components are stored outside the recommended range, the resulting change in physical properties will adversely affect its performance (Table 7.5).

Generally, multi-part adhesives have a shelf life of 1 year and single component adhesives 4 to 6 months, when the

**Table 7.5** Typical Storage Temperatures

| Storage location | Temperature | Component type |
|---|---|---|
| Room temperature | 15–27°C (60–80°F) | Multi-part |
| Refrigeration | 1–4°C (35–40°F) | One- and multi-part |
| Freezing | −18–12°C (0–10°F) | One-part |

manufacturer's recommendations for storage are followed. If there is any doubt as to the recommended storage condition, refrigeration should be performed.

When removing adhesives from storage it is important to allow them to reach the worksite temperature before opening the container, to prevent condensation on or around the adhesive. Depending upon the adhesive chemistry, the moisture may combine with or dilute the uncured material, resulting in poor or undesired performance. Light-sensitive adhesives require, at a minimum, storage in a dark place to reduce the potential for accidental exposure.

## CURE METHODS

The type of curing process required for an adhesive is dependent on its form and chemistry. The initial curing of an adhesive to sufficiently support the materials in any given application is known as its "green strength." This varies for each adhesive and application but is an important parameter in considering the required fixturing and manufacturing processing times.

One-part adhesives require some action to initiate or complete the chemical reaction necessary to finalize or accelerate the curing process. These are: elevated temperature, oxygen removal, moisture, activator, accelerator, primer, light cure, and pressure. The requirement for these conditions are dependent on the adhesive. Elevated temperatures (oven cure) are required to accelerate the reaction process and is required by single-part epoxy adhesives. These types of epoxy adhesives often have a limited shelf life because the catalytic reaction is occurring at room temperature but at a significantly reduced rate. Their shelf life can be extended by placing them in cold storage and further reducing the storage temperature. Elevated temperatures are also required in hot melt adhesives as the name implies. These are stored in solid form at room temperature and dispensed after being melted.

Adhesives which cure in the absence of oxygen are called anaerobic adhesives. In addition to oxygen removal, the adhesive requires active metal ions to complete the room temperature curing process. These adhesives are best known for their use as threadlockers, which provide the necessary environment to properly cure.

The presence of moisture is required for three commonly used adhesives: cyanoacrylates, silicones, and polyurethanes. All these require moisture in the air or on the surface to initiate the room temperature curing process. Although high temperature and heat can be used to accelerate the curing process, it is not typical.

Many acrylic adhesives typically require an activator to initiate the curing process. It is applied directly to one of the substrates, while the adhesive is applied to the mating substrate. The activator is required to "flash-off," during which time it dries to a desired state prior to assembly. Adhesives requiring an activator work best when surface areas are sufficient to allow the activator to spread uniformly, preventing edge effects which can result in uneven or excessive activator.

An accelerator is a surface treatment used to increase the cure rate and improve gap-filling capabilities. It is typically used with cyanoacrylate instant adhesives to cure exposed adhesive found at the exterior of the bond line to allow immediate use.

Primers are low viscosity fluids applied to promote adhesion of the adhesive to the desired substrate. They do so by either thoroughly wetting the substrate so that a uniform coating of a material that the adhesive will readily bond to is present, or to protect the surface finish from undesired oxidation or contamination until the adhesive is applied.

UV light cured adhesives cure by exposure to high intensity light of a specific wavelength which catalyzes the polymer. This requires the adhesive to be accessible by the light source when assembled. Although these types of adhesives will eventually cure in ambient light the resulting bond is of poor quality. This curing process was originally developed for acrylic-based adhesives but has since been expanded to include cyanoacrylates and epoxies, allowing the application of these unique adhesive properties to a greater range of engineering solutions.

Pressure-sensitive adhesives are one-part adhesives that cure when assembled between two substrates using moderate pressure. The intimate contact and pressure are sufficient to initiate the curing process.

Two-part adhesives produce room temperature bonds that are initiated by mixing, producing an exothermic curing reaction. Two-part epoxies are the most common form of this type of adhesive.

## CLASSIFICATIONS

### Reactive Acrylics

Thermosetting acrylic adhesives, sometimes called reactive or modified acrylics, provide tough, flexible bonds when used on a wide range of materials. They are available in three curable forms: surface catalyst or activator, thermal, and ultraviolet (UV). The most common form is the two-part, activator cured adhesive, which requires the activator be placed on one adherend and the adhesive on the other and then assembled. The other two forms, thermal and UV, require a secondary operation to complete the curing process. For thermally cured forms, 300°F (149°C) for ten minutes is typical, but care must be taken to design a bond joint that minimizes the possibility of losing adhesive prior to cure at the elevated temperature. This curing method typically produces higher bond strengths and shorter cure times.

### Mechanical

In general, reactive acrylics have good mechanical properties through a wide temperature range (−160 to 300°F; −107 to 147°C). Tensile lap shear strengths ranging from 3000 to 5000 psi (20.7 to 34.7 MPa) for metals and over 1000 psi (6.9 MPa) for plastics are not uncommon. These adhesives also have better than average impact and peel strengths and are known for their ability to adhere to materials that have received minimal surface preparation. Metals with residual oils from handling and untreated plastics have little effect on the final bond strength, when using acrylics. Even though fully cured bonds exhibit only small amounts of shrinkage, they have limited gap-filling capabilities. Because of their need for controlled application on both adherents, reactive acrylics do not lend themselves to automated dispensing.

## Bonding Issues

Typically, adherend materials that have a reasonably smooth, continuous surface that will not absorb the adhesive components work best because they allow the activator to be spread evenly and produce a uniform bond line. Surface area is an important consideration for using the activator-based acrylic because of the need for the activator to spread uniformly. Smaller surface areas have a greater proportion of edge effects (pooling at the edge due to the activator surface tension) that results in an excess of activator compared with what is required for the bond, preventing it from properly curing. Although acrylics do not need to be rigorously proportioned, the effect on small surface is pronounced enough to reduce the bond strength.

## Surface Preparation

The surface preparation necessary to obtain bonds of reasonable strength is not as rigorous as demanded by other adhesives. Limited amounts of surface contaminants, such as oil or dirt can be tolerated without significantly reducing the bond strength. Although this consideration is attractive, it is always best to properly clean all bond line surfaces before applying the adhesive components.

Two-component systems require the adhesive be applied to one surface and the activator to the other followed by joining the two to initiate the cross-linking curing process. These systems have rapid fixture cure times ranging from 30 sec to 10 min depending on the formulation. This makes them attractive for automated production processing. Full cure is typically 24 h. The cure process is also affected by the type of activator used and the quantity applied prior to bonding. Solvent-based activators have limited sensitivity to the quantity that is applied, as they are allowed to flash-off, while solvent-free activators must be applied with care to ensure that a correct amount is applied in a uniform manner. Application of an excessive amount of a solventless activator can result in a poor bond, owing to the adhesive's inability to successfully adhere.

Single-component types cured using either heat or a UV light source are attractive because they do not require the application of an activator. However, this convenience is offset by limitations from secondary operations. If heat is used to cure the adhesive the assembly must be capable of being placed in an oven, which not only imposes size restrictions but also adds process time. The use of a UV-cure requires a bond that is visible and limited in thickness. The adhesive must be accessible to UV-exposure which may require the adherend to be clear or opaque. Otherwise, the bond must be shaped similar to a filet weld, leaving the adhesive fully exposed for the curing process.

## Performance

Acrylics perform well at high humidity but exhibit low strength at elevated temperatures. They resist aging from direct sunlight but are not suited for continued exposure to hydrocarbon-based materials, such as fuels and lubricants. They are also highly flammable in their uncured forms. Additionally, many consider their odor objectionable, which makes them more suited for automated applications rather than manual. There have been advances in developing formulations with monomers that produce less odor in an attempt to make manual more desirable.

## Anaerobic

These single-part adhesives cure in the absence of oxygen when applied to active material surfaces and are most commonly used for threadlocker adhesives for securing fasteners. They are also used for gasketing or sealing of structural components which have close-fitting surfaces. Anaerobic adhesives typically provide gap-filling up to 0.03 in. (0.8 mm) with some formulations providing up to 0.05 in. (1.3 mm). There are also low viscosity formulations that are used for sealing the porosity found in castings and welds, as well as for bearing retension.

## Mechanical

Because of the curing method of anaerobic adhesives, they generally fill the void between two structural members. They must transfer the applied load in a captive manner, while not being subjected to peel or shear forces. In general, they are not rated for these types of loading conditions. Threadlockers are categorized by their use, either for securing serviceable or non-serviceable fasteners, and by fastener diameter. They are formulated in viscosities ranging from 12 cP to thixotropic compounds. They are also offered for high-temperature applications, bonding the fastener to the extent that temperatures exceeding the maximum service temperature is required to disassemble it.

Retaining compounds are another category of anaerobics that are formulated for use in precision assemblies. They are typically low-viscosity forms that help retain bearings and shafts that are assembled with close fits but available in greater viscosities and thixotropic forms to aid in the installation of looser fitting parts. These adhesives are rated for shear loading capability because of their intended use, but their typical shear strengths are similar across product offerings.

Anaerobics formulated for gasketing or sealing are available and are best used in assemblies with machined surfaces, providing close-fitting surfaces and small gaps. They are formulated in varying viscosities and material properties ranging from rigid to flexible with gap-filling capabilities. Others with slow curing characteristics allow assembly and adjustment.

## Bonding Issues

Anaerobic adhesives are generally used with metals because of their curing requirements but there are formulations for plastic fasteners. Careful consideration and testing during the selection process can identify incompatible combinations.

## Surface Preparation

The need for extremely clean surfaces is not critical to the success of most forms of anaerobic adhesives because of the application environment. The adhesive is generally captive, providing not only bonding of materials but also interference because of the gap-filling nature of their application.

Anaerobics can be used to bond a wide variety of materials, but may require different cure methods to do so effectively. Material surfaces are grouped into three types: active, inactive, and inhibiting. Active surfaces are those of clean metals and some thermoset polymers which provide the desired oxygen-free environments for rapid curing in the presence of the appropriate metal ions. Inactive surfaces also include metals and plastics that lack the necessary ions and do not promote rapid curing, which results in extended cure times lasting hours. Inhibiting surfaces are those that actually prevent curing, requiring primers or heat to cure the adhesive. These surfaces include certain platings such as cadmium and zinc, chromate conversion coatings, metal oxide finishes and certain aluminum anodize finishes.

## Performance

Typical formulations provide service temperatures of 300°F (149°C) and special formulations up to 450°F (232°C). The threadlocker, retaining, and gasketing forms have good resistance to moisture and solvents.

## Cyanoacrylate

Cyanoacrylates are single component adhesives that cure rapidly at room temperature while in the presence of minute amounts of moisture. These adhesives became widely known by consumers in the 1970s as the glue that could instantly repair broken china or lift large loads with a single drop. They were marketed under the trade names "Super Glue" and "Krazy Glue."

## Mechanical

There are two common types of cyanoacrylates in use, monomer-based methyl or ethyl ester groups. Methyl cyanoacrylates are preferred when bonding metals or other rigid surfaces because they provide greater strength and impact-resistance than ethyl cyanoacrylates. More compliant materials, such as plastic or rubber, benefit from using ethyl cyanoacrylates by creating stronger bonds.

## Bonding Issues

The strength of the cured bond is influenced by several factors. Although cyanoacrylates cure because of the presence of moisture, high relative humidity or excessive surface moisture can prevent it from properly curing, resulting in a useless bond. The thickness of the bond can also result in incomplete curing because of the inability of the water to reach the center of the bond. Because of this, cyanoacrylates are poor gap-fillers and perform best when bond lines are thin.

Generically, cyanoacrylates are of low viscosity and brittle upon curing, resulting in poor peel and impact strengths. There are formulations which improve upon both of these properties. Thixotropic gel forms address the applications requiring higher viscosity and elastomerically toughened ethyl cyanoacrylates for those applications involving shock, vibration, or thermal cycling.

Materials which are stiff and nonporous bond best with cyanoacrylates because they provide the necessary thin bond line. Stiff, porous materials can be bonded using viscous gel forms and must be evaluated on a case by case basis. Bonding of dissimilar material combinations, such as plastic and metal, also produces mixed results and needs to be carefully evaluated. Cyanoacrylates have been used extensively by the military in the treatment of flesh wounds, because of their ability to rapidly bond human skin, closing wounds, and preventing blood loss.

## Surface Preparation

Extensive surface preparation is not typically required for cyanoacrylates. Lightly abrading the surface followed by cleaning with alcohol is generally sufficient for most applications.

Performance

Cyanoacrylates benefit from the use of accelerators and primers to control the cure process. Accelerators are primarily used to reduce fixture times and cure exposed bond lines or excess adhesive. Wire tacking on printed circuit boards is commonly performed this way. Primers are used to prepare adherents that are difficult to bond to such as polyolefins (polypropylene and polyethylene), fluoropolymers (HALAR © 300) and acetals (Delrin®). Adhesion can be improved up to 20 times the untreated bond strength, depending on the adherent properties.

In some curing environments, cyanoacrylates produce a "blooming" or "frosting" around the bond line. This is the result of the exposed monomer reacting with ambient moisture to produce a white residue which settles onto the adherent. There have been advances in cyanoacrylate chemistry to address this problem. Low-odor/low-bloom formulations have been produced in which the vapor pressure is lower than the standard forms so they do not react as readily with ambient moisture.

Upon curing, cyanoacrylates form rigid thermoplastic making them susceptible to creep and moisture degradation. They are also limited in their temperature-resistance and are not typically recommended for use above 160°F (71°C). There are high-temperature grades that survive exposure to 250°F (121°C).

## Epoxy

Epoxy adhesives are widely used because of their good wetting and superior adhesion characteristics. They are available in a wide array of formulations. The most common form is a room temperature cure, two component system. Single component systems requiring heat cure are also available. Although epoxies perform best when the adherents are properly prepared, they are more forgiving than most adhesives when it comes to cured strength, because they tolerate being applied in thick sections, helping overcome poor joint design.

Recent developments include the design of thermally removable epoxy adhesives that behave as cross-linked thermoset plastics below 140°F (60°C), but when heated above 194°F (90°C) they are no longer cross-linked. This behavior allows simplified disassembly of applications requiring access after testing or at the time of recycling.

## Mechanical

It is inappropriate to attempt to classify epoxies in a generic sense because their properties vary widely depending on the formulation. An unmodified cured epoxy is inherently brittle and has low shrinkage and high creep resistance, which makes them excellent materials in nonimpact or static applications. Fortunately, epoxies can be cured with various chemical agents and modified with numerous materials, all of which affects the resulting cured material. For example, diluents can be added to reduce the viscosity of epoxy potting compounds. Conversely, fillers such as fibers can be added to produce a thixotropic compound; minerals, to reduce cost by acting as an extender or improve electrical or thermal conductivity; and rubbers, to improve impact and shock-resistance. Formulated hybrids extend the offering by modifying the epoxy by combining it with other compatible chemical families such as phenolics, nylons, polysulfides, and vinyls, all of which alter the formulation to provide the desired material properties (uncured and cured).

## Bonding Issues

There are few materials that epoxies will not adhere to, which is why they are widely used. They form their strongest bonds with metals and ceramics because of the benefits of a thin bond line and lack of porosity. Other materials, including wood and paper-based products, plastics, and many plating and coatings also bond well with epoxies. Materials that promote limited adhesion include those with low surface energies such as PTFE, acetals, and specialized coatings, such as PTFE-impregnated anodize.

## Surface Preparation

Epoxies do not typically require extensive surface preparation to provide an effective bond. Use of solvent-based and nonsolvent-based degreasers are usually sufficient, unless significant surface deterioration is present that necessitates mechanical or chemical processes to remove it.

Performance

Curing of epoxies is accomplished using many different agents which produce different material properties, both before and after curing. They affect cohesive strength, hardness, and environmental performance such as temperature and solvent-resistance.

Single-form epoxies generally cure in one of two ways. One, from the release of insoluble curing agents upon the application of heat and two, by chemically inhibiting the process at room temperature, requiring heat to unblock the reaction. Although greater processing time is often the penalty for using single part epoxies, the improved crosslinking and resulting higher quality bond necessitiates their use.

Specialized single-forms include B-staged and UV-curable adhesives. B-staged epoxies are resins that are semi-cured and in a solid state, requiring only heat to complete the curing process. The advantage of using these forms is that there is reduced waste and better consistency, because the adhesive is in its final or near final form. Single and two component UV epoxies provide the ease of processing associated with UV acrylic adhesives, with the rigid, low shrinkage properties possible with single- or two-part epoxies at the expense of strength and some environmental resistance.

Two-component systems require mixing to initiate the curing process, which results in an exothermic reaction that produces no physical byproducts. Thorough mixing in the correct proportions are necessary. Manufacturers provide syringe type systems with mixing nozzles that simplify the process while improving the quality and consistency of the bonding process. High temperature formulations that require heat curing can resist oils, moisture and many solvents while offering thermal resistance to 200°F (98°C). Temperatures reaching 400°F (204°C) can also be applied but with a significant reduction in bond strength. UV exposure can also affect two part adhesives by discoloring them with some loss of material properties.

## Explosive Bonding

Bonding materials by explosive impact is an unconventional solution to a process typically considered simple and inexpensive. Explosive bonding forcibly joins materials by driving one into the other at high speed and pressure, forming a cohesive bond which is vacuum tight and resists thermal expansion forces. It is an expensive process which requires elaborate preparation and a high level of skill to complete successfully and without personal injury.

### Mechanical

The bonding properties are purely a function of the process because there is no intermediate adhesive material involved. Although the process involves high levels of pressure and heat, they are for such a short duration that original properties of the materials involved typically are retained. The resulting assembly requires additional processing because the bonding process does not produce finished dimensions. The bonded adherents are not flat, and require skilled machining to be made flat and to prevent damage to the bond.

### Bonding Issues

The adherents that can be bonded by this process are limited to metals because of the high levels of pressure and heat involved. Not unlike traditional bonding, it is also a process that does not lend itself to all combinations of adherents, requiring trial and error to determine compatibility. The following combinations are currently in use: stainless steel/aluminum, stainless steel/copper, aluminum/copper, and copper/molybdenum.

### Surface Preparation

The intense forces involved in the bonding process produce a plasma that cleans the surface ahead of the bonding process.

Because of this, it is not necessary to perform any pre-treatment process to prepare the adherent surfaces other than traditional degreasing. There is considerable time required to assemble and prepare for the explosive process which added considerable cost to the final assembly.

## Performance

The explosive mechanism that actually produces the bond generates contact speeds in excess of 1,600 ft/sec (488 m/s), resulting in contact pressures of nearly 600,000 psi (4,000 MPa) (Big Bang Bonding, Machine Design November 16, 2000, p. 67). Control of the plasma flow and detonation wave form is critical to the success of the bond. The resulting interface between the materials is a wavy bond line, because of the forces which produce viscous behavior at the material surfaces. The resulting bond is one that includes, at a minimum, diffusion and mechanical interlocking adhesion to secure the materials together.

The byproducts of this process of bonding are not of the usual form found in polymer-based adhesives. Explosive bonding produces obvious byproducts, such as residual blast materials that include undesirable chemicals from the explosive formulations. It also requires structures and equipment for blast-containment, which considerably increases the expense for the safety process and when combined with a high degree of skill necessary to be successful, lends itself to outsourcing.

## Polyimide

Polyimide adhesives are known for their ability to retain their mechanical properties at elevated temperatures in excess of 550°F (288°C). They find the most use in aerospace applications because of their performance capabilities and high cost. Polyimide adhesives are available in two forms: thermoplastics and thermosets.

## Mechanical

Applications requiring extreme temperature-resistance and structural strength often use thermoset polyimide adhesives.

They have higher operating temperatures than epoxies and phenolics. To obtain these properties require extended cure times at elevated temperatures while fixtured under pressure. Because of these process requirements and high resin costs, polyimide adhesives are found in demanding applications that cannot be satisfied by less costly adhesives. They are typically produced in resin, film, and coating forms.

## Bonding Issues

Thermoset polyimides are available in two forms as defined by their cure process: condensation reaction and addition reaction. The importance of the distinction is in their cure methods and resulting performance. Condensation reaction polyimides release moisture as part of their cure process, requiring sophisticated processing equipment to maintain fixturing pressure throughout the entire cure process. Addition reaction polyimides do not produce byproducts during cure and subsequently do not require fixturing for the entire process but do produce a slightly weaker bond.

There is a thermoplastic formulation that has found limited application in the electronic packaging industry. It does not require sophisticated processing but this is at the expense of thermal and mechanical properties.

## Surface Preparation

Surface preparation is important for this high-performance adhesive. The adherent surfaces must be exceptionally clean so that the best possible bond can be obtained. Solvent cleaning followed by abrading the surface and a final solvent cleaning is the minimum necessary obtain a quality bond. Surface etching or anodizing may be necessary in order to provide the desired adhesion properties, as measured by the bond strength.

## Performance

Polyimides excel in high-temperature applications, producing lap shear strengths of greater than 2700 psi (18.6 MPa) at 550°F (288°C) in very specific applications. Room temperature lap shear strengths of 2500 psi (17.2 MPa) are typical. Generally, condensation reaction bonds have better properties than addition reaction bonds. Short-term (minutes) high-temperature exposure for addition reaction bonds is 600°F (316°C) while for condensation reaction bonds it is nearly 1000°F (538°C).

High-temperature materials, such as metals and composites, most likely require polyimide adhesives. Plastics are less desirable candidates because their limited high-temperature properties do not possess the advantages available with polyimides.

Polyimides commonly cure while at high temperatures (350°F; 177°C) and under pressure. Their strength is better than that of epoxies, even after exposure to high moisture environments.

### Polyurethane

Polyurethane is a versatile polymer that is available in many forms. As an adhesive, it produces a flexible, tough interface between two adherents. Although it is suitable for bonding rigid materials, polyurethanes excel in bonds requiring flexibility as well as chemical and thermal stability. They also adhere quite well to materials with low surface energies such as PTFE. Additionally, polyurethanes are formulated for use in encapsulating and potting electronics, casting of mechanical components, and sealing. They cure to different hardnesses, ranging from soft elastomers (50 Shore A) to rigid, hard plastics (90 Shore D).

Polyurethanes are available in one- and two-part formulations, with viscosities ranging from self-leveling liquids to thixotropic paste. One-part forms cure at room temperature or with the application of heat. Two-part systems cure upon mixing at room temperature and can have pot lives ranging from 15 sec to 16 h.

### Mechanical

The lap shear strength of polyurethanes can be as high as 3000 psi (21 MPa) at room temperature and exceed 5000 psi

(35 MPa) at −100°F (−73°C). The room temperature bond is flexible and tough but with limited shear and tensile strength (300–500 psi; 2.1–3.5 MPa). These adhesives exhibit exceptional cryogenic temperature properties, exceeded only by silicones which cost considerably more. The hardness ranges from 50 Shore A to 90 Shore D at room temperature. At reduced temperatures the hardness increases, which leads to improved shear strength, as the once compliant adhesive becomes firm, resisting shear loads. Elevated temperature performance is not exceptional and is limited to 300°F (149°C). Thermal cycling has little effect on these properties.

Mechanical strength degradation becomes a significant concern in high humidity environments where the adhesive is prone to reversion (polymer breakdown). Sealant failure on F-111 fighter bombers during the Vietnam war were found to be the result of reversion phenomena.

## Bonding Issues

Polyurethanes provide the ideal bond for those applications requiring a compliant, tough adhesive to secure to adherents together. Their high surface energy allows them to wet many different surfaces including metals, glass, concrete, ceramics, wood, vulcanized rubbers, and many fabrics and plastics. They do not bond well to polyolefins (polyethylene, polypropylene).

## Surface Preparation

No special surface preparation is necessary. Removal of oxidation and other contaminants from metals and removal of mold-release from plastics and rubber is essential to proper bonding. Adhesion promoters or catalysts may be needed (under certain conditions).

## Performance

One-component formulations cure to a thermoset plastic at room temperature owing to ambient moisture, or require heat. Like silicones, moisture must reach the adhesive bond line or else it will not cure. Their cure depth is limited to about

0.38 inch (9.6 mm). Two-component systems cure due to mixing and the related chemical reaction and therefore can cure in any depth required. Both adhesives typically have a 6 month shelf life because of their hydroscopic tendencies when exposed during the dispensing process.

Generally, chemical- and environmental-resistance is not as good as epoxies or acrylics but there are some areas of excellent performance including superior performance in water, acids, and organic chemicals.

## Pressure-Sensitive Adhesives

The development of pressure-sensitive adhesives (PSAs) was born out of the need to apply a uniform strip of uncontaminated, solvent-free adhesive to a precise location with minimal pressure. PSAs require no mixing or activation energy and are not a cross-linked adhesive as supplied. They are most commonly available on rolls or flat sheets. There are three common base polymers used in the formulation of PSAs: rubber, acrylic, and silicone. The adhesive is applied to a carrier of some form, which may or may not remain with it after it is positioned, and then wound into a roll, or cut as a flat sheet. The most common carriers are polyester film, fabric, or foam, which can be coated on one or both sides. There are forms that do not require a carrier and essentially are made up of a strip of adhesive applied to a backing material or release liner and wound into a roll. PSAs are also known as adhesive transfer tapes.

## Mechanical

Although the base polymer has a significant influence regarding the PSAs performance, there are some generic mechanical properties that apply to all PSAs. Bond strength develops over a period of days rather than hours as is typical for other adhesive forms. Epoxy formulations commonly reach full cure within 24 h. Curing is also dependent on the critical wetting tension of the adherent. If the adhesive's surface tension is lower than the adherent's surface tension, the bond will be weaker than if the opposite were the case. The strength of some

PSAs can be improved by employing a secondary operation to induce cross-linking of the polymer, usually by UV radiation.

PSA shear or peel strengths are typically not provided quantitatively by their manufacturers because they may vary widely depending on the adherent. Qualitative measures are employed to give a sense of a PSA's performance relative to the adherent being evaluated by the user.

Bond failure is typically evaluated by peel tests rather than lap shear or pure tensile loading because PSAs are most often applied to a flexible adherent. The peel strength is dependent on ambient temperature and peel rate. Cohesive failure is observed at low peel rates while higher rates induce adhesion failure. A phenomena called "stick-slip," during which the adhesive failure is oscillatory, occurs when the required peel force is low compared to the peel rate. This behavior may be of concern if the PSA is to be removed from a delicate adherent.

Formulations have been developed that provide exceptional tack, but are done so at the expense of shear and peel strengths. These forms are typically of reduced strength, since they have low molecular weights, making them also prone to creep behavior. These forms are best used as labels or other non-loadbearing applications. Decreasing the tack reduces creep behavior to a point at which adhesion becomes so poor that bond strength begins to decrease, resulting in adhesion failure rather than cohesive failure.

## Bonding Issues

PSAs have unique properties that let them bond at room temperature with the application of light pressure. They must have a balance of tack and cohesive strength to be effective in developing an effective bond. Tack is a measure of the instantaneous wetting of an adherent by an adhesive, with little or no applied pressure, and is best provided by low molecular weight polymers. Cohesive strength is a measure of the load carrying capability of an adhesive and is best performed by a high molecular weight polymer because of the improved stiffness and strength. The blending of these two properties is the basis of the different formulations offered in this type of adhesive.

The adhesive's wettability onto the adherent is critical because of the low assembly forces involved. Also important is the adherent stiffness and the method of application of the adhesive. If the adherent is flexible, additional fixturing may be required, so the PSA is subjected to sufficient loading to properly adhere. If the adherent is extremely rigid and the fit and finish between them is not good (i.e., uneven surfaces, rough surface finish, etc.), then the resulting bond may be poor because of lack of sufficient adhesion. There are two approaches to correct this problem. One, improve the fit and finish sufficiently to allow intimate contact between the surfaces. A flatness less than half of the adhesive tape width is desirable. Two, use a product that has a foam carrier, which minimizes the effect of poor fit and finish. Foam carriers can improve bonding strength up to ten times.

## Surface Preparation

The surface prepartation required when using PSAs is no different than other adhesives, requiring the surface be free of contaminants and in some cases additional processing is necessary. These processes may include flame treatment, primers or topcoats to produce an acceptable surface. The surface energy of the material also plays a significant roll in the adhesion properties as it controls the "wetting" characteristic of the material. This characteristic defines the adhesive's ability to flow onto and into the surface when forming the bond.

Additional surface conditions including texture and contour play a significant role in the resulting performance of the adhesive. PSAs benefit from an abraded surface to improve adhesion. Cleaning the surface prior to abrading is required to prevent embedding contaminants into the surface that may inhibit adhesion. Surface contour can also affect the performance of a PSA depending on its thickness and formulation. Torturous contours can prevent the adhesive from sufficiently adhering depending on its interaction with the surface.

## Performance

As a general rule, strength properties of PSAs are affected by aging. A common example involves ordinary masking tape, which is made from a rubber-based polymer, it typically becomes brittle and loses its adhesion properties as it ages at room temperature. Other formulations do this to a lesser extent, but aging of PSAs must be considered when making their selection.

Rubber-based PSAs are known for their high initial tack or adhesion, good flexibility, and low cost when compared with acrylic and silicone PSAs. They are equally known for their poor aging and limited temperature-resistance. They are most commonly found as single-sided tape, in forms such as masking tape, with a paper carrier or with a polyester carrier, as used for plain office tape.

Acrylic-based PSAs are considerably more robust than rubber-based forms because of their greater mechanical strength and environmental-resistance properties. They have good shear strengths, UV- and chemical-resistance, and age well, much like reactive liquid acrylic adhesives. They do not have the initial tack or adhesion like rubber adhesives, exhibit poor creep behavior, and are more expensive than rubber PSAs. Their bonding ability is significantly affected by the pressure applied at assembly and joint fit and finish. They are commonly used in applications requiring near structural strength and excellent weatherability.

Silicone-based PSAs have exceptional chemical- and thermal-resistance which allows them to be used in a wide range of applications. They also have a low surface energy in the range of 21 mJ/m$^2$, which allows them to wet a wide range of adherents, more than rubber- or acrylic-based PSAs. These advantages are offset only by their higher costs and lower strength, making them the PSA of choice for environmentally demanding applications.

## Silicone

Adherents that have a low surface energy bond best using an adhesive which has an even lower surface energy. Because of this, silicone adhesives work best on most low surface energy materials, such as plastics and silicone-based adherents. There are three common forms of silicone adhesives: one-part liquids or paste, which cure by room temperature vulcanization (RTV); two-part liquids or pastes which cure when mixed; and pressure-sensitive adhesive (PSA) tape forms. Generally, these adhesives are more expensive than epoxies or polyurethanes that might be formulated with similar properties but are often worth the expense because of the greater adhesion and excellent property retention.

## Mechanical

One-part systems are commonly used for room temperature applications that have a bond thickness of 0.001 to 0.125 in. (0.25–3.18 mm). Bond thickness up to 0.5 in. (12.5 mm) can be obtained in some applications. Exposure of the bond line is important to curing of the silicone because the process requires ambient moisture. In consideration of this requirement, the joint should be designed such that the adhesive is never more than one inch from an exposed edge. Deeper sections will either take an exceedingly long period of time to cure or may never cure completely. Two-part systems do not have this limitation because they cure when mixed, which allows them to be used in thicker sections for potting electronic components or sealing.

## Bonding Issues

Silicones are first known for their thermal properties and second for their ability to form tough bonds with almost any adherent. Their bond strength is moderate with lap shear strengths around 500 psi (3.5 MPa) and resulting hardness ranging from 30 to 60 Shore A. Silicone adhesives also have low shrinkage and good gap-filling capabilities which makes them more forgiving with regard to joint design and fit.

Like epoxies, silicones can be formulated in so many numerous ways that their application uses are quite extensive. Single-part forms can be formulated for bonding metals, glass, ceramics, plastics, and vulcanized rubbers. Two-part forms are more common for bonding concrete and wood products because of the thicker sections.

## Surface Preparation

Surface cleanliness is important for silicones because of their dependence on the wettability theory to adhere. Solvent cleaning is preferred but cleaning with soap and water has also been found to be effective. Abrading the surface and then cleaning again improves the mechanical interlocking of the adhesive to the adherent. Use of primers for materials that are difficult to bond is also recommended.

## Performance

High temperatures are the domain for which silicones are most recognized with surface temperature ranging from $-150$ to 600°F ($-101$ to 316°C). Curing time of between 5 and 60 mins can be achieved at 50% relative humidity, with thick sections taking up to 72 h. (Bond strength continues to increase for 1 to 2 weeks after the initial application.) This occurs because the rate of cross-linking is slower at the adherent surface than within the silicone itself and so, depending on the adherent surface properties, the adhesion properties improve over time. Two-part systems cure in several hours and can be accelerated by using heat.

One-part silicones cure due to the presence of moisture and produce byproducts during the process. Some formulations produce acetic acid during the curing process, which can prove to be corrosive to adherents such as copper or brass. These forms have greater unprimed adhesion and longer shelf-life. To eliminate this problem there are formulations which release alcohols (methanol or ethanol) as a byproduct of the cure process but at the expense of some reduced performance. A third form of materials, known as oxime curing silicones, have properties which rival acidic producing forms but do not corrode ferric adherents. These have found widespread use in the automotive industry as gaskets and sealants. Two component systems do not release byproducts when curing.

## Solvent Bonding

The ability to fuse two like materials together, rather than bond them with an intermediate material, such as an adhesive, is preferred in most applications because of the improved strength and reduced risk of bonding irregularities. Solvent bonding or solvent welding of polymers is much like welding of metallic materials, it joins the materials by allowing them to freely mix and form a nearly homogeneous joint. The nature of this bonding process requires that the polymers be amorphous, which allows them to liquefy and intermix. Semi-crystalline polymers do not lend themselves to this process because of their high degree of order and precise melting points.

Although these properties are attractive, the process is limited in its ability to precisely control the fixturing and joining of the adherents and applies to a relatively small subset of materials. Solvents are the traditional means of forming the bond but they have restrictions due to their VOC. Formulations of doped solvents or solvent cements, as they are also called, are available and have little or no VOCs while producing bonds of similar strength. The commercially available solvents are of the following three basic forms:

1. Pure solvents — acetone, methylene chloride, toluene, styrene, and MEK.
2. Doped solvents — also known as solvent cements, contain solutions of the plastic being bonded that help fill gaps in ill-fitting parts. These are capable of bonding more types of materials than straight solvents.
3. Monomer and polymerization solvents—these are doped solvents that contain catalysts and promoters to produce polymerization at RT or below the softening temperature.

Polycarbonates are often solvent bonded to form a high quality assembly that has all the strength of the base material. Solvent bonding of plastics affords a similar structural benefit as welding does for ferrous metals.

Mechanical

The mechanical strength of a solvent bonded joint is typically 85 to 100% of the base material, similar to that found in welded metallics. Also like welded metals, there can be unwanted side effects such as stress-cracking or crazing of the adherents at or near the bond line. This is most often the result of residual stress-relief and can be controlled or avoided by annealing the adherents at a temperature below the glass transition temperature. Properties that are relevant when using an adhesive, such as thermal expansion, are not so for solvent bonds because the adherents retain nearly all their prebonded strength.

## Bonding Issues

The need for the material to be amorphous rather than semi-crystalline limits those suitable for the process (Table 7.6).

Other materials such as PVC can be joined using solvent cements but require a primer to prepare the surface. Materials such as nylon and acetal polymers can be bonded but with limited success. The appropriate solvent and process are best validated using the actual items.

Solvent must be applied in a controlled manner, using a syringe or brush. Uneven application can result in variable performance because of the excessive softening that occurs which can result in limited bond strength. The time between application of the solvent or cement and joining of the adherents is critical. Too much solvent and the adherents become too soft and cannot withstand the pressure necessary to fixture the parts, while too little results in a poor bond. Solvents are also more likely to cause stress-corrosion cracking than adhesives.

The solvent bonding process was originally intended for bonding like materials. It has matured to a point that bonding of unlike materials is possible when using the correct solvent cement. Choosing the correct formulation is important and is best done with the help of a qualified supplier (Table 7.7).

**Table 7.6**   Solvent Bonding Capabilities

| Works with | Not suitable with |
| --- | --- |
| Polystyrene | Polyethylene (PE) |
| ABS | Polypropylene (PP) |
| Polyphenylene oxide (PPO) | Polyamide |
| Vinyl | Polyacetal |
| Acrylic | PTFE |
| Polycarbonate | Or most thermosets |

**Table 7.7**  Common Solvents Used to Bond Selected Plastics

| Plastic | Common solvent |
| --- | --- |
| ABS | Methyl ethyl ketone (MEK) |
| Acrylics | Methylene chloride, trichloroethylene |
| Nylons | None |
| PVC | Acetone, tetrahedrofuran, methyl ethyl ketone (MEK) |
| Polycarbonate | Methylene chloride, trichloroethylene, methyl ethyl ketone (MEK) |
| Polystyrene | Methylene chloride, ethylene ketone, ethylene dichloride, trichloroethylene, toluene |

## Surface Preparation

Surface preparation is important to any bond and solvent bonding is no exception. It is necessary to remove all mold release or other contaminants and dry the surfaces before joining the adherents. Failure to do this may limit the ability of the adherents to intermix after the application of the solvent while also affecting the drying process.

## Performance

The cure process for solvent bonded materials is different from what might be considered traditional adhesive curing. There is no cross-linking owing to a chemical reaction, as with epoxies, or a high degree of wetting required, as with PSAs. The solvent is applied to both surfaces using a syringe or brush and then assembled using light pressure to fixture their position. The bonding process is complete once the joint hardens back to that of the original adherent, which may take up to 24 h depending on the polymers involved. The drying rate of the solvent or solvent cement determines the necessary fixture time for the assembly. Elevated temperatures can be used to accelerate the process. The manufacturer's recommendation and experimental evaluation will ultimately determine the appropriate fixture time.

The most significant drawback of solvent bonding are the VOCs produced by the process. The more recent forms of solvent cements address this problem but do not eliminate it, because the process requires the polymer partially dissolve the adherent and solidify by evaporation of the volatile organic solvents. The resulting joint is both chemically and thermally resistant as the base material.

## Hot Melts

The ability to bond materials rapidly, at low cost, and without solvents is the primary attraction to selecting hot melt adhesives. The food and beverage industries have long used this form of bonding for packaging because of its simple application methods and environmentally safe formulations. Common hot melt adhesives obtain their strength from resolidification after the desired adherents are joined. There is a pressure-sensitive form of hot melt adhesive that has the unique characteristic of remaining tacky after application to an adherent, allowing for future assembly.

### Mechanical

Hot melt adhesives are applied using a heat gun or similar heating element to melt a thermoplastic material and dispense it onto the desired adherents prior to their assembly. The materials typically used as hot melts are polyamides (nylons), polyolefins (polypropylene and polyethylene), polyesters, urethanes, and ethylene vinyl alcohols (EVAs). They are commonly provided in rod or pelt form, making handling and storage a much simpler task than required by other adhesives. The most significant limitation of this form of bonding is the equipment necessary to melt the thermoplastic material, for which high-volume applications can be costly.

To obtain the desired mechanical strength, hot melts are formulated using semi-crystalline rather than amorphous polymers. The molecular weight of thermoplastics are proportional to their viscosity and their melting temperature are also proportional to their molecular weight semi-crystalline polymers provide the necessary strength, while also having a low melt viscosity, which is important for several reasons.

First, this allows the adhesive to have good wetting characteristics at relatively low melt temperatures, which is important because excessive time at temperature will degrade the adhesive, diminishing its material properties. Wetting can be improved by increasing the application temperature because the melt viscosity is inversely proportional to temperature, but at the risk of adhesive degradation because of the higher dispensing temperatures.

Second, the thermal stability of the adhesive is also better at lower temperatures, so a low melt temperature means the processing equipment operating at lower temperatures and the likelihood of the adhesive remaining at a high temperature for an extended time and degrading is minimized.

Hot melts have poor creep characteristics because of their low molecular weight. The higher the molecular weight, the greater the density of polymer chains, minimizing creep. Generally, polyamide forms are strongest, followed by polyolefins and EVAs. Polyurethane hot melts develop lap shear strengths ranging from 2,500 to 4,000 psi (17.2–27.6 MPa) and elongations of 500 to 700%. These are notably stronger because once applied they react with moisture and form a stronger, thermoset material.

## Bonding Issues

Many materials can be bonded by hot melt adhesives, with the most common being wood-based products, ranging from furniture to cardboard boxes. Other, less porous materials can be bonded if the process is properly controlled, including ceramics, rubbers, and plastics. Metals can be bonded with hot melts but it is not common due to the high thermally conductive nature of the material which causes rapid solidification. As with any adhesive, its ability to wet a surface is critical to obtaining a quality bond and with metals, the rapid cooling afforded by the highly conductive surface hinders the adhesive's ability to flow or wet-out on either of the adherends.

## Surface Preparation

A clean, dry, porous material is an ideal candidate for bonding with a hot melt adhesive. No special surface preparations are necessary for bonding with these thermoplastic adhesives because their primary method of adhesion is by a mechanical interlocking of the plastic with the adherent. Dirt and moisture will reduce the bond strength of the adhesive regardless of the adherent.

## Performance

Hot melts have maximum temperatures of 250°F (121°C) with some formulations exceeding 300°F (149°C). They do not "cure" in a traditional manner. Unlike all other adhesives which cure as a result of some form of chemical reaction, hot melts cure only because of a change in temperature. Because they are thermoplastic, there is no cross-linking or other alterations to the adhesive when it cures, which does allow it to be reapplied or removed with the application of heat.

Holt melt adhesives do not perform well when exposed to solvents such as acetone, unleaded gasoline, and isopropyl alcohol. Because they are thermoplastics, they become rigid with improved shear strength when exposed to cold temperatures and reduced shear strength at elevated temperatures. The polyamide adhesive have the greatest strength, followed by polyolefins and EVAs.

The maximum time the adhesive will still adequately bond after being dispensed from the heat gun is known as the "open" or "range" time and is different for every combination of adhesive and adherent. This time ranges from 30 to 90 sec for most formulations and goes up to 180 sec for specially modified ones. Because the wetting of an adherent is instrumental to its success, minimizing a specific application's open time is in the best interest of obtaining a high-quality bond.

A significant benefit of hot melt adhesives is their quick setting characteristics. They develop nearly full strength in a matter of minutes and are fully cured in less than 24 h. This lends them to high-volume applications requiring minimal fixturing.

A special form of hot melt is one that never fully cures, remaining tacky after dispensing and cooling. These are known as hot melt pressure-sensitive adhesives or PSAs and have found use in bonding magazine inserts.

## Solvent-Based Thermoplastics

Contact cements are the most common form of solvent-based thermoplastic adhesives and are primarily formulated using phenolic resins. They are most often used commercially to bond wood products and rubber materials. Solvent-based adhesives are applied by brush or spray methods and allowed to dry prior to assembly. In specialized applications they are alloyed for high-temperature resistance. Alloyed formulations include nitrile-phenolics, vinyl-phenolics, and neoprene-phenolics. These adhesive forms are becoming less popular because of their chemical makeup, which includes petroleum-based solvents such as acetone, toluene, and methylethyl ketone (MEK).

### Mechanical

The mechanical properties of a solvent-based thermoplastic bond are highly dependent on the adherents involved. When bonding wood products, lap shear strength can be greater than 400 psi (2.8 MPa) at room temperature. The adhesive viscosity ranges from 50 to 500 cps and must be considered carefully when evaluating the joint requirements. If the adhesive is of low viscosity and applied to a porous adherent, too much of it may flow into the pores and not leave sufficient adhesive at the surface, resulting in a poor bond. Compared with other adhesive families, the viscosity is much lower because of the need to wet the porous surfaces to which they usually bond. Knowing this, applications usually require a generous coating to provide sufficient buildup and a proper bond.

Nitrile-phenolic formulations provide high temperature lap shear strengths exceeding 1700 psi (11.7 MPa) at 300°F (149°C). These are typically used to bond aircraft structures because of these properties and their ability to dampen vibrations.

### Bonding Issues

Application of wood veneers to plywood or particle board is commonly done using solvent-based adhesives. Because of their porous nature they are ideal for this form of adhesive. Other porous materials such as cloth, synthetic upholstery, and foams also bond well. Nonporous applications involving elastomers and metals are also common and bonded using alloyed formulations. These contain additives to increase the flexibility of the adhesive, to improve its shock and vibration-resistance. Some metals can be adversely affected by the carrier solvents to the extent of corroding them. Careful selection and thorough evaluation is necessary to avoid complications once the assembly is exposed to its application environment.

## Surface Preparation

Surface preparation is important and is best done using a solvent-based cleaner. Use of solvent-based adhesives is less attractive because of the controls required to support the preparation process. Removal of dust, dirt, oil, grease, wax, or loose contamination is needed to provide an adequate bonding surface.

## Performance

Solvent-based adhesives require ambient temperatures above 65°F (18°C) to evaporate the carrier solvents. For porous materials they are applied using a brush or by spraying them generously onto both adherent surfaces. Nonporous surfaces do not benefit from the extra adhesive because it will not be absorbed as with the porous materials. Adhesive quantities are quoted in coverage per unit area such as grams per cubic millimeter. Typical dry times range from 3 to 5 min or until the adhesive surface is no longer tacky with assembly to follow immediately. Once again, thorough evaluation is recommended for each application so that the appropriate process is validated.

Temperatures above 180°F (82°C) typically reduce lap shear bond strength by 50% or more for typical formulations. Reduced temperatures improve the strength but at the expense of toughness because the adhesive becomes more brittle.

The carrier solvent involved in this form of adhesive is less than a desirable attribute because of the environmental issues. The concern over fire, explosion, and health hazards limits their use. Because of this there is a need for greater control over the application and storage, which increases the operating cost of this adhesive family to the point that reduced performance of another adhesive is often acceptable.

## Water-Based Thermoplastics

The most notable water-based thermoplastic adhesive is "white" glue. It is a polyvinyl acetate latex emulsion that has no volatile organic compounds (VOC), is not flammable and poses no health hazard. It can bond a variety of wood and paper-based materials and is commonly found in the packaging industry as the adhesive used to assemble cardboard boxes and paper bags. Some other forms include small amounts of solvents such as toluene or methanol to improve wetting and curing. Also added are neoprene and other elastomers to improve material properties. Generally, these adhesives are inexpensive and require no special handling.

### Mechanical

The strength properties of water-based thermoplastic adhesives are similar to their solvent-based counterparts, only reduced. This is due to the nature of the adhesive's curing process which generally does not involve cross-linking to improve the strength of the bond. These adhesives are also available in a range of viscosities to allow for the varying adherents to which they are applied.

### Bonding Issues

The materials bonded with these adhesives are most commonly wood- or paper-based. They include wood furniture products that require an inexpensive, nontoxic adhesive to retain fitted wood parts. Because fixturing is not critical to this application, the longer cure time is not detrimental to the bond's success. Water-based adhesives are also used extensively in the construction of paper bags and cardboard boxes because they are inexpensive and nontoxic and will not harm the products they are intended to transport.

## Surface Preparation

There is no special surface preparation required when bonding with water-based adhesives other than applying good practice in cleaning the adherents. Use of a nonsolvent-based citrus cleaner is sufficient.

## Performance

For most common water-based adhesives, curing is brought about by the evaporation of the water carrier. This results in longer curing times than those associated with solvent-based adhesive but is often offset by the environmentally friendly process. There are some special formulations that can be cross-linked by application of heat or a catalyst.

When used in the contact adhesive form there is often a time limit for which to assemble the adherents, otherwise the adhesive dries too thoroughly and will not bond adequately. This can be remedied by recoating the surface to reactivate the adhesive so that the adherents can be assembled. Once joined, the materials have adequate strength to be processed immediately.

Water-based adhesives generally have poor moisture-resistance but are unaffected by oil and grease. The environmental impact of this form of adhesive is minimal because it is odorless, tasteless, nonflammable, and nontoxic, making its use and storage a simple matter.

## UV-Curable

The development of light curable adhesives has paralleled the growth of the electronics and fiber optics industries because of their importance in providing a method of final assembly. The electronic industries require component potting to stabilize assemblies and prevent vibration and shock damage, as well as provide a moisture barrier. They also require a method of tacking surface mount components prior to wave soldering.

The fiber optics industries require optically clear and index matched adhesives for assembling complex transmission equipment which requires the system components be positioned and then bonded, something UV adhesives do quite well.

Rapid curing from a controlled UV light source make this adhesive suitable for high-volume production applications which require moderate strength. Cure times range from 2 to 60 sec, with cure depths in excess of 0.5 inch (12.5 mm). The adhesive types available most commonly include acrylics, cyanoacrylates, epoxies, and silicones. The acrylics and silicones are single component forms, while the epoxy is a two-part component adhesive. The cyanoacrylates are available in either one- or two-part forms. Because all or most of the bond must be visible to allow curing, inspection of the bond is generally easily done visibly, improving the quality of the bonded joint. Some formulations include a fluorescent additive that improves the ability to inspect the joint visually or with black lamp detection and observe the adhesive coverage, an added advantage for those critical applications requiring thorough inspection of the bond. There are also UV-curable PSAs which combines the benefit of the high tack of a PSA for initial assembly with the rapid curing of a UV adhesive.

Mechanical

The mechanical properties of UV-curable adhesives vary widely because they originate from several different families of polymers. In most cases their lap shear tensile strength, viscosities, and hardness properties parallel their non-UV-curable forms with a few notable exceptions. UV-curable cyanoacrylates generally have improved properties over their non-UV-curable counterparts because of greater cross-linking that occurs as a result of the additional curing mechanism. Cured strength and hardness can also be increased by extending the UV exposure and/or intensity but usually at the expense of toughness. For example, UV-curable acrylics can have lap shear strengths up to 5000 psi (34.5 MPa) but they become brittle when cured to provide that strength and therefore are not generally considered to be structural adhesives.

## Bonding Issues

Polymers, metals, glass, and ceramics all can be bonded using UV-curable adhesives with consideration given for irradiating the adhesive. Clear materials, such as glass or polycarbonates, provide for the greatest number of joint configurations because directly exposing the adhesive is possible. Nontransparent materials such as steels and ceramics can be tacked together using a fillet or edge bond that can be exposed to UV light, but not with a lap joint because there is no way to cure the adhesive in the center. There is a phenomenon called shadow curing, which involves the curing of an adhesive that is not directly exposed to the UV radiation. It is a valuable characteristic for those joints that do not allow the entire bond line to be exposed.

## Surface Preparation

No special surface preparation is required when using UV-curable adhesives. Working with clean materials such as electronic integrated circuits may not require a separate cleaning process while molded plastic medical devices may require surface degreasing to remove old release agents to produce acceptable bonds.

## Performance

UV adhesives are cured by exposure to ultraviolet light of wavelength ranging from 100 to 400 nm. Controlling the combination of UV intensity and exposure time determines how rapidly the adhesive cures and the resulting material properties. Underexposure can result in reduced strength or premature failure because the adhesive is too soft, while overexposure makes the adhesive brittle. Depending on the polymer family of the UV adhesive, there may be additional catalysts required to complete curing. UV acrylics replace the liquid catalyst with light. Single-component epoxies replace the need for heat with UV radiation while two-part systems are exposed to UV light to rapidly gel the adhesive and the hardener to complete the process. Cyanoacrylates and silicones require moisture as well as the UV exposure to completely cure. Some can be cured in combination with activators or exposed to heat to meet specific application needs.

When used in thicker sections such as potting, UV adhesive generally can be cured to a depth of 0.38 inch (9.5 mm). UV intensity varies depending on the amount and type of adhesive being cured. It can range from 25–30 to 80–100 mW/cm$^2$ intensity. The distance of the UV source also affect the intensity level. The inverse square rule applies, intensity decreases in inverse proportion to the square of the distance to the source.

The lack of VOCs is a significant factor in the use of UV adhesives. They lend themselves to use in clean rooms and indoor production areas without the need for expensive ventilation recovery systems. Their chemical-resistance once again parallels their polymer family non-UV-cured counterparts without any significant differences.

## JOINT DESIGN

The joint that provides the structural integrity required by the application must also accommodate the material properties of the adhesive. Adhesive selection is more than matching the adhesive properties to the requirements of the application environment, it must also include consideration for the bonded joint assembly. For example, selection of a cyanoacrylate to repair broken pottery because it cures quickly and bonds well to nonporous surfaces is only a portion of the considerations needed to obtain a successful joint. If the pottery has large gaps because of missing fragments or damaged edges, the ability of the cyanoacrylate to fill these is limited and will adversely affect the bond strength. Also, if the pottery is to be used for holding water for extended periods, this adhesive is not likely to work well, either. If moisture contact is incidental or for limited periods, then the adhesive should perform as intended. Critical evaluation of the application and designing the joint for the worst case conditions is the best approach to selecting an adhesive and developing a joint that will perform as desired (Figure 7.16).

Select adhesive based on application extremes

| Applied condition | Extreme condition for load case |
|---|---|
| Tension | High temperature<br>Rapid loading<br>Cyclic loading |
| Compression | Impact<br>Bending |
| Shear | High temperature<br>Joint flexure |
| Peel | High & low temperatures<br>Cyclic loading |
| Cleavage | High & low temperatures<br>Cyclic loading |
| Impact | Low temperature<br>Cyclic loading |
| Vibration | Large amplitude |
| Solvents | Strongest |

**Figure 7.16** Worst case conditions. Determining the extremes of the application provides a way to bound the design requirement, allowing a more accurate review of the candidate adhesives. Using only the nominal values to make an adhesive selection leaves the design engineer open to possible premature joint failure.

Selecting the most appropriate adhesive requires consideration of not only loading and environmental conditions, but must also include the form of the joint that is to hold the adhesive in place during curing and subsequently support the applied load. The shape and orientation of the joint during assembly and cure will dictate what adhesive viscosity is required. Joints that have poor fit, unavoidably large gaps or high porosity need a thixotropic adhesive, which can tolerate these conditions to perform successfully. Refinement of the joint geometry by improved machining or some design change is desirable if it reduces the bond thickness, because thicker bonds are generally not as strong as a comparable thinner bonds. Those that have exceptional fit and are assembled with excessive pressure result in bonded assemblies that have a thin or starved bond. This condition is also undesirable because it makes for a rigid assembly with a low tolerance for impact or vibration loads owing to stress concentrations that occur as a result of the limited joint compliance. Typically, joints which produce a bond thickness of 0.002 to 0.008 in. (0.05–0.20 mm) provide the greatest strength and minimal stress concentrations.

The orientation, magnitude, and frequency of the applied load must be determined prior to designing the joint. There are five basic loading conditions that all bonded joints produce with usually more than one form contributing to the overall loading condition (Figure 7.17).

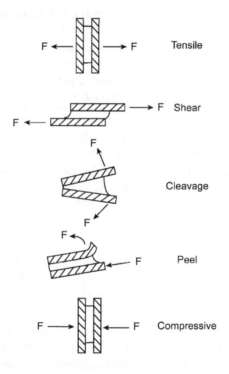

**Figure 7.17** Applied loading. Understanding the nature of the joint load is another critical consideration that must be included in the adhesive selection process. Failing to accurately predict the applied load magnitude and orientation can lead to catastrophic failure with little or no warning. Unlike bolted or welded joints that exhibit degradation through corrosion and cracking, adhesive joints typically do make joint deterioration difficult to observe.

1. Tensile — application of a tensile load perpendicular to the major joint plane.
2. Shear — application of the load parallel to the major joint plane.
3. Cleavage — application of a prying force along one or both edges of the joint.
4. Peel — application of a load along one edge of a flexible adherend.
5. Compressive — application of a compressive load perpendicular to the major joint plane.

Tensile loading is generally considered the strongest mode when evaluating a monolithic part, such as a steel bolt or flange. Adhesive joints respond similarly when subjected to a purely axial load with the applied stress measured as a function of load divided by bonded area and reported in pounds per square inch (psi) or Newton meter squared (MPa). Unfortunately, most applications produce some off-axis loading, owing to factors outside the design engineers control. These include bonding surfaces being machined slightly less than perpendicular to the load axis, off-axis transient loading inadvertently initiated by the equipment or the user or even the result of an uneven bond line. As a general rule, tensile loading of bonded joints should be avoided.

Shear loading is the most common and best performing load condition for adhesives because the forces are easily transferred between the adhesive and adherend by their relative interference as well as by other adhesion forces. Shear stress is also measured like tensile stress, as a function of the load and bonded area. The benefits of mechanical interlocking are realized in a lap shear joint and are not significantly influenced by off-axis loading conditions. Properly filleting the adhesive to the adherend helps minimize the stress concentration that exists at ends perpendicular to the applied force. Flexible adherends will affect the lap shear joint performance due to their bending and peel loading resulting in reduced performance.

Cleavage loading involves rigid substrates and should be avoided whenever possible. It is similar to off-axis tensile loading but with the distinction that the load is applied along one or both edges, tending to pry the adhesive from the substrate. For example, this condition would exist when mounting a transformer to a vertical surface, where the center of gravity is significantly removed from the bonding surfaces. Peel loading is similar but involves at least one flexible substrate rather than two rigid ones, also resulting in a loading condition that induces high stresses at the edge of the bond which most often initiates joint failure. Both of these loading conditions should be avoided when possible.

Compressive loading rarely produces a failure in bonded joints because the adhesives are sufficiently compliant to resist being crushed. They can crack or cold flow (creep) as a result of prolonged application of a compressive load. Designing the joint to limit compression or creep can be done by use of shims, glass threads, or microballoons to prevent compression as well as preserve the required gap, preventing adhesive starvation (Figure 7.18).

After identifying the load condition of the proposed joint and the load range it will be subjected to, it is necessary to calculate the required surface area of the bond. Manufacturer's strength data for the adhesive candidate should be used as a guide to determine the required area for the applied load.

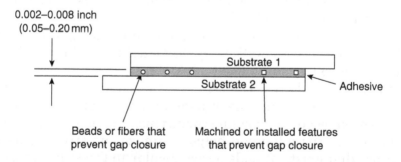

**Figure 7.18**   Joint gap preservation. Preserving the gap between the adherends to 0.002–0.008 in. is desirable when optimizing the joint design. This can be controlled by the design of the joint by including a step or some form of a hard stop, or it can be done by including a spacing mechanism, involving glass beads or fibers.

It may be necessary to apply additional safety factors to the calculation to compensate for the load condition and their effect on the overall joint performance. Because of the unique nature of every joint design and application, there are no set methods for applying safety factors to their design, and therefore, it is best for design engineers to work closely with the adhesive manufacture's application engineers to develop the most appropriate values for the specific application.

Because adhesive joints inherently reduce stress concentrations and absorb vibration, they are often selected over rivets or bolts for use in cyclic loading conditions. These latter fastening methods produce stress concentrations that can lead to early failure. An adhesively bonded joint minimizes stress concentrations by spreading the load across the entire width of the joint, rather than at discrete points as found with rivets or bolts. The life of an adhesive joint is inversely proportional to the range of the cyclic loading: if it is small, the joint will last longer than if it is large. Properly designed joints subjected to cyclic loading will most often have adherend failure before there is adhesive or cohesive failure. Knowing the peak loads and their durations can provide valuable insight into how best to design the joint for increased life by configuring it to minimize stress concentrations and dampen the peak loads.

The materials being bonded need to be considered from not only their adhesion characteristics, but also their stiffness. Plastics, from which many adhesives have their origin, are generally 100 times more flexible than steels. Flexible substrates require flexible adhesives to allow them to conform as the joint distorts. Use of a rigid adhesive with a flexible substrate will result in early failure as the adhesive attempts to conform to it and fails by cracking and separation from the adherend. Rigid substrates are more tolerant of rigid adhesives because their elastic moduli or stiffnesses are similar enough that excessive stress concentrations do not develop readily along the bond line. Flexible adhesives can be used when bonding rigid materials but their load carrying ability is often severely compromised, while they introduce excessive flexture of the joint. Applications which involve thermal cycling, vibration, or shock loading require this compliance, making it an important selection criteria for the adhesive.

Thermal expansion characteristics of the joint must be considered when selecting an adhesive. Matching the expansion behavior of the adhesive to the adherends will improve the joint performance by preventing thermal fatigue. If the expansion cannot be matched, then use of thicker bond lines and more compliant adhesive can reduce the thermally induced stress, improving joint life.

Successful joint design must take advantage of the increased area provided by this method to reduce stress concentrations at the material interface. Unlike bolted or riveted joints which have high local stresses where the fastener contacts the substrate, adhesive bonding avoids this problem by spreading out the contact surface over the entire joint, effectively reducing the unit loading. The joint can be modified further by using a more flexible adhesive (lower elastic modulus), reduced overlap or increased joint width and increased bond thickness. Because failure is initiated at high stress locations, minimizing stress concentrations is important to the design of successful joints.

## Joint Configurations

Determining the appropriate construction of an adhesive joint involves many considerations and compromises in regard to the substrate selection, joint geometry, joint loading, and adhesive. The substrate being joined must not only provide the required strength but also have the desired adhesion characteristics. This requires that the design engineer consider the substrate strength concurrently with the required bond strength to produce a design that includes compatible materials. Postponing the bonding issues until the design is too mature to alter the substrate can result, at a minimum, in increased manufacturing costs and worse, premature failure of a poorly designed joint. Joint geometry and how it will be loaded in the final assembly needs also to be considered when selecting the substrate as they can often be used to offset reduced adhesion or low material strength. For example, it may be possible to increase the joint surface area to compensate for limited

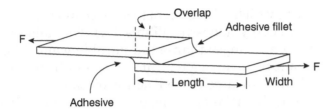

**Figure 7.19** Joint geometry. Greater joint width is desirable over increased adherend overlap, because the overlap induces greater bond line stresses. Including a generous fillet is also beneficial in improving the transition between the adherend and adhesive.

adhesion properties of a material and thus avoiding costly surface preparation.

The basic geometry of an adhesively bonded joint involves the thickness of the adherends, the width of the joint, adherend overlap, and adhesive thickness (Figure 7.19). Generally, joints perform best when the adherends are thick, joint width is maximized, and the overlap and bond line minimized. They should also maximize shear and compressive loading and minimize peel, cleavage, and tensile loading.

An adhesive joint is dependent on five essential factors for its strength:

1. Strength of the materials to be bonded.
2. Strength of the adhesive bond to the adherends.
3. Strength of the adhesive material.
4. Geometry of bonded area.
5. Adhesive thickness.

These are all important to the success of a joint and must be considered collectively when designing a successful joint. The substrates used to construct a joint have a significant impact on the resulting performance of it. The issues involved can be generalized by product family and include metallics, plastics, elastomers, and glass/ceramics.

Metallics

Adhesive bonds involving metals are typically stronger than those formed with other substrates, because of their higher surface energy, which allows them to adhere more readily with lower surface energy adhesives (Table 7.4). The adhesive bond may initially be stronger than a comparable welded joint but not as resistant to adverse environments, possibly resulting in reduced life due to accelerated aging of the adhesive. The greater material stiffness of metals is also an advantage because it allows for use of more rigid adhesives, which generally form a stronger bond, without concern for cracking or otherwise overstressing them from flexing of the substrate.

All metals have some form of oxide covering their surface which can have an impact on the adhesion properties and is best removed or stabilized before bonding. Carbon steels require removal of a weak oxide layer while stainless steels and aluminums have strong oxide layers, which allow for good adhesion even without surface treatments. Copper alloys provide good initial bond strength but have a weak oxide layer that does not hold up well in moist environments. Because all metals have some form of oxide present that must be removed or stabilized by mechanical or chemical means, it is important to seal the surface by primers or conversion coatings to protect the substrate prior to bonding.

## Polymers

Joining polymers is challenging because their material properties can change significantly during the life of the application. Unlike metallic materials, which are typically stable and decay slowly over the life of the application, polymers can lose their adhesion capabilities due to changing surface conditions that are beyond the design engineer's control. Additives such as prepolymers and plasticizers can migrate to the surface as the material ages, reducing the adhesion and weakening the original bond. The substrate flexibility can also have a significant effect on the performance of the joint and requires careful adhesive selection.

Joint requirements vary between thermoset and thermoplastics polymers. Thermoset polymers are those that are cross-linked and so do not have melting temperature and do not dissolve. Their surface energies are generally higher than thermoplastics and therefore require minimal surface preparation. They are best prepared by solvent degreasing and abrading, followed by a solvent wash with one of many environmentally compatible solvents now available from prominent suppliers. Previously used chemicals including acetone, toluene, and methyl ethyl ketone (MEK) which were not only hazardous to the user but also had undesirable environmental impact such as ozone depletion.

Thermoplastics are those polymers that are not cross-linked and soften when exposed to heat or solvents used for solvent bonding. Upon cooling or evaporation of the solvent, the polymer hardens. Thermoplastics usually have lower surface energies than thermosets, which leads to problems in finding an adhesive with a lower surface energy as is required to obtain a quality bond. To limit this problem by increasing the surface energy, treatments in the form of oxidation by electrical discharge or UV exposure have proven to be effective.

## Elastomers

The compliant nature of elastomers makes bonding difficult and joint design critical. They are highly compounded materials that can significantly impact their performance and as with polymers, make their aging difficult to predict. Designing a joint with a deformable substrate must require minimal assembly pressure to prevent generating internal stresses that would be detrimental to the bond. Joints requiring the least amount of pressure to achieve close contact are those likely to perform best when the appropriate flexible adhesive is selected.

## Ceramics/Glass

Joints for extremely rigid materials, such as ceramics and glasses, require careful consideration of the environment they are to endure. They are materials with high surface energy and therefore easily wet with most adhesives, making them appear to be less sensitive to adhesive selection. Adhesive shrinkage and thermal expansion differentials are the two most important properties that must be accommodated but are often overlooked. Most adhesives, with few exceptions, lose varying amounts of volume upon curing. Those adhesives, such as epoxies, which produce an exothermic reaction during the curing process, will expand thermally during the cross-linking process and shrink upon cooling and curing. Carefully reading the adhesive's specification and working closely with the supplier are extremely important when designing a joint for rigid materials. Choosing an adhesive that has minimal shrinkage is key.

Although the materials being joined are rigid, the adhesive should not be so, as it needs to be reasonably compliant to minimize the stress present at the bond line. These stresses can be from thermal expansion of the adherend or from handling. Maintaining a thin bond line is important to the strength of the joint but it must be balanced with the need for some joint compliance. Cyanoacrylates are popular and effective in bonding ceramics when the surface is properly solvent-cleaned. Abrading or more elaborate surface preparation is not generally required or practical but should be considered for demanding applications because the added mechanical interlocking can improve joint strength.

## Joints

Determining the appropriate joint geometry will be driven by the design requirements which should include a maximum load capability, loading profile, operating environment, and manufacturing limitations such as cure time and fixturing. The joint configuration that is most practical for meeting the requirements must include consideration of the adhesive and its curing methodology (Figure 7.20).

Butt — One of the simplest joints but also the weakest.

Lap — The most common joint type and the strongest when axially loaded.

Angle — Common for vertical or wall mounted design. Two forms, L-shaped or T-shaped.

Strap — A very simple and strong joint not requiring machining.

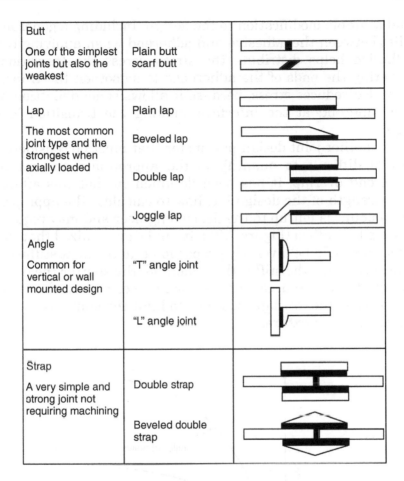

| Butt | | |
|---|---|---|
| One of the simplest joints but also the weakest | Plain butt<br>scarf butt | |

| Lap | | |
|---|---|---|
| The most common joint type and the strongest when axially loaded | Plain lap<br><br>Beveled lap<br><br>Double lap<br><br>Joggle lap | |

| Angle | | |
|---|---|---|
| Common for vertical or wall mounted design | "T" angle joint<br><br>"L" angle joint | |

| Strap | | |
|---|---|---|
| A very simple and strong joint not requiring machining | Double strap<br><br>Beveled double strap | |

**Figure 7.20** Common joint designs. The joint geometries shown here are found in some form or another in almost all joint designs. Note that minor alterations to a joint can often improve its strength, making it more efficient and robust.

*Source*: Ellsworth Adhesives Catalog, 1997, p. 271. Used with permission.

Subtle changes in joint design can make a significant difference in the stress distribution of the joint. Providing for a smooth transition between the adhesive and adherend is instrumental in reducing the joint stress and can often be accomplished with care in applying the adhesive to the joint and a simple modification to the design. Including a generous fillet between the adhesive and adherend when applying the adhesive helps distribute the stress across the transition. Tapering the ends of the adherends to smoothen the transition also reduces stress, because it allows some deflection of the adherend at the interface, making the transition less rigid.

Another joint design parameter that must be considered but is difficult to quantify is the appropriate amount of adherend overlap. It has been documented that this affects the strength of the design but how to calculate the appropriate amount is left up to the design engineer and may require testing to confirm (Figure 7.21). It can be generalized that the overlap should be inversely proportional to the joint stiffness, which is to say the stiffer the adherend, the shorter the overlap needs to be and, conversely, the more compliant the joint, the greater the overlap should be to limit the joint stress and accommodate flexture.

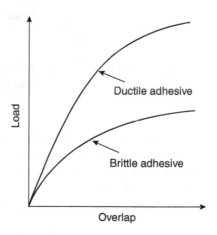

**Figure 7.21**   Effect of joint overlap. Increasing the overlap of the substrates in a lap joint improves its strength but does so with diminishing returns and is affected by the adhesive hardness. Increasing the width of the joint provides greater strength because it does not provide the high bond line stresses as observed with greater overlap.

## Joint Enhancements

Augmenting an adhesive joint to improve its manufacturability and performance is often necessary to gain full benefit of the adhesive joint. When considering the strength requirements of a joint it is often found to be more than adequate to support the desired load. What often prevents implementation is the concern over assembly times during manufacture and survivability, including environmental aging and overload effects. Many applications can benefit from using a combination of adhesive and mechanical fastening to meet the processing and load requirements (Figure 7.22). Consider a joint that is designed to use RTV to join two-sheet metal components. RTV is chosen because it can accommodate the flexible substrates while providing the required strength. Because the typical cure time for this type of adhesive is 24 h with limited green strength, fixturing and the long processing time (waiting for cure to allow handling) can be eliminated by the use of a limited number of mechanical fasteners such as rivets or sheet metal screws. This allows the added benefit that adhesives

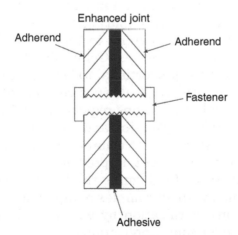

**Figure 7.22** Enhanced joints. Complementing an adhesive joint using a mechanical fastener to assist in reducing assembly time while providing redundancy is often desirable because it may not only provide reduced assembly time but also added safety.

provide such as vibration damping and uniform stress distribution without sacrificing manufacturability or safety.

## Preparation

Surface preparation is a critical element of any adhesive application and must be considered concurrently during the selection of the substrate and adhesive so that they may complement each other. The objective is to remove oils, greases, paints, oxides, films, dust, mold release agents, rust inhibitors, mill scale, rust, and other similar undesired materials and chemicals. Often, the substrate is predetermined, owing to design factors that demand materials of specific properties, leaving the adhesive selection a nontrivial exercise. All substrates such as metals, plastics, elastomers, and ceramics require some form cleaning or surface modification to increase the surface energy which promotes wetting of the adhesive and subsequent adhesion. A surface that produces the best bond is one that is clean or free of contamination, is uniform and continuous in finish, and stable with high surface energy. Surface preparation has a serious impact on the resulting bond performance and must not be overlooked.

In its simplest form, preparing a substrate for bonding requires only sufficient cleaning to remove surface contaminants such as oils and greases. Generally, the quality of the surface preparation is proportional to both the complexity and cost (Figure 7.23). Although all surfaces benefit from proper preparation, not all applications require extensive preparation to be adequate. Solvent cleaning and mechanical abrasion are two of the simplest and most common preparation processes that can provide significant improvement when applied to almost any substrate. Confirming the surface cleanliness can also be performed without complex equipment or processes for most applications. Verification by wiping with a white cloth can be adequate for some applications.

Removing oils, greases, and mold release forms of surface contamination should always be done first, using a clean, white cloth and an appropriate solvent. Continuing to clean the surface until the cloth no longer picks up dirt or oils is a

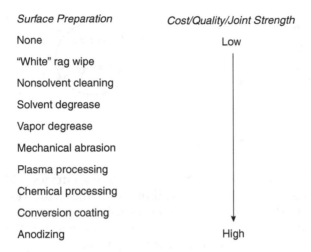

| Surface Preparation | Cost/Quality/Joint Strength |
|---|---|
| None | Low |
| "White" rag wipe | |
| Nonsolvent cleaning | |
| Solvent degrease | |
| Vapor degrease | |
| Mechanical abrasion | |
| Plasma processing | |
| Chemical processing | |
| Conversion coating | |
| Anodizing | High |

**Figure 7.23** Surface preparation is critical to the success of a adhesively bonded joint. The effort applied to preparation is proportional to the benefit and therefore must be match with the application.

must if the surface is to be properly prepared because any subsequent abrasive operation will force the residual contamination into the surface, reducing the effectiveness of the cleaning process. Following the cleaning process by surface abrasion will improve the adhesion by creating a micro-finish that promotes mechanical interlocking as well as increasing the effective surface area (Figure 7.24). Following the abrasive process with solvent cleaning using a white cloth will assure the removal of unwanted contaminants, leaving the surface ready for immediate bonding or application of primer. To simply confirm the surfaces are reasonably clean, place a few drops of water on them. If the surfaces are sufficiently clean, the water will spread to form a continuous film. If it forms beads, a second cleaning and retest are required until the surfaces is clean. It is best to immediately apply the adhesive or primers to freshly cleaned surfaces to avoid recontamination prior to assembly or during production delays.

The resulting bond strength can be observed in the joint performance and is a function of both the surface cleanliness and preparation. Cleaning is the initial component of the

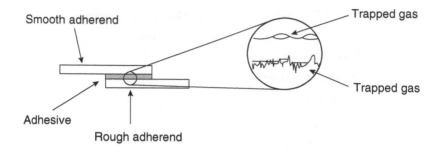

**Figure 7.24**  Surface roughness. Abrading a surface will typically improve its adhesion by providing mechanical interlocking and increased surface area. Excessive abrasion can actually be detrimental if it inhibits the wetting process by preventing the adhesive from flowing across the surface, creating voids.

preparation process. Surface treatments that alter or remove material and are followed by cleaning provides the best possible adhesion.

There are numerous methods of surface preparation to choose from but many manufacturers have specific requirements for their adhesives (Table 7.8). Understanding the processing and cost these will have on the final product is important to determine at the time of selection rather than at the point of design completion.

Although manufacturers can provide general surface preparation requirements, they rarely provide the required preparation based on the materials being bonded. The required preparation can be broken down by material families, including metals (Table 7.9), plastics (Table 7.10), and several miscellaneous materials (Table 7.11).

## Curing

Allowing for proper curing is an important factor in the successful formation of an adhesive joint. If the joint is handled too soon after it is assembled the adhesive will not have had time to set up sufficiently to support the adherends, possibly resulting in separation of the joint or unseen separation of the adhesive within the joint. Allowing it to remain fixtured for

**Table 7.8**  Surface Preparation Processes Commonly Used in Adhesive Bonding Applications

| Process | Description |
| --- | --- |
| Mechanical abrasion | Abrasive blasting, sanding (medium grit, 180 to 325), wire brush. Abraded surface must be small enough to promote capillary action into the microstructure. |
| Degreasing | Several types: pressure washing, mechanical agitation, vapor degreasing, mechanical abrasion in solvent solution and best is ultrasonic agitation. Does not increase surface energy. Degreasing method must be evaluated with consideration of the materials involved. |
| Acid etching | Three common types: chromic, sulfuric and phosphoric. Removes organic contamination, providing a clean, micro-etched surface. |
| Gas plasma treatments | Generated by high frequency and high voltage in low-pressure chamber in the presence of different gases. It produces an ionized gas which oxidizes the surface, improving adhesion. Very expensive. |
| Primers | Low viscosity liquids applied directly to the substrate to promote adhesion and prevent oxidation prior to application of the adhesive. |
| Conversion coatings | Application of an acidic solution to the surface of the metal to produce a smooth uniform organic coating that promotes adhesion and prevents corrosion. Typically used on aluminums but not for structural bonding. |
| Corona treatment | Most popular. Same as gas plasma (above) but done at room temperature and ambient atmosphere. Applied to thin sheets. Oxidizes surface to improve adhesion. |
| Flame treatment | Second most popular. Oxidizes surface which creates one that is better suited for wetting. Done using a blue flame from an oxyacetylene or propane torch. Apply until surface becomes glossy or shiny. Minimize heat to reduce distortion. Wash using water and soap, rinse and dry. Applied to polypo and polyeth. Its effectiveness can be limited by the variable nature of the process where air–gas mixture ratios can be subjective. |
| Ionizing treatment | Used on nylons to alter surface crystallinity, promoting adhesion. |
| Anodizing | Chromic and phosphoric types. Protective layer and micro porosity that lends to adhesion. Aluminum and titanium. |
| UV exposure | Surface oxidation. |
| Laser treatments | Removal of surface contamination/weak boundary layer and roughening. From excimer lasers, oxidize the surface. |

**Table 7.9**  Surface Preparations for Metallic Materials

| Substrate | Solvent cleaning | Intermediate cleaning | Chemical treatment or "other" |
|---|---|---|---|
| Aluminum, aluminum alloys | Immerse, spray, or wipe with chlorinated solvent, ketone or mineral spirits of vapor degrease with chlorinated solvent | Detergent scrub | Sulfuric acid (96%) 77.8 pbw<br>Sodium dichromate 22.2<br>Solution temperature: 25°C; immersion time: 20 min; rinse: tap water followed by distilled water; dry: 30 min at 70°C |
| Beryllium | Immerse, spray, or wipe with chlorinated solvent, ketone or mineral spirits or vapor degrease with chlorinated solvent | Wet abrasive blast | Sodium hydroxide 20.0 pbw<br>Distilled water 80.0<br>Solution temperature: 79°–85°C; immersion time: 3–4 min; rinse: tap water followed by distilled water; dry: 65°–82°C |
| Beryllium/copper | Immerse, spray, or wipe with chlorinated solvent, ketone or mineral spirits or vapor degrease with chlorinated solvent | Wet abrasive blast | — |
| Cadmium | — | — | Electroplate with nickel |
| Chromium | Immerse, spray, or wipe with chlorinated solvent, ketone or mineral spirits or vapor degrease with chlorinated solvent | Dry abrasion or dry abrasive blast | Hydrochloric acid (37%) 46.0 pbw<br>Distilled water 54.0<br>Solution temperature: 90°–95°C; immersion time: 1–5 min; rinse: tap water followed by distilled water; dry: hot air |

| Material | Solvent cleaning | Abrasive treatment | Chemical treatment |
|---|---|---|---|
| Copper, copper alloys | Immerse, spray, or wipe with chlorinated solvent, ketone or mineral spirits or vapor degrease with chlorinated solvent | Dry abrasion or wire brushing | Nitric acid (69%) 12.4 pbw<br>Ferric chloride 6.2<br>Distilled water 81.4<br>Solution temperature: 21°–32°C; immersion time: 1–2 min; rinse: tap water followed by distilled water; dry: 65°C maximum |
| Germanium | Vapor degrease with chlorinated solvent | Dry or wet abrasive blast | — |
| Gold, platinum, silver | Vapor degrease with chlorinated solvent | Dry abrasion | — |
| Lead, solders | Immerse, spray, or wipe with chlorinated solvent or vapor degrease with chlorinated solvent | Dry abrasion or wet or dry abrasive blast or wet abrasive scour | — |
| Magnesium, magnesium alloys | Immerse, spray, or wipe with chlorinated solvent, ketone or mineral spirits or vapor degrease with chlorinated solvent | Dry abrasion | Chromic acid 20.0 pbw<br>Distilled water 80.0<br>Solution temperature: 71°–88°C; immersion time: 10 min; rinse: tap water followed by distilled water; dry: below 60°C |
| Nickel | Vapor degrease with chlorinated solvent | — | Nitric acid (69%) 100.0 pbw<br>Solution temperature: 20°C; immersion time: 5 sec; rinse: tap water followed by distilled water; dry: 40°C |

**Table 7.9** (*Continued*)

| Substrate | Solvent cleaning | Intermediate cleaning | Chemical treatment or "other" |
|---|---|---|---|
| Steel (stainless) | Immerse, spray, or wipe with chlorinated or aromatic solvent or vapor degrease with chlorinated solvent | Heavy duty alkaline cleaner | Nitric acid (69%) 20.0 pbw<br>Distilled water 80.0<br>Solution temperature: 21°–32°C;<br>immersion time: 25–35 min;<br>rinse: tap water followed by<br>distilled water; dry: 65°C<br>maximum |
| Steel (mild) | Immerse, spray, or wipe with chlorinated or aromatic solvent or vapor degrease with chlorinated solvent | Heavy duty alkaline cleaner | Ethyl alcohol (denatured) 66.7 pbw<br>Orthophosphoric acid (85%) 33.3<br>Solution temperature: 60°C;<br>immersion time: 10 min; rinse:<br>tap water followed by distilled<br>water; dry: 60 min at 120°C |
| Tin | Immerse, spray, or wipe with chlorinated solvent or vapor degrease with chlorinated solvent | Dry abrasion | — |
| Titanium, titanium alloys | Immerse, spray, or wipe with ketone or aromatic solvent | Mild alkaline cleaner or wet abrasive scour | Nitric acid (69%) 28.8 pbw<br>Hydrofluoric acid (60%) 3.4<br>Distilled water 67.8<br>Solution temperature: 38°–52°C; |

| Material | Solvent method | Abrasion method | Chemical method |
|---|---|---|---|
| Tungsten, tungsten alloys | Immerse, spray, or wipe with chlorinated solvent, ketone or mineral spirits or vapor degrease with chlorinated solvent | | immersion time: 10–15 min; rinse: tap water followed by distilled water; dry: 15 min at 71–82°C; brush off carbon residue with nylon brush while rinsing<br><br>Nitric acid (69%) 30.0 pbw<br>Sulfuric acid (96%) 50.0<br>Hydrofluoric acid (60%) 5.0<br>Distilled water 15.0<br>Hydrogen peroxide few drops<br>Solution temperature: 20°C; immersion time: 1–5 min; rinse: tap water followed by distilled water; dry: 15 min at 71–82°C |
| Zinc, zinc alloys | Immerse, spray, or wipe with chlorinated solvent, ketone or mineral spirits or vapor degrease with chlorinated solvent | Dry abrasion or wet or dry abrasive blast | Hydrochloric acid (37%) 20.0 pbw<br>Distilled water 80.0<br>Solution temperature: 20°C; immersion time: 2–4 min; rinse: tap water followed by distilled water; dry: 30 min at 66°–71°C |

*Source:* Ellsworth Adhesives Catalog, 1997, Table 1, p. 273. Used with permission.

**Table 7.10** Surface Preparation for Plastic Materials

| Substrate | Solvent cleaning | Intermediate cleaning | Chemical treatment or "other" |
|---|---|---|---|
| Acetal (copolytner) | Immerse, spray, or wipe with ketone solvent | Dry abrasion or wet or dry abrasive blast | Sulfuric acid (96%) 88.5 pbw Potassium dichromate 4.4 Distilled water 7.1 Solution temperature: 25°C; immersion time: 10 sec; rinse: tap water followed by distilled water; dry: room temperature |
| Acetal (homopolymer) | Immerse, spray, or wipe with ketone solvent | Dry abrasion | 1,4-Dioxane 3.00 pbw Perchloroethylene 96.85 p-Toluene sulfonic acid 0.05 Hydrophilic fumed silica 0.10 Solution temperature: 80°–120°C; immersion time: 5.30 sec; rinse: hot water (place in oven at 120°C for 1 min prior to rinsing); dry: 120°C |
| ABS or methyl pentene | Immerse, spray, or wipe with ketone solvent | Dry abrasion or wet or dry abrasive blast | Sulfuric acid (96%) 65.0 pbw Potassium chromate 7.5 Distilled water 27.5 Solution temperature: 60°C; immersion time: 20 min (ABS)/ 60 min (methyl pentene); rinse: tap water followed by distilled water; dry: warm air |
| Cellulosics | Immerse, spray, or wipe with alcohol | Dry abrasion or wet or dry abrasive blast | Dry plastic at 100°C for 60 min, apply adhesive before plastic cools to room temperature |

| Material | Solvent cleaning | Mechanical treatment | Chemical treatment |
|---|---|---|---|
| Diallyl phthalate | Immerse, spray, or wipe with ketone solvent | Dry abrasion or wet or dry abrasive blast | — |
| Epoxy, phenolics | Immerse, spray, or wipe with ketone solvent | Dry abrasive blast or wet abrasive scour or detergent scrub | — |
| Ethylene vinyl acetate | Immerse, spray, or wipe with alcohol | — | Prime with epoxy adhesive and fuse into the surface by heating for 30 min at 100°C |
| Furane, ionomer, melamine resins, SAN, polysulfone or vinyl (rigid) | Immerse, spray, or wipe with ketone solvent | Dry abrasion or wet or dry abrasive blast | — |
| Polyamide | Immerse, spray, or wipe with chlorinated, aromatic, or ketone solvent | Dry abrasion | Phenol 80.0 pbw<br>Distilled water 20.0<br>Solution temperature: 70°–90°C; immersion time: brush on; rinse: none; dry: 20 min at 60°–71°C |
| Polycarbonate | Immerse, spray, or wipe with alcohol | Dry abrasion or wet or dry abrasive blast | — |
| Fluorocarbons | Immerse, spray, or wipe with ketone solvent or alcohol | — | Naphthalene 12.3 pbw<br>Sodium 2.2<br>Tetrahydrofuran 85.5<br>Solution temperature: 70°–90°C; immersion time: 60–120 sec; rinse: ketone + aromatic + distilled water; dry: 65°C maximum |

**Table 7.10** (*Continued*)

| Substrate | Solvent cleaning | Intermediate cleaning | Chemical treatment or "other" |
|---|---|---|---|
| Polyester (Mylar) | Immerse, spray, or wipe with ketone solvent | — | Sodium hydroxide 20.0 pbw<br>Distilled water 80.0<br>Solution temperature: 70°–95°C; immersion time: 10 min; rinse: hot water; dry: hot air |
| Polyethylene, polypropylene, or chlorinated polyether | Immerse, spray, or wipe with ketone solvent | — | Sulfuric acid (96%) 88.5 pbw<br>Sodium dichromate 4.4<br>Distilled water 7.1<br>Immersion time: 60 sec at 70°C (PE, PP); 5–10 min at 70°C (CPE)<br>Expose surface to gas burner flame (or oxyacetylene oxidizing flame) until the substrate is glossy |
| Polyimide or Polymethylmethacrylate | Immerse, spray, or wipe with ketone alcohol or chlorinated solvent | Dry abrasion or wet or dry abrasive blast or wet abrasive scour | — |
| Polyphenylene oxide | Immerse, spray, or wipe with alcohol | — | May be primed with an adhesive containing xylene |
| Polystyrene | Immerse, spray, or wipe with alcohol | Dry abrasion | — |
| Polyurethane | Immerse, spray, or wipe with alcohol | — | — |
| Vinyl (flexible) | Immerse, spray, or wipe with ketone solvent | — | — |

*Source:* Ellsworth Adhesives Catalog, 1997, Table 2, p. 274. Used with permission.

**Table 7.11  Surface Preparation for Miscellaneous Materials**

| Substrate | Solvent cleaning | Intermediate cleaning | Chemical treatment or "other" |
|---|---|---|---|
| Brick | Immerse, spray, or wipe with ketone solvent | Wire brushing | — |
| Carbon graphite | Immerse, spray, or wipe with ketone solvent | Dry abrasion | — |
| Glass (nonoptical) | Immerse, spray, or wipe with ketone solvent | Wet abrasive blast | Sulfuric acid (96%) 96.8 pbw<br>Sodium dichromate 1.7<br>Distilled water 1.7<br>Solution temperature: 20°C; immersion time: 10–15 min; rinse: tap water followed by distilled water; dry: 65°C maximum |
| Glass (optical) | Immerse, spray, or wipe with chlorinated solvent | | Clean in ultrasonically agitated detergent bath; rinse: tap water followed by distilled water; dry: 38°C maximum |
| Concrete, granite, stone | Immerse, spray, or wipe with chlorinated solvent | Wire brushing | Hydrochloric acid (15%) 100.0 pbw<br>Solution temperature: 20°C; immersion time: until effervescence stops; rinse: tap water until neutral, then rinse with 1% ammonia and water; dry: 38°C maximum |
| Ceramics | Immerse, spray, or wipe with ketone solvent | — | Sulfuric acid (96%) 96.6 pbw<br>Sodium dichromate 1.7<br>Distilled water 1.7<br>Solution temperature: 20°C; immersion time: 15 min; rinse: tap water followed by distilled water; dry: 65°C maximum |
| Wood, plywood | — | Dry abrasion | |
| Painted surface | — | Detergent scrub or dry | Remove paint by solvent or abrasion; pretreat exposed base |

*Source:* Ellsworth Adhesives Catalog, 1997, Table 3, p. 275. Used with permission.

too long may unnecessarily prolong the assembly process and increase manufacturing time and expense.

Adhesives typically have what are known as a "set" time and a "cure" time. The set time is the duration required for the adhesive to develop sufficient strength to support the bonded assembly. Once the adhesive has set, the assembly fixture can be removed and the joined substrates can be moved to the next level of assembly or into place in preparation for use. The cure time is the duration required to obtain full strength or near full strength. This time is commonly 24 h but can be as long as 7 days for some adhesives and is dependent on cure temperature, relative humidity, joint geometry, and bond thickness.

Accelerating the curing process is most commonly achieved by elevating the cure temperature above ambient, with higher temperatures producing the shortest cure times. The reverse is also applicable, reducing the cure temperature slows the reaction and at lower temperatures halts it entirely. This suspended state is known as "B-staging" an adhesive and is often used on laminate and sheet forms. Reducing the cure temperature can also extend the time from when the adhesive is first mixed to when it sets, typically called the "working life" of the adhesive.

Manufacturers often provide a range of curing options or can tailor adhesive formulations to provide required set and cure times. Generally, the strongest bonds are obtained by elevated temperature curing and the weakest by rapid room temperature curing. An oven-cured epoxy will be stronger than a 90 sec, room temperature cured epoxy because of the improved cross-linking provided by the elevated temperature-curing process.

## Evaluation

The quality of the bonded assembly is significantly influenced by the processes used to create it. Adhesives are unique from other fastening methods because it is difficult to determine the condition of the bonded assembly without destroying it. Other assembly methods, such as a threaded bolt, allows for verification by inspection using of a torque wrench, easily confirming the joint is properly assembled. Producing test coupons that simulate the bonded joint and are prepared at the same time, by the same technician and evaluated using the appropriate ASTM test method for the application is commonly done to prove the quality of the joint. Lap shear tests are often the easiest samples to prepare and test, making them the most popular. Tests of peel, cleavage, fatigue, impact, creep, and environmental responses and properties are also performed as the design and application requires. The most common faults revealed by mechanical testing are summarized in Table 7.12.

Inspecting the bonded assembly without simultaneously destroying it would be preferred, but is often difficult because it does not provide the desired performance data. Consider again a bolted assembly: making a direct measurement of the applied torque confirms the joint preload from which its strength and integrity can be inferred from prior testing of the joint. An adhesive joint does not provide for this type of direct measurement to confirm the strength and integrity after assembly. Use of qualitative nondestructive test (NDT) methods such as visual inspection are performed but these methods lack the desired quantitative measure. A trained assembler or inspector can inspect a joint and observe obvious flaws, such as insufficient adhesive or joint misalignment, as well as more subtle problems, such as lack a of bond line uniformity or a soft, improperly cured adhesive. Although this information is useful and provides a reasonable estimate of the bond integrity, experience has shown that it by no means guarantees the bond performance.

Other forms of NDT include such methods as proof-testing, ultrasonic and X-ray inspection as well as several forms of thermal transmission tests. Proof-testing provides a quantitative level of strength without testing until failure, preserving the joint. Ultrasonic and X-ray methods both involve ways of imaging the joint to confirm the presence of adhesive but does not provide direct measurement of strength other than what can be inferred by the presence of the adhesive and correlated to previous testing. Thermal testing is a newer form of

**Table 7.12**   Bonding Problems, Causes and their Corrective Actions

| Bonding problem | Cause | Corrective action |
|---|---|---|
| Uneven or thick bond line | Insufficient clamping pressure | Increase pressure. Confirm clamp method is sufficiently rigid for the application |
| | Low curing temperature | Confirm curing temperature is within manufacturer's specification |
| | Adhesive has exceeded shelf life | Use new or adhesive with known age and storage history |
| Exposed adhesive contains bubbles | Entrained air from mixing | Vacuum degas adhesive during mixing or reduce mixing agitation |
| | Adhesive out gassing | Consult manufacture on proper usage |
| Voids in failed joint | Improper surface preparation | Consult manufactures recommendations. Cause likely is a result of one or more of the following: contamination, low surface energy, lack of surface abrasion |
| | Contaminated adhesive | Replace |
| | Improper or uneven bond thickness | Establish design criteria that provides for uniform bond line of appropriate thickness for the adhesive to be used |
| Adhesive softening when exposed to solvents or heat | Improperly cured | Confirm mixing ratios, cure time and temperature are per the manufacture's specification |

evaluation that provides some insight into not only the presence of the adhesive but also of subsurface anomalies, as detected by the variation of measure heat flow. Some helpful sources involving material and process specifications are summarized in Table 7.13.

**Table 7.13**  Sources for Material and Process Specifications

| Source | Website |
|---|---|
| American Society for Testing and Materials (ASTM) | http://www.astm.org |
| American Society for Nondestructive Testing | http://www.asnt.org |
| United States Military Specifications | http://dodssp.daps.dla.mil |
| Pressure Sensitive Tape Council | http://www.pstc.org |
| Technical Association for the Pulp and Paper Industries (TAPPI) | http://www.tappi.org |
| Association of European Adhesive Manufactures (FEICA) | http://www.feica.com |

## SELECTION CONSIDERATIONS

Adhesive selection is one of the most challenging tasks a design engineer can undertake because of the large number of variables that affect an adhesive's performance and the catastrophic nature of the failure mode. Adhesives typically are thought of as an all-or-nothing proposition because they usually either work or do not, failing completely and often without warning. This is in stark contrast to mechanical fasteners, such as bolts or rivets, which are assembled to a preset torque or load, providing measurable information regarding the assembly process. They also are readily inspected for loss of preload or pending failure, allowing them to be either retorqued or replaced as necessary. Adhesives offer no such convenience, making visual or quantitative inspection a challenge.

Considering the complete lifecycle of an adhesive is important to its success and is often overlooked during the selection process. There are four primary considerations that must be addressed during the selection process: adhesive technology, design criteria, manufacturing, and operation/service.

The adhesive technology refers to the basic chemistry of the adhesive which needs to be the starting point in the selection process and is heavily influenced by the performance requirements. Making the selection based only on the strength requirements, as is often the case, can result in failure. Recall that the best bonds are produced when the critical wetting tension of the adhesive is greater than the surface energy of adherend. Although this is the basis for the adhesive selection and recommendations of adhesive manufacturers, it is not commonly listed in their product literature. Critical bonding applications may require investigating the joint properties in detail, including the critical wetting tension and surface energy.

The adhesive technology not only needs to be appropriate for the materials involved, it must also fit within the manufacturing environment and related process capabilities. For example, selecting an oven-cured single component epoxy for bonding two halves of a picture frame may provide the required strength but would be inappropriate for the high-volume manufacturing processes available. A cyanoacrylate would be a more appropriate selection for the close-fitting parts of a picture frame as well as the available manufacturing processes. Other considerations regarding such properties as clarity, creep, viscoelastic behavior, and toughness are also highly dependent on the adhesive polymer chemistry.

The selection must also consider the adhesive technology as it relates to the intended operational environment. The performance and life expectancy should be considered in the context of the extremes of the operational environment to avoid unexpected degradation and resulting failures. Although time and costs tend to limit the scope of environmental testing of specific bonds, making a thorough evaluation difficult or impractical, it does not prevent considering the effects of the extremes on the adhesive chemistry as a means to obtain greater understanding of potential behaviors. Some polymers react with moisture resulting in what is called hydrolysis or reversion, which degrades the adhesive by reducing the $T_g$, hardness, and cohesive strength. Swelling often precedes the complete reversion, causing bond deformation and subsequent failure. The worst case conditions are high temperature and high humidity, causing some to transform to a fluid. Many potted electronic devices suffered this fate during the Vietnam War, causing electronic failures in aircraft operating in the intense jungle heat and humidity. Hydrophobic polymers have been developed which are designed to resist this form of failure and should be considered for applications involving high heat and humidity.

Developing the appropriate design criteria can be a challenge for even the most experienced engineer because of the numerous variables that are involved in adhesive bonding. The most common approach to adhesive selection is to consider a narrow set of properties such as lap shear strength or hardness along with maximum temperature. These properties may meet the immediate design needs of the engineer but ignore the other important elements of the selection process. The design criteria needs to address not only nominal property performance such as strength or hardness, but also include the full range of operational exposure.

Generally, initially considering a short list of properties as they apply to the design requirements will improve the focus of the selection to include the important adhesive properties necessary for a successful bonded joint. The list should include at a minimum the following: lap shear strength, temperature range (high, low and $T_g$), thermal expansion, thermal conductivity, electrical conductivity, chemical, humidity, viscosity, and pot life.

The adhesive bond strength, as measured by the lap shear strength, is the most common criteria for selection but should not be the only one. Consideration for the operational temperature range should also be made because of the detrimental effects they can have on bond performance. Elevated temperatures can oxidize the adhesive, reducing its stress and impact-resistance and accelerate thermal aging. Also, requiring an adhesive to operate near its glass transition temperature, $T_g$, may result in bond deformation and loss of cohesive strength. Low temperatures most often result in lowering of elastic modulus and corresponding toughness, making the bond brittle. Additionally, thermal expansion and conductivity properties need to be considered for those applications that require thermal management, because the appropriate selection can reduce thermal-induced stresses and control the thermal characteristics of the bond.

The effects of an adhesive when exposed to chemicals and excessive moisture can be devastating to the long-term performance of the bond. Because these are surface phenomena, they typically take time to manifest themselves into an observable problem such as adhesive degradation or substrate corrosion. Adhesives affected by chemical or water exposure often become more flexible which may actually improve impact resistance and cyclic loading but at a loss of cohesive strength. This type of exposure can also result in the adhesive swelling, upsetting the joint geometry sufficiently to accelerate or initiate failure by increasing interference or joint loads.

Exposure to either chemicals or water can also result in corrosion of the adherend by wicking of the liquid between the adhesive and adherend, resulting in the formation of oxide compounds and corresponding loss of adhesion. This may be minimized or prevented entirely by surface preparation in the form of either pretreatments such as anodizing or application of primers. They not only provide a protective layer, but also fill in surface voids that may have otherwise been present due to incomplete wetting of the adherent, reducing the opportunity for wicking.

Designing for worst case conditions includes designing for adherend failure, rather than for bond failure. Although joint geometry may limit the ability to do this, it is desirable for the joint to involve thick substrates rather than thin, in order to avoid peel or cleavage loading. A conflicting requirement is often the need to reduce costs, which are typically the result of the joint geometry, preparation, and cure schedule, while still meeting the requirements. When possible, perform only the minimum of surface preparation necessary to be successful, such as removing the weak boundary layer and solvent cleaning by hand. Some common errors that design engineers make are presented in Table 7.14.

Determining end life is possibly the most daunting task when considering the use of adhesives, more so in harsh or critical applications. Most products are designed with a required life expectancy and need to be subjected to testing to verify the requirement. Time and costs tend to limit the scope of testing of specific bonds, making a thorough evaluation difficult or impractical. Fully defining the extremes of the intended operating range and then testing for them is highly

**Table 7.14** Common Errors in Selection, Evaluation and Use of Adhesives

| Application considerations | Error |
|---|---|
| Adhesive technology | Overlooking important factors: surface preparation; cure time; operating temperatures |
| | Assuming strongest is best, regardless of cost or processing issues |
| | Poor testing/evaluation procedures |
| Design consideration | Improper joint selection |
| | Incorrect load assumptions |
| | Overlooking service temperatures or chemical resistance |
| | Lack of consideration of thermal expansion issues of joined materials |
| | Using a heat cure when the parts being bonded cannot handle it |
| | Over-design due to lack of understanding of the application |
| Production-line problems | Improper surface preparation |
| | Failing to validate the process and the resulting joint integrity (does it meet the design specification) |
| | Using product that has exceeded the manufactures shelf life |
| | Improper storage of adhesive |

desirable because it quantifies the robustness of the design and may possibly expose its weaknesses. In order to obtain this data quickly, accelerated tests are often performed. These types of tests are especially difficult to perform with adhesives due to their polymeric nature. For example, differences have been noted between stressed-aged samples versus non-stressed-aged samples, indicating a relationship between the samples that then needs to be considered in the context of the requirements. If the stressed-aged sample was subjected to a stress higher than that to be seen in operation, the data may be inappropriate to represent the production design.

## Performance Requirements

Defining the necessary joint performance can be a ambiguous task if not performed in a manner consistent with the system requirements. It is important to consider adhesively bonded joints as an extension of the substrates which they join, so they must survive the same conditions. Based on the material properties, adhesives should be chosen much the same way but with greater consideration for the manufacturing and life issues.

The performance requirements should be broken into two distinct categories, those that the adhesive must have and those that are desired, but not absolutely required. The most common requirements of an adhesive are maximum load or stress capability (as it relates to the specific substrate) and chemical-resistance. These are often the easiest to quantify and the ones for which manufacturer's data is most readily available. Additional requirements that should be specified but often are overlooked are electrically or thermally conductive or resistant properties as well as cyclic loading or effects due to average constant stress.

Inherently, adhesives are not thermally or electrically conductive because of their polymer chemistry but can be formulated to have varying degrees of both properties with the general consideration that they are not mutually exclusive properties. This allows a design engineer to influence a system's thermal and electrical characteristics by selecting an adhesive with the appropriate properties to improve heat transfer. Greater heat dissipation may not only keep the system elements cooler, but the joint will likely see a reduced thermal gradient which will limit internal joint stresses. Distortion from thermal expansion can also be controlled by attempting to match the adhesive's expansion coefficient to that of the substrates. Electrical conductivity can also be controlled to either improve it or eliminate it as necessary. Potting compounds are an example of an electrically insulated formulation that both seals and supports the electrical components, preventing damage.

Those properties that are often desirable but not required include a rapid or modified cure schedule, clarity or some form of color and shelf life. Generally, cure time and adhesive properties are proportional, meaning those that require a longer or more involved cure process, including fixturing and elevated temperatures, produces a cured adhesive with better properties than one that cures faster at room temperature. This is contrary to what is most desirable for cost-effective manufacturing, namely, no fixturing and nearly zero cure time at room temperature. Also, partial curing or B-staging an adhesive may be beneficial for some applications.

Adhesives are also selected based on appearance. Optically clear adhesives provide the much needed property of assembly of light-based systems, including position sensors or vision systems. Other applications may require the color of the adhesive to be similar to the adherends to promote a cosmetically appealing design.

## Adhesive Review

Adhesives selected for use in engineering applications are generally selected for their structural capabilities or for their sealing capabilities or the unique combination of both. Structural adhesives are defined as those that can withstand operating conditions for an extended period without degradation. When properly matched with the materials to be bonded a joint similar to a welded or brazed metal joint is formed, which is one that uniformly transfers the stress across the common bonded area.

There are six distinct advantages of adhesives over conventional fastening or joining methods. These include:

1. Uniform stress distribution. Rivets, bolts, spot welds and similar joining methods induce stress concentrations that necessitate thicker or heavier gauges than would otherwise be required. Adhesives allow the use of material thicknesses that are selected based on whole sections without stress-risers, resulting in lighter structures.
2. Bonding of dissimilar materials. The ability to bond dissimilar materials allows the fabrication of higher

strength and improved performance assemblies. The flexible layer of adhesive can be selected to be an insulator or conductor, minimize the affects of thermal expansion and provide a barrier to bimetallic corrosion.

3. Retain material integrity. Other than the required material preparation prior to bonding, alteration to the base material is not significant as might be required by the use of mechanical fasteners or welding. These mechanical processes require finishing operations that can extend processing time.

4. Greater fatigue-resistance. The relative flexibility of the adhesive provides an energy absorption element to the assembled product, extending the typical fatigue life by several orders of magnitude over bolted, riveted, or spot welded assemblies.

5. Continuous contact. Because adhesives perform best when they are in complete contact with the materials being bonded, they are also excellent sealants and thermal or electrical conductors or insulators.

6. Improved design efficiency. The use of adhesives allows for reduced weight and more rapid product assembly time. To gain a reduction in assembly time may require significant tooling costs to provide proper fixturing of the parts being joined.

There are five mandatory steps that must be followed when using adhesives in any application. They are the following:

1. Proper joint design. Selection and application of the best joint design for the design plays a significant role for its success. Consult with adhesive manufacturer's for those joint designs that work best with their adhesives.

2. Proper surface preparation. The most important element of adhesive bonding is the surface preparation. If the surfaces are not properly prepared, regardless of joint design, the assembly will fail.

3. Appropriate application method. Adhesives come in numerous forms which allow them to be adapted to most design requirement. Establishing a close working

relationship with your supplier is mandatory to maximize the performance of the adhesive.

4. Use of proper curing method. Each application will determine the most appropriate curing method, whether it is at room or elevated temperature.
5. Joining/assembly method. Application of uniform assembly pressure is paramount to a successful bond because it promotes the distribution of the adhesive, while maintaining the necessary bond line thickness.

With regard to the last item, the application of contact pressure at the time of assembly is an important factor in the success of the joint (Figure 7.25). Because of the variations in surface energy and roughness of the materials involved, the time required to allow the adhesive to come into proper contact, or to "flow" into the surface, varies and is best determined by experiment. It is also important to note that the contact pressure will also influence the bond line thickness.

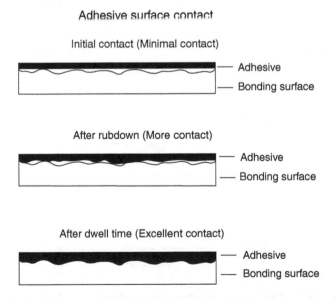

**Figure 7.25** For the best performance of an adhesive bond the proper pressure is required at assembly to ensure that the adhesive "flows" into the mating surfaces and forms the necessary bond line.

Maintaining a uniform and controlled thickness is also necessary to consistently achieve the optimum performance.

An adhesive's initial contact adhesion can vary significantly and, therefore, is an important parameter in the selection process. For example, the initial adhesion of a rubber-based adhesive is significantly greater than that afforded by an acrylic-based adhesive (Figure 7.26). This property can be an advantage in applications requiring immediate holding strength, but not in those applications which may require temporary removal for positioning or alignment. For these applications the acrylic not only offers lower initial tack, it ultimately forms a stronger bond than the rubber-based adhesive.

Operating temperature of adhesives cover a wide range and follow the general trend of reduced performance at higher temperatures (Table 7.15). Being a viscoelastic material, they exhibit a declining performance as the ambient temperature increases, requiring a close review of this data when using an adhesive above room temperature. It is always best to perform extensive application testing as part of the evaluation process and review the results with the manufacture. This review should obtain confirmation of the repeatability of the outcome as it could vary based on formulation variations and other processes that are beyond the end user's control.

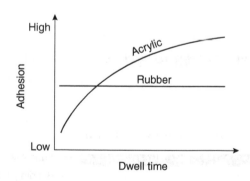

**Figure 7.26**   The effect of adhesion as function of time an important factor in the selection process. Rubber and acrylic adhesives offer significantly different performance as a function of contact or "dwell" time.

**Table 7.15** A General Comparison of the Maximum Operating Temperature while Under Load

| Adhesive type | Maximum operating temperature at load °F (°C) |
|---|---|
| Silicones | 500 (260) |
| Polyimide | 500 (260) |
| Cyanoacrylates | 475 (246) |
| Epoxies | 400 (204) |
| Acrylics | 300 (149) |
| Anaerobics | 300 (149) |
| UV Curable | 300 (149) |
| Polyurethane | 250 (121) |
| Hot Melts | 250 (121) |
| Solvent Bonding | Base material property |

Shrinkage of adhesives should be considered during the selection process, as it may result in loss of integrity of the bond for the intended application. For example, if the adhesive must not only secure two parts together but also provide a seal between them, shrinkage and joint design must be reviewed and tested to confirm the selection.

A general review of common requirements and the anticipated performance of the adhesive family is an excellent place to start the selection process (Table 7.16). A quick review of this type of information can help provide direction as to which adhesive may perform best for the designed application. But, as stated before, there is no substitute for rigorous tests and evaluation. Translating these performance characteristics to a specific design is, then, the challenge that must be undertaken with great attention to detail. Published performance values are often very idealized and do not take into account the processing variations that can have a significant influence on the end result.

There are four basic types of curing methods for adhesives: room temperature cures for one-part formulations; room temperature cures for two-part formulations; elevated

**Table 7.16**  Selection Chart for Determining Appropriate Adhesive Family

| | Viscosity (thickness) | Void filling | Heat resistance | Cold resistance | Flexibility | Chemical resistance | Humidity resistance | Work time | Cure time | Metal bonding (steel, alum.) | Plastic bonding (abs, styrene) | Polyolefin bonding | Wood | Paper cardboard |
|---|---|---|---|---|---|---|---|---|---|---|---|---|---|---|
| Acrylic | M | G | G | G | G | G | G | FT–M | M–FT | G | VG | F | NS | NS |
| Anaerobic | L | P–F | G | G | G | G | G | M | M | F | P | NS | NS | NS |
| Cyanoacrylate | L | P–F | F | F | P–F | G | F | FT | FT | G | VG | G | NS | NS |
| Epoxy | M–T | VG | G | F | F | G | G | M–S | S | G | F | P | G | NS |
| Hot melt | T | VG | P–F | F | F–G | F | G | FT | FT | F | F | P | VG | VG |
| Polyurethane | M | G | F | G | G | G | F | M–S | M | G | VG | G | VG | VG |
| Polysulfide | T | VG | G | G | G | VG | G | M | M | G | F | NS | NS | NS |
| Silicone | T | VG | VG | VG | VG | VG | VG | S–M | M | F | F | F | NS | NS |
| Solvent-base | L–M | F | G | G | G | G | G | S–M | M | G | F | F | G | G |
| Water-base | L–M | P | F | F | P | P | P | M | M | P | P | P | VG | VG |
| UV | L–M | L–M | F | G | G | F | G | S | FT | G | G | F | F | F |

F Fair; FT Fast; G Good; L Low; M Medium; P Poor; S Slow; T Thick; VG Very Good; NS Not Suggested

*Source:*  Ellsworth Adhesives Catalog, 1997, p. 6. Used with permission.

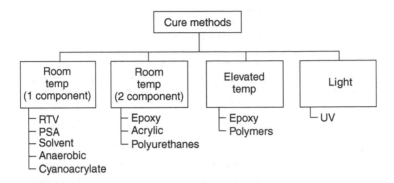

**Figure 7.27** Four basic types of cure methods are used to initiate the chemical reaction to complete the curing process.

temperature cure formulations; and UV light cure formulations (Figure 7.27). There are often implied requirements for the curing process due to the nature of the application. For example, an adhesive required for field repair would not likely be of the elevated curing type because of the lack of controlled thermal processing capabilities.

The time required to achieve sufficient strength such that the bond can physically support the components involved in the application is known as the "fixture time." Adhesive families can be classified in this manner as it is often important in the selection process to know what supporting equipments may be required during processing (Figure 7.28). Adhesives such as epoxies can be formulated to cure at predetermined rates but with the penalty of reduced mechanical performance.

Bond line and gap-filling capabilities are opposing requirements, as the resulting joint's performance will be very different. Generally, the thinner the bond line the better-performing the joint will be because it is not significantly influenced by the adhesive modulus. Gap-filling is the ability of the adhesive to support its own weight while curing, such that it does not run out of the joint while doing so. Adhesives can be sorted by this behavior and it can be used as a means of selection, based on known joint geometries (Figure 7.29).

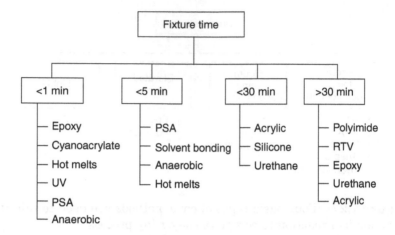

**Figure 7.28**   The required fixture time is a function of the adhesive chemistry and the time it takes to sufficiently cure so that it can sport the application.

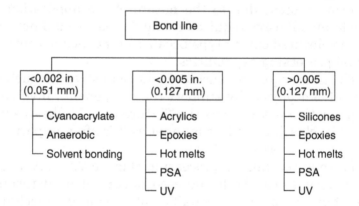

**Figure 7.29**   Bond line and gap-filling are unique properties of each adhesive formulation. Thinner bond lines are best for structural applications while gap-filling is generally for potting or filling applications.

## Selection Methodologies

The selection methodologies employed in determining the appropriate adhesive for a given application are similar to those used for metals or plastics. Not only are there numerous families with similar performance characteristics, there are custom formulations possible within many of them. For example, epoxies can be tailored to the application by varying the cure method and time, altering the hardness by using fillers and selecting from a range of viscosities. The resulting performance of the formulation will certainly be affected by these selections and will need to be evaluated for the specific application. All these options make the selection process a challenging one that is highly influenced by the material application process rather than directly by the material performance, which is contrary to the typical selection of metals and plastics.

The mechanical performance is a primary selection criteria, as it is with other materials with the added consideration for joint design. Adhesives perform best when the influence of the material modulus is limited, as it is when the bond line is at its thinnest. Other applications may require a thick bond line to act not only as an adhesive to join the materials, but also to fill larger gaps, providing a capability for sealing.

The primary selection criteria must also include consideration for the product form (one or two part) and the cure method because they are closely related to the mechanical strength of the adhesive. Generally, two-part adhesives provide better performance than single-part adhesives and elevated curing typically results in a better bond than those obtained at room temperature. Tackiness can also be considered a primary criteria for selection, because many applications may require rapid adhesion to limit the required fixturing but these bonds are often not structural in design.

Secondary considerations include adhesive viscosity and surface preparation methods. Adhesive formulations are generally available in several, if not many, viscosities, allowing for considerable latitude in its selection. The selection of the appropriate viscosity is dependent on several factors including the joint design and cure method. The ability of the adhesive to properly fill the joint prior to and during curing is important to the overall joint qualtiy. The surface preparation requirement is driven by both a design factor and the need to control costs. The time and effort required to produce a specific surface finish is proportional to its cost (see Figure 7.23), requiring careful review and testing of the design to evaluate the required level of preparation necessary to support the design requirements.

Using lap shear strength for solvent-cleaned carbon steel as a primary selection criteria can be useful in providing a relative measure of performance (Figure 7.30). In reviewing the strength values it must be stressed that an adhesives performance is significantly affected by the adherent, surface finish, and surface preparation before bonding. Varying any one of these will influence the actual performance significantly, making the selection process difficult and certainly require rigorous testing before implementation.

Adhesives can also be generally considered for their lap shear strength at varying temperatures. Because adhesives are polymers, they exhibit many of the same characteristics as plastics and elastomers, including their thermal performance. It can generally be stated that elastomers are plastics that are above their glass transition temperature ($T_g$) when evaluated at room temperature, and that plastics exhibit elastomeric behavior above their $T_g$. Adhesives behave in a similar manner, in that

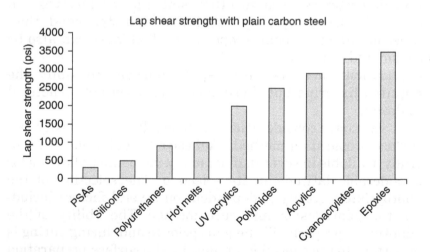

**Figure 7.30** Adhesive selection based on lap shear strength at room temperature for solvent-cleaned plain carbon steel.

they become stiffer at temperatures below the ambient and softer at those higher than ambient. Of course, some forms resist this notion in varying degrees, with silicone the most robust adhesive, when temperature performance is paramount. It does lack the ability to support high lap shear loads and large gaps but as with any material selection, these variations must be taken into account during the selection and evaluation process.

The process flow for making an informed selection need not be overly complicated, even though there are numerous families and formulations of adhesives to choose from (Figure 7.31). As with any material selection, the designer must develop a requirements list and then determine the primary and secondary performance criteria. Strength is most commonly a primary consideration for structural adhesives regardless of the required value because they are selected for some specific joining property. Alternatively, adhesives that are not required to provide some form of structural support are considered sealants.

It is quite likely that the selection between a single- and two-part adhesive will be made early in the process, because, as a general rule, they differentiate between the possible adhesion capabilities of the adhesive. One-part adhesives generally do not offer the higher strength afforded by two-part systems, but this does not automatically rule them out of selection for those higher performing applications. Altering the joint design to provide greater contact area can provide the required load carrying capability, while still allowing a one-part adhesive to be selected.

Once the issue of one- or two-part form is determined, the desired cure method of the adhesive must be selected. Which method is appropriate is again defined by the end application and whether the cure method can be implemented within the constraints imposed by the design. For example, UV-curing requires that the joint be "line-of-sight" or involve transparent materials, otherwise it is difficult to achieve full cure of the adhesive. This issue has been addressed in recent formulations that allow for "shadow" curing, or the ability to initiate curing using UV light on the visible portion of the joint, which then carries through to the nonvisible portion.

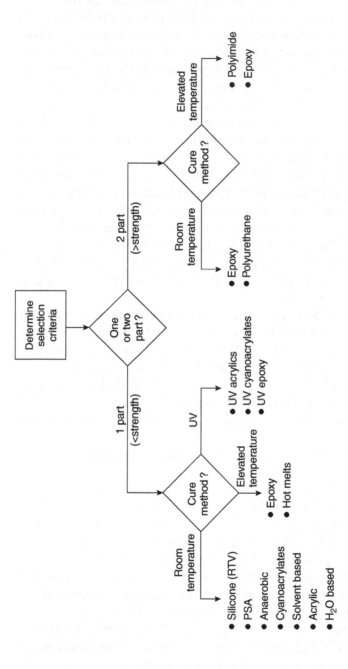

**Figure 7.31** Adhesive selection flow chart. Consideration for adhesive strength and processing complexity is required as part of the selection process. Many additional considerations must be taken into account, including the composition of the adherent material, surface preparation, bond line thickness (gap fill) and joint design.

Once one or several candidates have been selected it is then imperative that extensive application testing be performed to determine affects of joint performance from the point of view of both production assembly and product life. Variations in production methods can often result in off-nominal performance that limits the capability of the adhesive. Complete process specifications should be developed and include the appropriate training to those involved in the application and handling of the adhesive products. The resulting processes should then be audited both during manufacture and with the use of test coupons. Random component level testing should also be performed to assure that the highest quality product is provided to the customer.

# Appendices

## GENERAL ENGINEERING CONVERSION FACTORS

The following table gives conversion factors from various units of measure to SI units. It is reproduced from NIST Special Publication 811, *Guide for the Use of the International System of Units (SI)*. The table gives the factor by which a quantity expressed in a non-SI unit should be multiplied in order to calculate its value in the SI. The SI values are expressed in terms of the base, supplementary, and derived units of SI in order to provide a coherent presentation of the conversion factors and facilitate computations (see the table "International System of Units" in this section). If desired, powers of ten can be avoided by using SI Prefixes and shifting the decimal point if necessary.

Conversion from a non-SI unit to a different non-SI unit may be carried out by using this table in two stages, for example,

$$1 \text{ cal}_{th} = 4.184 \text{ J}$$
$$1 \text{ Btu}_{IT} = 1.055056 \text{ E} + 03 \text{ J}$$

Thus,

$$1 \text{ Btu}_{IT} = (1.055056 \text{ E} + 03 \div 4.184) \text{ cal}_{th} = 252.164 \text{ cal}_{th}$$

Conversion factors are presented for ready adaptation to computer readout and electronic data transmission. The factors are written as a number $\geq 1$ and $<10$ with six or fewer decimal places. This number is followed by the letter E (for exponent), a plus or a minus sign, and two digits which indicate the power of 10 by which the number must be multiplied to obtain the correct value. For example,

$$3.523\,907 \text{ E} - 02 \text{ is } 3.523\,907 \times 10^{-2}$$

or

0.035 239 07

Similarly,

3.386 389 E + 03 is 3.386 389 × 10³

or

3 386.389

A factor in boldface is exact; that is, all subsequent digits are zero. All other conversion factors have been rounded to the figures given in accordance with accepted practice. Where less than six digits after the decimal point are shown, more precision is not warranted.

It is often desirable to round a number obtained from a conversion of units in order to retain information on the precision of the value. The following rounding rules may be followed:

1.  If the digits to be discarded begin with a digit <5, the digit preceding the first discarded digit is not changed.

    Example: 6.974 951 5 rounded to 3 digits is 6.97

2.  If the digits to be discarded begin with a digit >5, the digit preceding the first discarded digit is increased by one.

    Example: 6.974 951 5 rounded to 4 digits is 6.975

3.  If the digits to be discarded begin with a 5 and at least one of the following digits is >0, the digit preceding the 5 is increased by 1.

    Example: 6.974 851 rounded to 5 digits is 6.974 9

4.  If the digits to be discarded begin with a 5 and all of the following digits are 0, the digit preceding the 5 is unchanged if it is even and increased by one if it is odd. (Note that this means that the final digit is always even.)

    Examples: 6.974 951 5 rounded to 7 digits is 6.974 952
              6.974 950 5 rounded to 7 digits is 6.974 950

## REFERENCE

Taylor, B. N., *Guide for the Use of the International System of Units (SI)*. NIST Special Publication 811, 1995 Edition, Superintendent of Documents, U.S. Government Printing Office, Washington, DC 20402, 1995.

| To convert from | To | Multiply by | |
|---|---|---|---|
| abampere | ampere (A) | **1.0** | **E + 01** |
| abcoulomb | coulomb (C) | **1.0** | **E + 01** |
| abfarad | farad (F) | **1.0** | **E + 09** |
| abhenry | henry (H) | **1.0** | **E − 09** |
| abmho | siemens (S) | **1.0** | **E + 09** |
| abohm | ohm (Ω) | **1.0** | **E − 09** |
| abvolt | volt (V) | **1.0** | **E − 08** |
| acceleration of free fall, standard ($g_n$) | meter per second squared (m/sec$^2$) | **9.806 65** | **E + 00** |
| acre (based on U.S. survey foot)[a] | square meter (m$^2$) | 4.046 873 | E + 03 |
| acre foot (based on U.S. survey foot)[a] | cubic meter (m$^3$) | 1.233 489 | E + 03 |
| *ampere hour* (A · h) | coulomb (C) | **3.6** | **E + 03** |
| ångström (Å) | meter (m) | **1.0** | **E − 10** |
| ångström (Å) | nanometer (nm) | **1.0** | **E − 01** |
| *are* (a) | square meter (m$^2$) | **1.0** | **E + 02** |
| astronomical unit (AU) | meter (m) | 1.495 979 | E + 11 |
| atmosphere, standard (atm) | pascal (Pa) | **1.013 25** | **E + 05** |
| atmosphere, standard (atm) | kilopascal (kPa) | **1.013 25** | **E + 02** |
| atmosphere, technical (at)[b] | pascal (Pa) | **9.806 65** | **E + 04** |
| atmosphere, technical (at)[b] | kilopascal (kPa) | **9.806 65** | **E + 01** |
| bar (bar) | pascal (Pa) | **1.0** | **E + 05** |
| bar (bar) | kilopascal (kPa) | **1.0** | **E + 02** |
| barn (b) | square meter (m$^2$) | **1.0** | **E − 28** |
| barrel [for petroleum, 42 gallons (U.S.)](bbl) | cubic meter (m$^3$) | 1.589 873 | E − 01 |
| barrel [for petroleum, 42 gallons (U.S.)](bbl) | liter (L) | 1.589 873 | E + 02 |
| biot (Bi) | ampere (A) | **1.0** | **E + 01** |
| British thermal unit$_{IT}$ (Btu$_{IT}$)[c] | joule (J) | 1.055 056 | E + 03 |
| British thermal unit$_{th}$ (Btu$_{th}$)[c] | joule (J) | 1.054 350 | E + 03 |
| British thermal unit (mean) (Btu) | joule (J) | 1.055 87 | E + 03 |
| British thermal unit (39°F) (Btu) | joule (J) | 1.059 67 | E + 03 |
| British thermal unit (59°F) (Btu) | joule (J) | 1.054 80 | E + 03 |
| British thermal unit (60°F) (Btu) | joule (J) | 1.054 68 | E + 03 |
| British thermal unit$_{IT}$ foot per hour square foot degree Fahrenheit [Btu$_{IT}$ · ft/(h · ft$^2$ · °F)] | watt per meter kelvin [W/(m · K)] | 1.730 735 | E + 00 |

| To convert from | To | Multiply by |
|---|---|---|
| British thermal unit$_{th}$ foot per hour square foot degree Fahrenheit [Btu$_{th}$ · ft/(h · ft$^2$ · °F)] | watt per meter kelvin [W/(m · K)] | 1.729 577  E + 00 |
| British thermal unit$_{IT}$ inch per hour square foot degree Fahrenheit [Btu$_{IT}$ · in/(h · ft$^2$ · °F)] | watt per meter kelvin [W/(m · K)] | 1.442 279  E − 01 |
| British thermal unit$_{th}$ inch per hour square foot degree Fahrenheit [Btu$_{th}$ · in/(h · ft$^2$ · °F)] | watt per meter kelvin [W/(m · K)] | 1.441 314  E − 01 |
| British thermal unit$_{IT}$ inch per second square foot degree Fahrenheit [Btu$_{IT}$ · in/(sec · ft$^2$ · °F)] | watt per meter kelvin [W/(m · K)] | 5.192 204  E + 02 |
| British thermal unit$_{th}$ inch per second square foot degree Fahrenheit [Btu$_{th}$ · in/(sec · ft$^2$ · °F)] | watt per meter kelvin [W/(m · K)] | 5.188 732  E + 02 |
| British thermal unit$_{IT}$ per cubic foot (Btu$_{IT}$/ft$^3$) | joule per cubic meter (J/m$^3$) | 3.725 895  E + 04 |
| British thermal unit$_{th}$ per cubic foot (Btu$_{th}$/ft$^3$) | joule per cubic meter (J/m$^3$) | 3.723 403  E + 04 |
| British thermal unit$_{IT}$ per degree Fahrenheit (Btu$_{IT}$/°F) | joule per kelvin (J/k) | 1.899 101  E + 03 |
| British thermal unit$_{th}$ per degree Fahrenheit (Btu$_{th}$/°F) | joule per kelvin (J/k) | 1.897 830  E + 03 |
| British thermal unit$_{IT}$ per degree Rankine (Btu$_{IT}$/°R) | joule per kelvin (J/k) | 1.899 101  E + 03 |
| British thermal unit$_{th}$ per degree Rankine (Btu$_{th}$/°R) | joule per kelvin (J/k) | 1.897 830  E + 03 |
| British thermal unit$_{IT}$ per hour (Btu$_{IT}$/h) | watt (W) | 2.930 711  E − 01 |
| British thermal unit$_{th}$ per hour (Btu$_{th}$/h) | watt (W) | 2.928 751  E − 01 |
| British thermal unit$_{IT}$ per hour square foot degree Fahrenheit [Btu$_{IT}$/(h · ft$^2$ · °F)] | watt per square meter kelvin [W/(m$^2$ · K)] | 5.678 263  E + 00 |

| To convert from | To | Multiply by |
|---|---|---|
| British thermal unit$_{th}$ per hour square foot degree Fahrenheit [Btu$_{th}$/(h · ft$^2$ · °F)] | watt per square meter kelvin [W/(m$^2$ · K)] | 5.674 466 E + 00 |
| British thermal unit$_{th}$ per minute (Btu$_{th}$/min) | watt (W) | 1.757 250 E + 01 |
| British thermal unit$_{IT}$ per pound (Btu$_{IT}$/lb) | joule per kilogram (J/kg) | **2.326** **E + 03** |
| British thermal unit$_{th}$ per pound (Btu$_{th}$/lb) | joule per kilogram (J/kg) | 2.324 444 E + 03 |
| British thermal unit$_{IT}$ per pound degree Fahrenheit [Btu$_{IT}$/(lb · °F)] | joule per kilogram kelvin (J/(kg · K)) | **4.1868** **E + 03** |
| British thermal unit$_{th}$ per pound degree Fahrenheit [Btu$_{th}$/(lb · °F)] | joule per kilogram kelvin [J/(kg · K)] | **4.184** **E + 03** |
| British thermal unit$_{IT}$ per pound degree Rankine [Btu$_{IT}$/(lb · °R)] | joule per kilogram kelvin [J/(kg · K)] | **4.1868** **E + 03** |
| British thermal unit$_{th}$ per pound degree Rankine [Btu$_{th}$/(lb · °R)] | joule per kilogram kelvin [J/(kg · K)] | **4.184** **E + 03** |
| British thermal unit$_{IT}$ per second (Btu$_{IT}$/sec) | watt (W) | 1.055 056 E + 03 |
| British thermal unit$_{th}$ per second (Btu$_{th}$/sec) | watt (W) | 1.054 350 E + 03 |
| British thermal unit$_{IT}$ per second square foot degree Fahrenheit [Btu$_{IT}$/(sec · ft$^2$ · °F)] | watt per square meter kelvin [W/(m$^2$ · K)] | 2.044 175 E + 04 |
| British thermal unit$_{th}$ per second square foot degree Fahrenheit [Btu$_{th}$/(sec · ft$^2$ · °F)] | watt per square meter kelvin [W/(m$^2$ · K)] | 2.042 808 E + 04 |
| British thermal unit$_{IT}$ per square foot (Btu$_{IT}$/ft$^2$) | joule per square meter (J/m$^2$) | 1.135 653 E + 04 |
| British thermal unit$_{th}$ per square foot (Btu$_{th}$/ft$^2$) | joule per square meter (J/m$^2$) | 1.134 893 E + 04 |
| British thermal unit$_{IT}$ per square foot hour [(Btu$_{IT}$/(ft$^2$ · h)] | watt per square meter (W/m$^2$) | 3.154 591 E + 00 |

| To convert from | To | Multiply by |
|---|---|---|
| British thermal unit$_{th}$ per square foot hour [Btu$_{th}$/(ft$^2$ · h)] | watt per square meter (W/m$^2$) | 3.152 481 E + 00 |
| British thermal unit$_{th}$ per square foot minute [Btu$_{th}$/(ft$^2$ · min)] | watt per square meter (W/m$^2$) | 1.891 489 E + 02 |
| British thermal unit$_{IT}$ per square foot second [(Btu$_{IT}$/(ft$^2$ · sec)] | watt per square meter (W/m$^2$) | 1.135 653 E + 04 |
| British thermal unit$_{th}$ per square foot second [Btu$_{th}$/(ft$^2$ · sec)] | watt per square meter (W/m$^2$) | 1.134 893 E + 04 |
| British thermal unit$_{th}$ per square inch second [Btu$_{th}$/(in$^2$ · sec)] | watt per square meter (W/m$^2$) | 1.634 246 E + 06 |
| bushel (U.S.) (bu) | cubic meter (m$^3$) | 3.523 907 E − 02 |
| bushel (U.S.) (bu) | liter (L) | 3.523 907 E + 01 |
| calorie$_{IT}$ (cal$_{IT}$)$^c$ | joule (J) | **4.1868     E + 00** |
| calorie$_{th}$ (cal$_{th}$)$^c$ | joule (J) | **4.184     E + 00** |
| calorie (cal) (mean) | joule (J) | 4.190 02   E + 00 |
| calorie (15°C) (cal$_{15}$) | joule (J) | 4.185 80   E + 00 |
| calorie (20°C) (cal$_{20}$) | joule (J) | 4.181 90   E + 00 |
| calorie$_{IT}$, kilogram (nutrition)$^d$ | joule (J) | **4.1868     E + 03** |
| calorie$_{th}$, kilogram (nutrition)$^d$ | joule (J) | **4.184     E + 03** |
| calorie (mean), kilogram (nutrition)$^d$ | joule (J) | 4.190 02   E + 03 |
| calorie$_{th}$ per centimeter second degree Celsius [cal$_{th}$/(cm · sec · °C)] | watt per meter kelvin [W/(m · K)] | **4.184     E + 02** |
| calorie$_{IT}$ per gram (cal$_{IT}$/g) | joule per kilogram (J/kg) | **4.1868     E + 03** |
| calorie$_{th}$ per gram (cal$_{th}$/g) | joule per kilogram (J/kg) | **4.184     E + 03** |
| calorie$_{IT}$ per gram degree Celsius [cal$_{IT}$/(g · °C)] | joule per kilogram kelvin [J/(kg · K)] | **4.1868     E + 03** |
| calorie$_{th}$ per gram degree Celsius [cal$_{th}$/(g · °C)] | joule per kilogram kelvin [J/(kg · K)] | **4.184     E + 03** |
| calorie$_{IT}$ per gram kelvin [cal$_{IT}$/(g · K)] | joule per kilogram kelvin [J/(kg · K)] | **4.1868     E + 03** |
| calorie$_{th}$ per gram kelvin [cal$_{th}$/(g · K)] | joule per kilogram kelvin [J/(kg · K)] | **4.184     E + 03** |
| calorie$_{th}$ per minute (cal$_{th}$/min) | watt (W) | 6.973 333 E + 02 |
| calorie$_{th}$ per second (cal$_{th}$/sec) | watt (W) | **4.184     E + 00** |

| To convert from | To | Multiply by | |
|---|---|---|---|
| calorie$_{th}$ per square centimeter (cal$_{th}$/cm$^2$) | joule per square meter (J/m$^2$) | **4.184** | **E + 04** |
| calorie$_{th}$ per square centimeter minute [cal$_{th}$/(cm$^2$ · min)] | watt per square meter (W/m$^2$) | 6.973 333 | E + 02 |
| calorie$_{th}$ per square centimeter second [cal$_{th}$/(cm$^2$ · sec)] | watt per square meter (W/m$^2$) | **4.184** | **E + 04** |
| candela per square inch (cd/in$^2$) | candela per square meter (cd/m$^2$) | 1.550 003 | E + 03 |
| carat, metric | kilogram (kg) | **2.0** | **E + 04** |
| carat, metric | gram (g) | **2.0** | **E + 01** |
| centimeter of mercury (0°C)$^e$ | pascal (Pa) | 1.333 22 | E + 03 |
| centimeter of mercury (0°C)$^e$ | kilopascal (kPa) | 1.333 22 | E + 00 |
| centimeter of mercury, conventional (cmHg)$^e$ | pascal (Pa) | 1.333 224 | E + 03 |
| centimeter of mercury, conventional (cmHg)$^e$ | kilopascal (kPa) | 1.333 224 | E + 00 |
| centimeter of water (4°C)$^e$ | pascal (Pa) | 9.806 38 | E + 01 |
| centimeter of water, conventional (cmH$_2$O)$^e$ | pascal (Pa) | **9.806 65** | **E + 01** |
| centipoise (cP) | pascal second (Pa · s) | **1.0** | **E − 03** |
| centistokes (cSt) | meter squared per second (m$^2$/s) | 1.0 | E − 06 |
| chain (based on U.S. survey foot) (ch)$^u$ | meter (m) | 2.011 684 | E + 01 |
| circular mil | square meter (m$^2$) | 5.067 075 | E − 10 |
| circular mil | square millimeter (mm$^2$) | 5.067 075 | E − 04 |
| clo | square meter kelvin per watt (m$^2$ · K/W) | 1.55 | E − 01 |
| cord (128 ft$^3$) | cubic meter (m$^3$) | 3.624 556 | E + 00 |
| cubic foot (ft$^3$) | cubic meter (m$^3$) | 2.831 685 | E − 02 |
| cubic foot per minute (ft$^3$/min) | cubic meter per second (m$^3$/sec) | 4.719 474 | E − 04 |
| cubic foot per minute (ft$^3$/min) | liter per second (L/sec) | 4.719 474 | E − 01 |
| cubic foot per second (ft$^3$/sec) | cubic meter per second (m$^3$/sec) | 2.831 685 | E − 02 |
| cubic inch (in$^3$)$^f$ | cubic meter (m$^3$) | 1.638 706 | E − 05 |
| cubic inch per minute (in$^3$/min) | cubic meter per second (m$^3$/s) | 2.731 177 | E − 07 |
| cubic mile (mi$^3$) | cubic meter (m$^3$) | 4.168 182 | E + 09 |
| cubic yard (yd$^3$) | cubic meter (m$^3$) | 7.645 549 | E − 01 |
| cubic yard per minute (yd$^3$/min) | cubic meter per second (m$^3$/s) | 1.274 258 | E − 02 |

| To convert from | To | Multiply by |
|---|---|---|
| cup (U.S.) | cubic meter (m$^3$) | 2.365 882 E − 04 |
| cup (U.S.) | liter (L) | 2.365 882 E − 01 |
| cup (U.S.) | milliliter (mL) | 2.365 882 E + 02 |
| *curie* (Ci) | becquerel (Bq) | **3.7          E + 10** |
| darcy[g] | meter squared (m$^2$) | 9.869 233 E − 13 |
| *day* (d) | second (s) | **8.64          E + 04** |
| day (sidereal) | second (s) | 8.616 409 E + 04 |
| debye (D) | coulomb meter (C · m) | 3.335 641 E − 30 |
| *degree* (angle) (°) | radian (rad) | 1.745 329 E − 02 |
| *degree Celsius* (temperature) (°C) | kelvin (K) | $T/K = t/°C +$ **273.15** |
| *degree Celsius* (temperature interval) (°C) | kelvin (K) | **1.0          E + 00** |
| degree centigrade (temperature)[h] | degree Celsius (°C) | $t/°C ≈ t/\text{deg. cent.}$ |
| degree centigrade (temperature interval)[h] | degree Celsius (°C) | 1.0          E + 00 |
| degree Fahrenheit (temperature) (°F) | degree Celsius (°C) | $t/°C =$ $(t/°F − \mathbf{32})/\mathbf{1.8}$ |
| degree Fahrenheit (temperature) (°F) | kelvin (K) | $T/K = (t/°F +$ **459.67**$)/\mathbf{1.8}$ |
| degree Fahrenheit (temperature interval)(°F) | degree Celsius (°C) | 5.555 556 E − 01 |
| degree Fahrenheit (temperature interval)(°F) | kelvin (K) | 5.555 556 E − 01 |
| degree Fahrenheit hour per British thermal unit$_{IT}$ (°F · h/Btu$_{IT}$) | kelvin per watt (K/W) | 1.895 634 E + 00 |
| degree Fahrenheit hour per British thermal unit$_{th}$ (°F · h/Btu$_{th}$) | kelvin per watt (K/W) | 1.896 903 E + 00 |
| degree Fahrenheit hour square foot per British thermal unit$_{IT}$ (°F · h · ft$^2$/Btu$_{IT}$) | square meter kelvin per watt (m$^2$ · K/W) | 1.761 102 E − 01 |
| degree Fahrenheit hour square foot per British thermal unit$_{th}$ (°F · h · ft$^2$ /Btu$_{th}$) | square meter kelvin per watt (m$^2$ · K/W) | 1.762 280 E − 01 |
| degree Fahrenheit hour square foot per British thermal unit$_{IT}$ inch [°F · h · ft$^2$/(Btu$_{IT}$ · in)] | meter kelvin per watt (m · K/W) | 6.933 472 E + 00 |

| To convert from | To | Multiply by |
|---|---|---|
| degree Fahrenheit hour square foot per British thermal unit$_{th}$ inch [°F · h · ft$^2$/(Btu$_{th}$ · in)] | meter kelvin per watt (m · K/W) | 6.938 112 E + 00 |
| degree Fahrenheit second per British thermal unit$_{IT}$ (°F · sec/Btu$_{IT}$) | kelvin per watt (K/W) | 5.265 651 E − 04 |
| degree Fahrenheit second per British thermal unit$_{th}$ (°F · sec/ Btu$_{th}$) | kelvin per watt (K/W) | 5.269 175 E − 04 |
| degree Rankine (°R) | kelvin (K) | $T/K = (T/°R)/1.8$ |
| degree Rankine (temperature interval) (°R) | kelvin (K) | 5.555 556 E − 01 |
| denier | kilogram per meter (kg/m) | 1.111 111 E − 07 |
| denier | gram per meter (g/m) | 1.111 111 E − 04 |
| dyne (dyn) | newton (N) | **1.0      E − 05** |
| dyne centimeter (dyn · cm) | newton meter (N · m) | **1.0      E − 07** |
| dyne per square centimeter (dyn/cm$^2$) | pascal (Pa) | **1.0      E − 01** |
| *electronvolt* (eV) | joule (J) | 1.602 177 E − 19 |
| EMU of capacitance (abfarad) | farad (F) | **1.0      E + 09** |
| EMU of current (abampere) | ampere (A) | **1.0      E + 01** |
| EMU of electric potential (abvolt) | volt (V) | **1.0      E − 08** |
| EMU of inductance (abhenry) | henry (H) | **1.0      E − 09** |
| EMU of resistance (abohm) | ohm (Ω) | **1.0      E − 09** |
| erg (erg) | joule (J) | **1.0      E − 07** |
| erg per second (erg/sec) | watt (W) | **1.0      E − 07** |
| erg per square centimeter second |1obrktξ1ru|/(cm$^2$ · sec)] | watt per square meter (W/m$^2$) | **1.0      E − 03** |
| ESU of capacitance (statfarad) | farad (F) | 1.112 650 E − 12 |
| ESU of current (statampere) | ampere (A) | 3.335 641 E − 10 |
| ESU of electric potential (statvolt) | volt (V) | 2.997 925 E + 02 |
| ESU of inductance (stathenry) | henry (H) | 8.987 552 E + 11 |
| ESU of resistance (statohm) | ohm (Ω) | 8.987 552 E + 11 |
| faraday (based on carbon 12) | coulomb (C) | 9.648 531 E + 04 |
| fathom (based on U.S. survey foot)[a] | meter (m) | 1.828 804 E + 00 |
| fermi | meter (m) | **1.0      E − 15** |
| fermi | femtometer (fm) | **1.0      E + 00** |

| To convert from | To | Multiply by |
|---|---|---|
| fluid ounce (U.S.) (fl oz) | cubic meter ($m^3$) | 2.957 353  E − 05 |
| fluid ounce (U.S.) (fl oz) | milliliter (mL) | 2.957 353  E + 01 |
| foot (ft) | meter (m) | **3.048      E − 01** |
| foot (U.S. survey) (ft)[a] | meter (m) | 3.048 006  E − 01 |
| footcandle | lux (lx) | 1.076 391  E + 01 |
| footlambert | candela per square meter ($cd/m^2$) | 3.426 259  E + 00 |
| foot of mercury, conventional (ftHg)[e] | pascal (Pa) | 4.063 666  E + 04 |
| foot of mercury, conventional (ftHg)[e] | kilopascal (kPa) | 4.063 666  E + 01 |
| foot of water (39.2°F)[e] | pascal (Pa) | 2.988 98   E + 03 |
| foot of water (39.2°F)[e] | kilopascal (kPa) | 2.988 98   E + 00 |
| foot of water, conventional ($ftH_2O$)[e] | pascal (Pa) | 2.989 067  E + 03 |
| foot of water, conventional ($ftH_2O$)[e] | kilopascal (kPa) | 2.989 067  E + 00 |
| foot per hour (ft/h) | meter per second (m/sec) | 8.466 667  E − 05 |
| foot per minute (ft/min) | meter per second (m/sec) | **5.08      E − 03** |
| foot per second (ft/sec) | meter per second (m/sec) | **3.048      E − 01** |
| foot per second squared ($ft/sec^2$) | meter per second squared ($m/sec^2$) | **3.048      E − 01** |
| foot poundal | joule (J) | 4.214 011  E − 02 |
| foot pound-force (ft · lbf) | joule (J) | 1.355 818  E + 00 |
| foot pound-force per hour (ft · lbf/h) | watt (W) | 3.766 161  E − 04 |
| foot pound-force per minute (ft · lbf/min) | watt (W) | 2.259 697  E − 02 |
| foot pound-force per second (ft · lbf/sec) | watt (W) | 1.355 818  E + 00 |
| foot to the fourth power ($ft^4$)[i] | meter to the fourth power ($m^4$) | 8.630 975  E − 03 |
| franklin (Fr) | coulomb (C) | 3.335 641  E − 10 |
| gal (Gal) | meter per second squared ($m/sec^2$) | **1.0      E − 02** |
| gallon [Canadian and U.K. (Imperial)] (gal) | cubic meter ($m^3$) | **4.546 09  E − 03** |
| gallon [Canadian and U.K. (Imperial)] (gal) | liter (L) | **4.546 09  E + 00** |
| gallon (U.S.) (gal) | cubic meter ($m^3$) | 3.785 412  E − 03 |

| To convert from | To | Multiply by | |
|---|---|---|---|
| gallon (U.S.) (gal) | liter (L) | 3.785 412 | E + 00 |
| gallon (U.S.) per day (gal/d) | cubic meter per second ($m^3$/sec) | 4.381 264 | E − 08 |
| gallon (U.S.) per day (gal/d) | liter per second (L/sec) | 4.381 264 | E − 05 |
| gallon (U.S.) per horsepower hour [gal/(hp · h)] | cubic meter per joule ($m^3$/J) | 1.410 089 | E − 09 |
| gallon (U.S.) per horsepower hour [gal/(hp · h)] | liter per joule (L/J) | 1.410 089 | E − 06 |
| gallon (U.S.) per minute (gpm) (gal/min) | cubic meter per second ($m^3$/sec) | 6.309 020 | E − 05 |
| gallon (U.S.) per minute (gpm) (gal/min) | liter per second (L/sec) | 6.309 020 | E − 02 |
| gamma ($\gamma$) | tesla (T) | **1.0** | **E − 09** |
| gauss (Gs, G) | tesla (T) | **1.0** | **E − 04** |
| gilbert (Gi) | ampere (A) | 7.957 747 | E − 01 |
| gill [Canadian and U.K. (Imperial)] (gi) | cubic meter ($m^3$) | 1.420 653 | E − 04 |
| gill [Canadian and U.K. (Imperial)] (gi) | liter (L) | 1.420 653 | E − 01 |
| gill (U.S.) (gi) | cubic meter ($m^3$) | 1.182 941 | E − 04 |
| gill (U.S.) (gi) | liter (L) | 1.182 941 | E − 01 |
| gon (also called grade) (gon) | radian (rad) | 1.570 796 | E − 02 |
| gon (also called grade) (gon) | degree (angle) (°) | **9.0** | **E − 01** |
| grain (gr) | kilogram (kg) | **6.479 891** | **E − 05** |
| grain (gr) | milligram (mg) | **6.479 891** | **E + 01** |
| grain per gallon (U.S.) (gr/gal) | kilogram per cubic meter (kg/$m^3$) | 1.711 806 | E − 02 |
| grain per gallon (U.S.) (gr/gal) | milligram per liter (mg/L) | 1.711 806 | E + 01 |
| gram-force per square centimeter (gf/$cm^2$) | pascal (Pa) | **9.806 65** | **E + 01** |
| *gram per cubic centimeter* (g/$cm^3$) | kilogram per cubic meter (kg/$m^3$) | **1.0** | **E + 03** |
| *hectare* (ha) | square meter ($m^2$) | **1.0** | **E + 04** |
| horsepower (550 ft · lbf/s) (hp) | watt (W) | 7.456 999 | E + 02 |
| horsepower (boiler) | watt (W) | 9.809 50 | E + 03 |
| horsepower (electric) | watt (W) | **7.46** | **E + 02** |
| horsepower (metric) | watt (W) | 7.354 988 | E + 02 |
| horsepower (U.K.) | watt (W) | 7.4570 | E + 02 |
| horsepower (water) | watt (W) | 7.460 43 | E + 02 |
| *hour* (h) | second (s) | **3.6** | **E + 03** |
| hour (sidereal) | second (s) | 3.590 170 | E + 03 |
| hundredweight (long, 112 lb) | kilogram (kg) | 5.080 235 | E + 01 |
| hundredweight (short, 100 lb) | kilogram (kg) | 4.535 924 | E + 01 |

| To convert from | To | Multiply by | |
|---|---|---|---|
| inch (in) | meter (m) | **2.54** | **E − 02** |
| inch (in) | centimeter (cm) | **2.54** | **E + 00** |
| inch of mercury (32°F)[e] | pascal (Pa) | 3.386 38 | E + 03 |
| inch of mercury (32°F)[e] | kilopascal (kPa) | 3.386 38 | E + 00 |
| inch of mercury (60°F)[e] | pascal (Pa) | 3.376 85 | E + 03 |
| inch of mercury (60°F)[e] | kilopascal (kPa) | 3.376 85 | E + 00 |
| inch of mercury, conventional (inHg)[e] | pascal (Pa) | 3.386 389 | E + 03 |
| inch of mercury, conventional (inHg)[e] | kilopascal (kPa) | 3.386 389 | E + 00 |
| inch of water (39.2°F)[e] | pascal (Pa) | 2.490 82 | E + 02 |
| inch of water (60°F)[e] | pascal (Pa) | 2.4884 | E + 02 |
| inch of water, conventional (inH$_2$O)[e] | pascal (Pa) | 2.490 889 | E + 02 |
| inch per second (in/sec) | meter per second (m/sec) | **2.54** | **E − 02** |
| inch per second squared (in/sec$^2$) | meter per second squared (m/sec$^2$) | **2.54** | **E − 02** |
| inch to the fourth power (in$^4$)[i] | meter to the fourth power (m$^4$) | 4.162 314 | E − 07 |
| kayser (K) | reciprocal meter (m$^{-1}$) | **1.0** | **E + 02** |
| *kelvin* (K) | degree Celsius (°C) | $t/°C =$ $T/K - 273.15$ | |
| kilocalorie$_{IT}$ (kcal$_{IT}$) | joule (J) | **4.1868** | **E + 03** |
| kilocalorie$_{th}$ (kcal$_{th}$) | joule (J) | **4.184** | **E + 03** |
| kilocalorie (mean) (kcal) | joule (J) | 4.190 02 | E + 03 |
| kilocalorie$_{th}$ per minute (kcal$_{th}$ /min) | watt (W) | 6.973 333 | E + 01 |
| kilocalorie$_{th}$ per second (kcal$_{th}$/sec) | watt (W) | **4.184** | **E + 03** |
| kilogram-force (kgf) | newton (N) | **9.806 65** | **E + 00** |
| kilogram-force meter (kgf · m) | newton meter (N · m) | **9.806 65** | **E + 00** |
| kilogram-force per square centimeter (kgf/cm$^2$) | pascal (Pa) | **9.806 65** | **E + 04** |
| kilogram-force per square centimeter (kgf/cm$^2$) | kilopascal (kPa) | **9.806 65** | **E + 01** |
| kilogram-force per square meter (kgf/m$^2$) | pascal (Pa) | **9.806 65** | **E + 00** |
| kilogram-force per square millimeter (kgf/mm$^2$) | pascal (Pa) | **9.806 65** | **E + 06** |

| To convert from | To | Multiply by |
|---|---|---|
| kilogram-force per square millimeter (kgf/mm²) | megapascal (MPa) | **9.806 65  E + 00** |
| kilogram-force second squared per meter (kgf · s²/m) | kilogram (kg) | **9.806 65  E + 00** |
| *kilometer per hour* (km/h) | meter per second (m/sec) | 2.777 778  E − 01 |
| kilopond (kilogram-force) (kp) | newton (N) | **9.806 65  E + 00** |
| *kilowatt hour* (kW · h) | joule (J) | **3.6  E + 06** |
| *kilowatt hour* (kW · h) | megajoule (MJ) | **3.6  E + 00** |
| kip (1 kip = 1000 lbf) | newton (N) | 4.448 222  E + 03 |
| kip (1 kip = 1000 lbf) | kilonewton (kN) | 4.448 222  E + 00 |
| kip per square inch (ksi) (kip/in²) | pascal (Pa) | 6.894 757  E + 06 |
| kip per square inch (ksi) (kip/in²) | kilopascal (kPa) | 6.894 757  E + 03 |
| *knot* (nautical mile per hour) | meter per second (m/sec) | 5.144 444  E − 01 |
| lambert[j] | candela per square meter (cd/m²) | 3.183 099  E + 03 |
| langley (cal$_{th}$/cm²) | joule per square meter (J/m²) | **4.184  E + 04** |
| light year (l.y.)[k] | meter (m) | 9.460 73  E + 15 |
| *liter* (L)[l] | cubic meter (m³) | **1.0  E − 03** |
| lumen per square foot (lm/ft²) | lux (lx) | 1.076 391  E + 01 |
| maxwell (Mx) | weber (Wb) | **1.0  E − 08** |
| mho | siemens (S) | **1.0  E + 00** |
| microinch | meter (m) | **2.54  E − 08** |
| microinch | micrometer (m) | **2.54  E − 02** |
| micron (μ) | meter (m) | **1.0  E − 06** |
| micron (μ) | micrometer (μm) | **1.0  E + 00** |
| mil (0.001 in) | meter (m) | **2.54  E − 05** |
| mil (0.001 in) | millimeter (mm) | **2.54  E − 02** |
| mil (angle) | radian (rad) | 9.817 477  E − 04 |
| mil (angle) | degree (°) | **5.625  E − 02** |
| mile (mi) | meter (m) | **1.609 344  E + 03** |
| mile (mi) | kilometer (km) | **1.609 344  E + 00** |
| mile (based on U.S. survey foot) (mi)[a] | meter (m) | 1.609 347  E + 03 |
| mile (based on U.S. survey foot) (mi)[a] | kilometer (km) | 1.609 347  E + 00 |
| *mile, nautical*[m] | meter (m) | **1.852  E + 03** |
| mile per gallon (U.S.) (mpg) (mi/gal) | meter per cubic meter (m/m³) | 4.251 437  E + 05 |

| To convert from | To | Multiply by |
|---|---|---|
| mile per gallon (U.S.) (mpg) (mi/gal) | kilometer per liter (km /L) | 4.251 437 E − 01 |
| mile per gallon (U.S.) (mpg) (mi/gal)[a] | liter per 100 kilometer (L/100 km) | divide 235.215 by number of miles per gallon |
| mile per hour (mi/h) | meter per second (m/sec) | **4.4704    E − 01** |
| mile per hour (mi/h) | kilometer per hour (km/h) | **1.609 344 E + 00** |
| mile per minute (mi/min) | meter per second (m/sec) | **2.682 24  E + 01** |
| mile per second (mi/sec) | meter per second (m/sec) | **1.609 344 E + 03** |
| millibar (mbar) | pascal (Pa) | **1.0      E + 02** |
| millibar (mbar) | kilopascal (kPa) | **1.0      E − 01** |
| millimeter of mercury, conventional (mmHg)[e] | pascal (Pa) | 1.333 224 E + 02 |
| millimeter of water, conventional (mmH$_2$O)[e] | pascal (Pa) | **9.806 65  E + 00** |
| *minute* (angle) (') | radian (rad) | 2.908 882 E − 04 |
| *minute* (min) | second (sec) | **6.0      E + 01** |
| minute (sidereal) | second (sec) | 5.983 617 E + 01 |
| oersted (Oe) | ampere per meter (A/m) | 7.957 747 E + 01 |
| *ohm centimeter* (Ω · cm) | ohm meter (Ω · m) | **1.0      E − 02** |
| ohm circular-mil per foot | ohm meter (Ω · m) | 1.662 426 E − 09 |
| ohm circular-mil per foot | ohm square millimeter per meter (Ω · mm$^2$/m) | 1.662 426 E − 03 |
| ounce (avoirdupois) (oz) | kilogram (kg) | 2.834 952 E − 02 |
| ounce (avoirdupois) (oz) | gram (g) | 2.834 952 E + 01 |
| ounce (troy or apothecary) (oz) | kilogram (kg) | 3.110 348 E − 02 |
| ounce (troy or apothecary) (oz) | gram (g) | 3.110 348 E + 01 |
| ounce [Canadian and U.K. fluid (Imperial)] (fl oz) | cubic meter (m$^3$) | 2.841 306 E − 05 |
| ounce [Canadian and U.K. fluid (Imperial)] (fl oz) | milliliter (mL) | 2.841 306 E + 01 |
| ounce (U.S. fluid) (fl oz) | cubic meter (m$^3$) | 2.957 353 E − 05 |
| ounce (U.S. fluid) (fl oz) | millimeter (mL) | 2.957 353 E + 01 |
| ounce (avoirdupois)-force (ozf) | newton (N) | 2.780 139 E − 01 |
| ounce (avoirdupois)-force inch (ozf · in) | newton meter (N · m) | 7.061 552 E − 03 |
| ounce (avoirdupois)-force inch (ozf · in) | millinewton meter (mN · m) | 7.061 552 E + 00 |

| To convert from | To | Multiply by | |
|---|---|---|---|
| ounce (avoirdupois) per cubic inch (oz/in$^3$) | kilogram per cubic meter (kg/m$^3$) | 1.729 994 | E + 03 |
| ounce (avoirdupois) per gallon [Canadian and U.K. (Imperial)] (oz/gal) | kilogram per cubic meter (kg/m$^3$) | 6.236 023 | E + 00 |
| ounce (avoirdupois) per gallon [Canadian and U.K. (Imperial)] (oz/gal) | gram per liter (g/L) | 6.236 023 | E + 00 |
| ounce (avoirdupois) per gallon (U.S.)(oz/gal) | kilogram per cubic meter (kg/m$^3$) | 7.489 152 | E + 00 |
| ounce (avoirdupois) per gallon (U.S.)(oz/gal) | gram per liter (g/L) | 7.489 152 | E + 00 |
| ounce (avoirdupois) per square foot (oz/ft$^2$) | kilogram per square meter (kg/m$^2$) | 3.051 517 | E − 01 |
| ounce (avoirdupois) per square inch (oz/in$^2$) | kilogram per square meter (kg/m$^2$) | 4.394 185 | E + 01 |
| ounce (avoirdupois) per square yard (oz/yd$^2$) | kilogram per square meter (kg/m$^2$) | 3.390 575 | E − 02 |
| parsec (pc) | meter (m) | 3.085 678 | E + 16 |
| peck (U.S.) (pk) | cubic meter (m$^3$) | 8.809 768 | E − 03 |
| peck (U.S.) (pk) | liter (L) | 8.809 768 | E + 00 |
| pennyweight (dwt) | kilogram (kg) | 1.555 174 | E − 03 |
| pennyweight (dwt) | gram (g) | 1.555 174 | E + 00 |
| perm (0°C) | kilogram per pascal second square meter [kg/(Pa · sec · m$^2$)] | 5.721 35 | E − 11 |
| perm (23°C) | kilogram per pascal second square meter [kg/(Pa · sec · m$^2$)] | 5.745 25 | E − 11 |
| perm inch (0°C) | kilogram per pascal second meter [kg/(Pa · sec · m)] | 1.453 22 | E − 12 |
| perm inch (23°C) | kilogram per pascal second meter [kg/(Pa · sec · m)] | 1.459 29 | E − 12 |
| phot (ph) | lux (lx) | **1.0** | **E + 04** |
| pica (computer) (1/6 in) | meter (m) | 4.233 333 | E − 03 |
| pica (computer) (1/6 in) | millimeter (mm) | 4.233 333 | E + 00 |
| pica (printer's) | meter (m) | 4.217 518 | E − 03 |
| pica (printer's) | millimeter (mm) | 4.217 518 | E + 00 |
| pint (U.S. dry) (dry pt) | cubic meter (m$^3$) | 5.506 105 | E − 04 |
| pint (U.S. dry) (dry pt) | liter (L) | 5.506 105 | E − 01 |
| pint (U.S. liquid) (liq pt) | cubic meter (m$^3$) | 4.731 765 | E − 04 |
| pint (U.S. liquid) (liq pt) | liter (L) | 4.731 765 | E − 01 |

| To convert from | To | Multiply by |
|---|---|---|
| point (computer) (1/72 in) | meter (m) | 3.527 778 E − 04 |
| point (computer) (1/72 in) | millimeter (mm) | 3.527 778 E − 01 |
| point (printer's) | meter (m) | 3.514 598 E − 04 |
| point (printer's) | millimeter (mm) | 3.514 598 E − 01 |
| poise (P) | pascal second (Pa · s) | **1.0        E − 01** |
| pound (avoirdupois) (lb)° | kilogram (kg) | 4.535 924 E − 01 |
| pound (troy or apothecary) (lb) | kilogram (kg) | 3.732 417 E − 01 |
| poundal | newton (N) | 1.382 550 E − 01 |
| poundal per square foot | pascal (Pa) | 1.488 164 E + 00 |
| poundal second per square foot | pascal second (Pa · s) | 1.488 164 E + 00 |
| pound foot squared (lb · ft²) | kilogram meter squared (kg · m²) | 4.214 011 E − 02 |
| pound-force (lbf)ᵖ | newton (N) | 4.448 222 E + 00 |
| pound-force foot (lbf · ft) | newton meter (N · m) | 1.355 818 E + 00 |
| pound-force foot per inch (lbf · ft/in) | newton meter per meter (N · m/m) | 5.337 866 E + 01 |
| pound-force inch (lbf · in) | newton meter (N · m) | 1.129 848 E − 01 |
| pound-force inch per inch (lbf · in/in) | newton meter per meter (N · m/m) | 4.448 222 E + 00 |
| pound-force per foot (lbf/ft) | newton per meter (N/m) | 1.459 390 E + 01 |
| pound-force per inch (lbf/in) | newton per meter (N/m) | 1.751 268 E + 02 |
| pound-force per pound (lbf/lb) (thrust to mass ratio) | newton per kilogram (N/kg) | **9.806 65   E + 00** |
| pound-force per square foot (lbf/ft²) | pascal (Pa) | 4.788 026 E + 01 |
| pound-force per square inch (psi) (lbf/in²) | pascal (Pa) | 6.894 757 E + 03 |
| pound-force per square inch (psi) (lbf/in²) | kilopascal (kPa) | 6.894 757 E + 00 |
| pound-force second per square foot (lbf · sec/ft²) | pascal second (Pa · s) | 4.788 026 E + 01 |
| pound-force second per square inch (lbf · sec/in²) | pascal second (Pa · s) | 6.894 757 E + 03 |
| pound inch squared (lb · in²) | kilogram meter squared (kg · m²) | 2.926 397 E − 04 |
| pound per cubic foot (lb/ft³) | kilogram per cubic meter (kg/m³) | 1.601 846 E + 01 |
| pound per cubic inch (lb/in³) | kilogram per cubic meter (kg/m³) | 2.767 990 E + 04 |
| pound per cubic yard (lb/yd³) | kilogram per cubic meter (kg/m³) | 5.932 764 E − 01 |

| To convert from | To | Multiply by |
|---|---|---|
| pound per foot (lb/ft) | kilogram per meter (kg/m) | 1.488 164  E + 00 |
| pound per foot hour [lb/(ft · h)] | pascal second (Pa · sec) | 4.133 789  E − 04 |
| pound per foot second [lb/(ft · sec)] | pascal second (Pa · sec) | 1.488 164  E + 00 |
| pound per gallon [Canadian and U.K. (Imperial)] (lb/gal) | kilogram per cubic meter (kg/m$^3$) | 9.977 637  E + 01 |
| pound per gallon [Canadian and U.K. (Imperial)] (lb/gal) | kilogram per liter (kg/L) | 9.977 637  E − 02 |
| pound per gallon (U.S.) (lb/gal) | kilogram per cubic meter (kg/m$^3$) | 1.198 264  E + 02 |
| pound per gallon (U.S.) (lb/gal) | kilogram per liter (kg/L) | 1.198 264  E − 01 |
| pound per horsepower hour [lb/(hp · h)] | kilogram per joule (kg/J) | 1.689 659  E − 07 |
| pound per hour (lb/h) | kilogram per second (kg/sec) | 1.259 979  E − 04 |
| pound per inch (lb/in) | kilogram per meter (kg/m) | 1.785 797  E + 01 |
| pound per minute (lb/min) | kilogram per second (kg/sec) | 7.559 873  E − 03 |
| pound per second (lb/sec) | kilogram per second (kg/sec) | 4.535 924  E − 01 |
| pound per square foot (lb/ft$^2$) | kilogram per square meter (kg/m$^2$) | 4.882 428  E + 00 |
| pound per square inch (*not* pound-force) (lb/in$^2$) | kilogram per square meter (kg/m$^2$) | 7.030 696  E + 02 |
| pound per yard (lb/yd) | kilogram per meter (kg/m) | 4.960 546  E − 01 |
| psi (pound-force per square inch) (lbf/in$^2$) | pascal (Pa) | 6.894 757  E + 03 |
| psi (pound-force per square inch) (lbf/in$^2$) | kilopascal (kPa) | 6.894 757  E + 00 |
| quad ($10^{15}$ Btu$_{IT}$)$^c$ | joule (J) | 1.055 056  E + 18 |
| quart (U.S. dry) (dry qt) | cubic meter (m$^3$) | 1.101 221  E − 03 |
| quart (U.S. dry) (dry qt) | liter (L) | 1.101 221  E + 00 |
| quart (U.S. liquid) (liq qt) | cubic meter (m$^3$) | 9.463 529  E − 04 |
| quart (U.S. liquid) (liq qt) | liter (L) | 9.463 529  E − 01 |
| *rad* (absorbed dose) (rad) | gray (Gy) | **1.0**     **E − 02** |
| *rem* (rem) | sievert (Sv) | **1.0**     **E − 02** |
| revolution (r) | radian (rad) | 6.283 185  E + 00 |
| revolution per minute (rpm) (r/min) | radian per second (rad/sec) | 1.047 198  E − 01 |

| To convert from | To | Multiply by | |
|---|---|---|---|
| rhe | reciprocal pascal second [(Pa · sec)⁻¹] | **1.0** | **E + 01** |
| rod (based on U.S. survey foot) (rd)[a] | meter (m) | 5.029 210 | E + 00 |
| *roentgen* (R) | coulomb per kilogram (C/kg) | **2.58** | **E − 04** |
| rpm (revolution per minute) (r/min) | radian per second (rad/sec) | 1.047 198 | E − 01 |
| *second* (angle) (″) | radian (rad) | 4.848 137 | E − 06 |
| second (sidereal) | second (sec) | 9.972 696 | E − 01 |
| shake | second (sec) | **1.0** | **E − 08** |
| shake | nanosecond (nsec) | **1.0** | **E + 01** |
| slug (slug) | kilogram (kg) | 1.459 390 | E + 01 |
| slug per cubic foot (slug/ft³) | kilogram per cubic meter (kg/m³) | 5.153 788 | E + 02 |
| slug per foot second [slug/(ft · sec)] | pascal second (Pa · sec) | 4.788 026 | E + 01 |
| square foot (ft²) | square meter (m²) | **9.290 304** | **E − 02** |
| square foot per hour (ft²/h) | square meter per second (m²/sec) | **2.580 64** | **E − 05** |
| square foot per second (ft²/sec) | square meter per second (m²/sec) | **9.290 304** | **E − 02** |
| square inch (in²) | square meter (m²) | **6.4516** | **E − 04** |
| square inch (in²) | square centimeter (cm²) | **6.4516** | **E + 00** |
| square mile (mi²) | square meter (m²) | 2.589 988 | E + 06 |
| square mile (mi²) | square kilometer (km²) | 2.589 988 | E + 00 |
| square mile (based on U.S. survey foot) (mi²)[a] | square meter (m²) | 2.589 998 | E + 06 |
| square mile (based on U.S. survey foot) (mi²)[a] | square kilometer (km²) | 2.589 998 | E + 00 |
| square yard (yd²) | square meter (m²) | 8.361 274 | E − 01 |
| statampere | ampere (A) | 3.335 641 | E − 10 |
| statcoulomb | coulomb (C) | 3.335 641 | E − 10 |
| statfarad | farad (F) | 1.112 650 | E − 12 |
| stathenry | henry (H) | 8.987 552 | E + 11 |
| statmho | siemens (S) | 1.112 650 | E − 12 |
| statohm | ohm (Ω) | 8.987 552 | E + 11 |
| statvolt | volt (V) | 2.997 925 | E + 02 |
| stere (st) | cubic meter (m³) | **1.0** | **E + 00** |
| stilb (sb) | candela per square meter (cd/m²) | **1.0** | **E + 04** |
| stokes (St) | meter squared per second (m²/sec) | **1.0** | **E − 04** |

| To convert from | To | Multiply by |
|---|---|---|
| tablespoon | cubic meter ($m^3$) | 1.478 676 E − 05 |
| tablespoon | milliliter (mL) | 1.478 676 E + 01 |
| teaspoon | cubic meter ($m^3$) | 4.928 922 E − 06 |
| teaspoon | milliliter (mL) | 4.928 922 E + 00 |
| tex | kilogram per meter (kg/m) | **1.0** **E − 06** |
| therm (EC)[q] | joule (J) | **1.055 06** **E + 08** |
| therm (U.S.)[q] | joule (J) | **1.054 804** **E + 08** |
| ton, assay (AT) | kilogram (kg) | 2.916 667 E − 02 |
| ton, assay (AT) | gram (g) | 2.916 667 E + 01 |
| ton-force (2000 lbf) | newton (N) | 8.896 443 E + 03 |
| ton-force (2000 lbf) | kilonewton (kN) | 8.896 443 E + 00 |
| ton, long (2240 lb) | kilogram (kg) | 1.016 047 E + 03 |
| ton, long, per cubic yard | kilogram per cubic meter (kg/$m^3$) | 1.328 939 E + 03 |
| *ton, metric* (t) | kilogram (kg) | **1.0** **E + 03** |
| tonne (called "metric ton" in U.S.) (t) | kilogram (kg) | **1.0** **E + 03** |
| ton of refrigeration (12,000 $Btu_{IT}$/h) | watt (W) | 3.516 853 E + 03 |
| ton of TNT (energy equivalent)[r] | joule (J) | **4.184** **E + 09** |
| ton, register | cubic meter ($m^3$) | 2.831 685 E + 00 |
| ton, short (2000 lb) | kilogram (kg) | 9.071 847 E + 02 |
| ton, short, per cubic yard | kilogram per cubic meter (kg/$m^3$) | 1.186 553 E + 03 |
| ton, short, per hour | kilogram per second (kg/s) | 2.519 958 E − 01 |
| torr (Torr) | pascal (Pa) | 1.333 224 E + 02 |
| unit pole | weber (Wb) | 1.256 637 E − 07 |
| *watt hour* (W · h) | joule (J) | **3.6** **E + 03** |
| *watt per square centimeter* (W/$cm^2$) | watt per square meter (W/$m^2$) | **1.0** **E + 04** |
| watt per square inch (W/$in^2$) | watt per square meter (W/$m^2$) | 1.550 003 E + 03 |
| *watt second* (W · sec) | joule (J) | **1.0** **E + 00** |
| yard (yd) | meter (m) | **9.144** **E − 01** |
| *year* (365 days) | second (s) | **3.1536** **E + 07** |
| year (sidereal) | second (s) | 3.155 815 E + 07 |
| year (tropical) | second (s) | 3.155 693 E + 07 |

Factors in **boldface** are exact.

[a] The U.S. survey foot equals (1200/3937) m. 1 international foot = 0.999998 survey foot.

[b] One technical atmosphere equals one kilogram-force per square centimeter (1 at = 1kgf/cm$^2$).

[c] The Fifth International Conference on the Properties of Steam (London, July 1956) defined the International Table calorie as 4.1868 J. Therefore the exact conversion factor for the International Table Btu is 1.055 055 852 62 kJ. Note that the notation for International Table used in this listing is subscript "IT." Similarily, the notation for thermochemical is subscript "th." Further, the thermochemical Btu, Btu$_{th}$, is based on the thermochemical calorie, cal$_{th}$, where cal$_{th}$ = 4.184 J exactly.

[d] The kilogram calorie or "large calorie" is an obsolete term used for the kilocalorie, which is the calorie used to express the energy content of foods. However, in practice, the prefix "kilo" is usually omitted.

[e] Conversion factors for mercury manometer pressure units are calculated using the standard value for the acceleration of gravity and the density of mercury at the stated temperature. Additional digits are not justified because the definitions of the units do not take into account the compressibility of mercury or the change in density caused by the revised practical temperature scale, ITS-90. Similar comments also apply to water manometer pressure units. Conversion factors for conventional mercury and water manometer pressure units are based on ISO 31–3.

[f] The exact conversion factor is 1.638 706 4 E − 05.

[g] The darcy is a unit for expressing the permeability of porous solids, not area.

[h] The centigrade temperature scale is obsolete; the degree centigrade is only approximately equal to the degree Celsius.

[i] This is a unit for the quantity second moment of area, which is sometimes called the "moment of section" or "area moment of inertia" of a plane section about a specified axis.

[j] The exact conversion factor is 10$^4$/$\pi$.

[k] This conversion factor is based on 1 d = 86 400 sec; and 1 Julian century = 36 525 d. (See *The Astronomical Almanac for the Year 1995*, page K6, U.S. Government Printing Office, Washington, DC, 1994).

[l] In 1964 the General Conference on Weights and Measures reestablished the name "liter" as a special name for the cubic decimeter. Between 1901 and 1964 the liter was slightly larger (1.000 028 dm$^3$); when one uses high-accuracy volume data of that time, this fact must be kept in mind.

[m] The value of this unit, 1 nautical mile = 1852 m, was adopted by the First International Extraordinary Hydrographic Conference, Monaco, 1929, under the name "International nautical mile."

[n] For converting fuel economy, as used in the United States, to fuel consumption.

[o] The exact conversion factor is 4.535 923 7 E − 01. All units that contain the pound refer to the avoirdupois pound.

[p] If the local value of the acceleration of free fall is taken as $g_n$ = 9.806 65 m/sec$^2$ (the standard value), the exact conversion factor is 4.448 221 615 260 5 E + 00.

[q] The therm (EC) is legally defined in the Council Directive of 20 December 1979, Council of the European Communities (now the European Union, EU). The therm (U.S.) is legally defined in the Federal Register of July 27, 1968. Although the therm (EC), which is based on the International Table Btu, is frequently used by engineers in the United States, the therm (U.S.) is the legal unit used by the U.S. natural gas industry.

[r] Defined (not measured) value.

# UNITS

Common engineering unit scales

| Prefix name | Symbol | Prefix value | Multiplication factor | Scientific notation |
|---|---|---|---|---|
| yotta | Y | one million million million times | 1 000 000 000 000 000 000 000 000 | $10^{24}$ |
| zetta | Z | one thousand million million million times | 1 000 000 000 000 000 000 000 | $10^{21}$ |
| exa | E | one million million million times | 1 000 000 000 000 000 000 | $10^{18}$ |
| peta | P | one thousand million million times | 1 000 000 000 000 000 | $10^{15}$ |
| tera | T | one million million times | 1 000 000 000 000 | $10^{12}$ |
| giga | G | one thousand million times | 1 000 000 000 | $10^{9}$ |
| mega | M | one million times | 1 000 000 | $10^{6}$ |
| kilo | k | one thousand times | 1 000 | $10^{3}$ |
| hecto[a] | h | one hundred times | 100 | $10^{2}$ |
| deka[a] | da | ten times | 10 | $10^{1}$ |
| unit name | | one time | 1 | |
| deci[a] | d | one tenth of | 0.1 | $10^{-1}$ |
| centi[a] | c | one hundredth of | 0,01 | $10^{-2}$ |
| milli | m | one thousandth of | 0.001 | $10^{-3}$ |
| micro | $\mu$ | one millionth of | 0.000 001 | $10^{-6}$ |
| nano | n | one thousandth millionth of | 0.000 000 001 | $10^{-9}$ |
| pico | p | one millionth millionth of | 0.000 000 000 001 | $10^{-12}$ |
| femto | f | one thousandth millionth millionth of | 0.000 000 000 000 001 | $10^{-15}$ |
| atto | a | one millionth millionth millionth of | 0.000 000 000 000 000 001 | $10^{-18}$ |
| zepto | z | one thousandth millionth millionth millionth of | 0.000 000 000 000 000 000 001 | $10^{-21}$ |
| yocto | y | one millionth millionth millionth millionth of | 0.000 000 000 000 000 000 000 001 | $10^{-24}$ |

[a] Avoid using these multiples and submultiples whenever possible. Prefixes representing steps of 1000 are recommended.

Order of magnitude conversion from English to SI

| in. | mm | microns | microns | in. |
|---|---|---|---|---|
| 1 | 25.4 | 25400 | 1,000,000 | 39.37 |
| 0.1 | 2.54000000 | 2540 | 100,000 | 3.937000000 |
| 0.01 | 0.25400000 | 254 | 10,000 | 0.393700000 |
| 0.001 | 0.02540000 | 25.4 | 1,000 | 0.039370000 |
| 0.0001 | 0.00254000 | 2.54 | 100 | 0.003937000 |
| 0.00001 | 0.00025400 | 0.254 | 10 | 0.000393700 |
| 0.000001 | 0.00002540 | 0.0254 | 1.0 | 0.000039370 |
| 0.0000001 | 0.00000254 | 0.00254 | 0.1 | 0.000003937 |

# HARDNESS CONVERSION

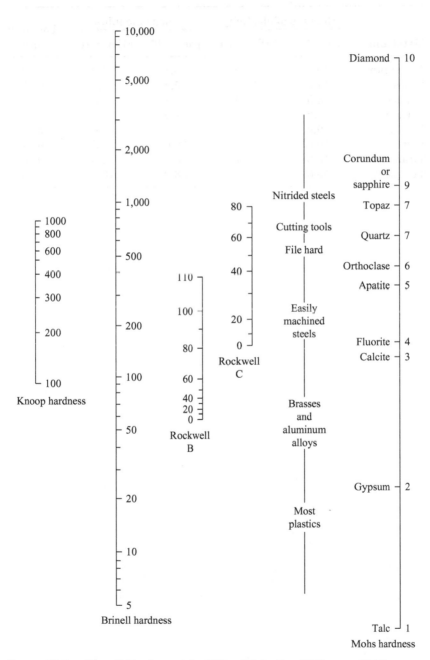

*Source*: Madou, Marc J. *Fundamentals of Microfabrication* (CRC press, 2002).

## METAL ALLOY PROPERTIES

| Metal alloy | Modulus of elasticity | | Shear modulus | | Poisson's ratio |
| --- | --- | --- | --- | --- | --- |
| | psi $\times 10^6$ | MPa $\times 10^4$ | psi $\times 10^6$ | MPa $\times 10^4$ | |
| Aluminum | 10.0 | 6.9 | 3.8 | 2.6 | 0.33 |
| Brass | 14.6 | 10.1 | 5.4 | 3.7 | 0.35 |
| Copper | 16.0 | 11.0 | 6.7 | 4.6 | 0.35 |
| Magnesium | 6.5 | 4.5 | 2.5 | 1.7 | 0.29 |
| Nickel | 30.0 | 20.7 | 11.0 | 7.6 | 0.31 |
| Steel | 30.0 | 20.7 | 12.0 | 8.3 | 0.27 |
| Titanium | 15.5 | 10.7 | 6.5 | 4.5 | 0.36 |
| Tungsten | 59.0 | 40.7 | 23.2 | 16.0 | 0.28 |

# MATERIAL PROPERTIES

## Physical properties of metals

| Material | Lattice | Density P, g cm³ | Specific heat $C_p$, cal sec g°C | Thermal conductivity K, cal sec cm°C | Thermal diffusivity $\alpha$ cm² sec | Melting temperature, °F | Modulus of elasticity E, psi × 10⁶ | Yield strength $S_y$, psi | Tensile strength $S_u$, psi | Elongation % | Reduction of area % | Heat of fusion, $\Delta H_f$ cal g |
|---|---|---|---|---|---|---|---|---|---|---|---|---|
| Steel; Ferrite | BCC | 7.86 | 0.11 | 0.15 | 0.173 | 2760 | 30 | 30,000 | 40,000 | 45 | 75 | 58 |
| Steel: 1020 | — | 7.86 | 0.11 | 0.15 | 0.173 | 2760 | 30 | 40,000 | 60,000 | 40 | 60 | 58 |
| Steel: 4340 | — | 7.84 | 0.120 | 0.08 | 0.085 | 2750 | 30 | 135,000 | 153,000 | 10 | 35 | 58 |
| Al: 99.996 | FCC | 2.70 | 0.22 | 0.53 | 0.89 | 1218 | 10 | 1,800 | 7,000 | 50 | — | 77 |
| Al: 2S-0 | FCC | 2.71 | 0.23 | 0.53 | 0.88 | 1215 | 10 | 5,000 | 13,000 | 45 | — | 77 |
| Zirconium | HCP | 6.5 | 0.066 | 0.07 | 0.16 | 3220 | 12 | 16,000 | 36,000 | 30 | — | f60 |
| Tantalum | BCC | 16.6 | 0.036 | 0.13 | 0.22 | 5425 | 27 | 23,000 | 50,000 | 40 | — | 38 |
| Columbium | BCC | 8.57 | 0.065 | 0.13 | 0.23 | 4330 | 15 | 30,000 | 40,000 | 30 | 80 | 69 |
| Beryllium | HCP | 1.82 | 0.516 | 0.385 | 0.41 | 2340 | 40 | 26,500 | 33,000 | 1 | — | 260 |
| Hafnium | HCP | 13.09 | 0.0351 | 0.0533 | 0.116 | 3100 | 20 | 33,600 | 64,800 | 23 | 37 | — |
| Molybdenum | BCC | 10.2 | 0.061 | 0.35 | 0.56 | 4760 | 50 | 60,000 | 150,000 | 10 | 50 | 70 |
| Copper | FCC | 8.96 | 0.092 | 0.94 | 1.14 | 1981 | 16 | 8,000 | 32,000 | 55 | 78 | 42 |
| 70-30 Brass | FCC | 8.5 | 0.09 | 0.25 | 0.33 | 1740 | 16 | 16,000 | 48,000 | 65 | 75 | — |
| 304 S.S. | FCC | 7.9 | 0.12 | 0.039 | 0.054 | 2600 | 25 | 33,000 | 86,000 | 60 | 75 | — |
| Titanium | HCP | 4.54 | 0.126 | 0.0685 | 0.120 | 3500 | 16.8 | 63,000 | 79,000 | 25.2 | 50 | 97 |
| Nickel | FCC | 8.9 | 0.105 | 0.22 | 0.236 | 2651 | 30 | 8,500 | 46,000 | 30 | 70 | 74 |
| Ni base alloy[a] | FCC | 8.25 | 0.108 | 0.575 | 0.645 | 2500 | 31.7 | 154,000 | 206,000 | 14 | 40 | — |
| Co base alloy[b] | FOC | 9.13 | 0.092 | 0.27 | 0.322 | 2500 | 32.9 | 70,000 | 150,000 | 65 | 40 | — |

Conversion factors: 1 cal = $3.968 \times 10^{-3}$ Btu; 1 cal/g = 1.8 Btu/lb; 1 cal/g °C = 1 Btu/lb °F.

[a] 19 Cr, 11 Co, 10 Mo, 3 Ti, Bal Ni.

[b] 10 Ni, 20 Cr, 15 W, Bal Co.

Source: Datsko, Joseph. Materials Selection for Design and Manufacturing (Marcel Dekker Inc., 1997).

Tensile properties of some metals[a]

| Material | Condition | Yield (ksi) | Tensile (ksi) | $\sigma_0$ (ksi) | m | $\varepsilon_f$ |
|---|---|---|---|---|---|---|
| | | **Strength** | | | | |

*Carbon and Alloy Steels[a]*

| Material | Condition | Yield (ksi) | Tensile (ksi) | $\sigma_0$ (ksi) | m | $\varepsilon_f$ |
|---|---|---|---|---|---|---|
| 1002 | 1500°F-1 h A 0.032 in. | 22.0 | 39.5 | 76.0 | 0.29 | 1.25 |
| 1002[b] | 1800°F-1 h A | 19.0 | 42.0 | 78.0 | 0.27 | 1.25 |
| 1008 DQ | as rec'd 0.024 in. | 25.0 | 39.0 | 70.0 | 0.24 | 1.20 |
| 1008 DQ | as above-trans | 27.0 | 43.0 | 70.0 | 0.24 | 1.10 |
| 1008 DQ | 1600°F-1 h A | 26.5 | 40.0 | — | — | — |
| 1010 | 0.024 in. CRS strip | 33.2 | 47.5 | 84.0 | 0.23 | 1.20 |
| 1010 | as above-trans | 36.8 | 48.5 | 88.0 | 0.26 | 1.00 |
| 1010 | 1600°F-1 h A | 28.6 | 44.2 | 82.0 | 0.23 | 1.20 |
| 1010 | as above-trans | 29.1 | 43.8 | 82.0 | 0.23 | 1.20 |
| 1018 | A | 32.0 | 49.5 | 90.0 | 0.25 | 1.05 |
| 1020 | HR | 42.0 | 66.2 | 115.0 | 0.22 | 0.90 |
| 1045 | HR | 60.0 | 92.5 | 140.0 | 0.14 | 0.58 |
| 1144 | A | 52.0 | 93.7 | 144.0 | 0.14 | 0.49 |
| 1144[c] | A | 50.0 | 93.7 | 144.0 | 0.14 | 0.05 |
| 1212 | HR | 28.0 | 61.5 | 110.0 | 0.24 | 0.85 |
| 4340 | HR | 132.0 | 151.0 | 210.0 | 0.09 | 0.45 |
| 52100 | spher A | 80.0 | 101.0 | 165.0 | 0.18 | 0.58 |
| 52100 | 1500°F A | 131.0 | 167.0 | 210.0 | 0.07 | 0.40 |

*Stainless Steels*

| Material | Condition | Yield (ksi) | Tensile (ksi) | $\sigma_0$ (ksi) | m | $\varepsilon_f$ |
|---|---|---|---|---|---|---|
| 18-8 | 1600°F-1 h A | 37.0 | 89.5 | 210.0 | 0.51 | 1.08 |
| 18-8 | 1800°F-1 h A | 37.5 | 96.5 | 230.0 | 0.53 | 1.38 |
| 302 | 1800°F-1 h A | 34.0 | 92.4 | 210.0 | 0.48 | 1.20 |
| 303 | A | 35.0 | 87.3 | 205.0 | 0.51 | 1.16 |
| 304 | A | 40.0 | 82.4 | 185.0 | 0.45 | 1.67 |
| 202 | 1900°F-1 h A | 55.0 | 105.0 | 195.0 | 0.30 | 1.00 |
| 17-4 PH | 1100°F aged | 240.0 | 246.0 | 260.0 | 0.01 | 0.65 |
| 17-4 PH | A | 135.0 | 142.0 | 173.0 | 0.05 | 1.20 |
| 17-7 PH | 1050°F aged | 155.0 | 185.0 | 225.0 | 0.05 | 0.90 |
| 17-7 PH | 900°F aged | 245.0 | 255.0 | 300.0 | 0.04 | 0.50 |
| 440 C | Solution HT | 63.5 | 107.0 | 153.0 | 0.11 | 0.36 |
| 440 C | A 1600°F–50°F/h | 67.6 | 117.0 | 180.0 | 0.14 | 0.12 |

*Aluminum Alloys*

| Material | Condition | Yield (ksi) | Tensile (ksi) | $\sigma_0$ (ksi) | m | $\varepsilon_f$ |
|---|---|---|---|---|---|---|
| 1100 | 900°F-1 h A | 4.5 | 12.1 | 22.0 | 0.25 | 2.30 |
| 3003 | 800°F-1 h A | 6.0 | 15.0 | 29.0 | 0.30 | 1.50 |
| 2024[c] | T-351 | 52.0 | 68.8 | 115.0 | 0.20 | 0.37 |
| 2024 | T-4 | 43.0 | 64.8 | 100.0 | 0.15 | 0.18 |
| 7075 | 800°F A | 24.3 | 33.9 | 61.0 | 0.22 | 0.53 |
| 7075 | T-6 | 78.6 | 86.0 | 128.0 | 0.13 | 0.18 |

Table (*Continued*)

| Material | Condition | Yield (ksi) | Tensile (ksi) | $\sigma_0$ (ksi) | m | $\varepsilon_f$ |
|---|---|---|---|---|---|---|
| 2011 | 800°F-1 h A | 7.0 | 25.2 | 41.0 | 0.18 | 0.35 |
| 2011 | T-6 | 24.5 | 47.0 | 90.0 | 0.28 | 0.10 |
| *Magnesium Alloys* | | | | | | |
| HK31XA | 800°F-1 h A | 19.0 | 25.5 | 49.5 | 0.22 | 0.33 |
| HK31XA | H24 | 31.0 | 36.2 | 48.0 | 0.08 | 0.20 |
| *Copper Alloys* | | | | | | |
| ETP Copper | 1000°F-1 h A | 4.7 | 31.0 | 78.0 | 0.55 | 1.19 |
| ETP Copper | 1250°F-1 h A | 4.6 | 30.6 | 72.0 | 0.50 | 1.21 |
| ETP Copper | 1500°F-1 h A | 4.2 | 30.0 | 68.0 | 0.48 | 1.26 |
| OFHC Copper | 1250°F-1 h A | 5.3 | 33.1 | 67.0 | 0.35 | 1.00 |
| 90–10 Brass | as rec'd[c] | 12.8 | 38.0 | 85.0 | 0.43 | — |
| 90–10 Brass | 1200°F-1 h A | 8.4 | 36.4 | 83.0 | 0.46 | — |
| 90–10 Brass | as above + 10% CW + 1200°F A | 6.9 | 35.0 | 87.0 | 0.51 | 1.83 |
| 80–20 Brass | 1200°F-1 h A | 7.2 | 35.8 | 84.0 | 0.48 | — |
| 80–20 Brass | as above + 10% CW + 1200°F A | 6.4 | 34.6 | 85.0 | 0.52 | 1.83 |
| 70–30 Brass | 1200°F-1 h A | 12.1 | 44.8 | 112.0 | 0.59 | — |
| 70–30 Brass | as above + 10% CW + 1200°F A | 10.7 | 43.4 | 107.0 | 0.59 | 1.62 |
| 70–30 Brass[b] | 1000°F-1 h A | 11.5 | 45.4 | 110.0 | 0.56 | 1.50 |
| 70–30 Brass[b] | 1200°F-1 h A | 10.5 | 44.0 | 105.0 | 0.52 | 1.55 |
| 70–30 Brass[b] | 1400°F-1 h A | 8.8 | 42.3 | 105.0 | 0.60 | 1.60 |
| 70–30 Leaded Brass | 1250°F-1 h A | 11.0 | 45.0 | 105.0 | 0.50 | 1.10 |
| Naval Brass[d] | 1350°F-1/2 h A | 17.0 | 54.5 | 125.0 | 0.58 | 1.00 |
| Naval Brass[d] | 1350°F-1/2 h Wq | 27.0 | 66.2 | 135.0 | 0.37 | 0.50 |
| Naval Brass[d] | 850°F-1/2 h A | 17.5 | 56.0 | 125.0 | 0.48 | 0.90 |
| Naval Brass[d] | 850°F-1/2 h WQ | 31.5 | 64.5 | 135.0 | 0.37 | 0.80 |
| Naval Brass[d] | 1500°F-3 h A | 11.0 | 48.0 | — | — | 0.74 |
| *Nickel Alloys*[e] | | | | | | |
| Nickel 200 | 1700°F-15 min WQ | 16.2 | 72.0 | 150.0 | 0.375 | 1.805 |
| Nickel 99.44%[e] | CD + A | 20.5 | 73.7 | 160.0 | 0.40 | 1.47 |
| Monel 400 | 1700°F-15 min WQ | 26.5 | 77.7 | 157.0 | 0.337 | 1.184 |
| Monel K500 | 1700°F-15 min WQ | 34.4 | 92.6 | 182.0 | 0.32 | 1.305 |
| Inconel 600 | 1700°F-15 min WQ | 46.6 | 102.5 | 201.0 | 0.3315 | 1.14 |

Table (*Continued*)

| Material | Condition | Strength Yield (ksi) | Tensile (ksi) | $\sigma_o$ (ksi) | m | $\varepsilon_f$ |
|---|---|---|---|---|---|---|
| Inconel 625 | 1700°F-30 min WQ | 77.1 | 139.7 | 297.0 | 0.395 | 0.75 |
| Inconel 718 | 1750°F-20 min WQ | 43.6 | 99.4 | 205.0 | 0.363 | 1.337 |
| Inconel 750 | 2050°F-45 min WQ | 36.4 | 106.5 | 230.0 | 0.415 | 1.27 |
| Incoloy 800 | 2050°F-2 h AC | 22.2 | 77.1 | 169.0 | 0.420 | 1.262 |
| Incoloy 825 | 1700°F-20 min WQ | 66.7 | 138.0 | 283.0 | 0.353 | 0.715 |
| Ni + 2% Be | 1800°F Sol. T. WQ | 41.0 | 104.0 | 222.0 | 0.39 | 1.00 |
| Ni + 2% Be | As above + 1070°F-2 h age | 140.0 | 195.0 | 300.0 | 0.15 | 0.18 |
| Ni + 15.8Cr + 7.2Fe | A | 36.0 | 90.0 | 203.0 | 0.45 | 0.92 |
| *Special alloys* | | | | | | |
| Cobalt Alloy[f] | 2250°F solution HT | 65.0 | 129.0 | 300.0 | 0.50 | 0.51 |
| As above | As above-trans[c] | 65.0 | 129.0 | 300.0 | 0.50 | 0.40 |
| Cobalt Alloy[e,g] | As rec'd (Ann'd) | 62.8 | 119.5 | 283.0 | 0.52 | 0.75 |
| As above | Machined + 2250°F sol HT | 48.0 | 112.5 | 283.0 | 0.62 | 0.70 |
| As above | 2250°F sol HT +925°F aged | 48.0 | 107.5 | 270.0 | 0.63 | 1.00 |
| Molybdenum | Extr'd A | 49.5 | 70.7 | 106.0 | 0.12 | 0.38 |
| Vanadium | A | 45.0 | 63.0 | 97.0 | 0.17 | 1.10 |

[a] All values are for longitudinal specimens except as noted. These are values obtained from only 1 or 2 different heats. The values will vary from heat to heat because of differences in composition. $\varepsilon_f$ may vary by 100%.
[b] 3/4 in. dia. bar.
[c] Tensile specimen machined from 4 in. dia. bar transverse to rolling direction.
[d] Specimens cut from 1/2 in. hot-rolled plate.
[e] 1/2 in. dia. Bar.
[f] HS25 or L605 alloy. 50 Co 20 Cr 15 W 10 Ni 3 Fe.
[g] Elgiloy: 50 Co 20 Cr 15 Ni 7 Mo 15 Fe.
*Source*: Datsko, Joseph. *Materials Selection for Design and Manufacturing* (Marcel Dekker Inc., 1997).

## Mechanical properties of some plastics

| Material | Strength Tensile (ksi) | Compression (ksi) | Tensile modulus ($10^5$ psi) | Elongation (%) | Rockwell hardness | Izod (ft lb/in.) |
|---|---|---|---|---|---|---|
| ABS | | | | | | |
| Medium impact | 6–8 | 10–12 | 3–4 | 15–25 | $R_R$ 108–115 | 2–5 |
| Very high impact | 4–6 | 8–11 | 2–3 | 15–40 | $R_R$ 85–105 | 7–8 |
| Acrylic | | | | | | |
| Cast | 6–12 | 11–19 | 3–5 | 5 | $R_M$ 80–102 | 0.5 |
| Molding grade | 9–11 | 12–19 | 3–5 | 5 | $R_M$ 85–95 | 0.4 |
| Epoxy | | | | | | |
| Cast rigid | 9–15 | 15–35 | 4–5 | 5 | $R_M$ 106 | |
| Molded | 8–20 | 20–40 | 15–25 | 5 | $R_B$ 75–80 | 1.4 |
| Glass cloth laminate | 30–40 | 30–60 | 30–39 | 3 | $R_M$ 115–117 | 10–30 |
| Filament-wound comp | 130–200 | 40–175 | 60–75 | 3 | $R_M$ 98–120 | 10–30 |
| Fluorocarbon | | | | | | |
| PTFE | 2–7 | 1–2 | 1 | 350 | $R_J$ 79085 | 3–6 |
| PVF | 5–7 | 8–9 | 1 | 100–300 | — | 3–4 |
| Nylons | | | | | | |
| Nylon 6 | 9–12 | 6–13 | 4 | 150 | $R_R$ 118 120 | 1–6 |
| Nylon 6/6 | 12 | 6–12 | 4–5 | 60–300 | $R_R$ 118 | 1–2 |
| Phenolic | | | | | | |
| Mineral/glass fiber | 5–12 | 30–40 | 1–2F | 1–2 | $R_E$ 50–90 | 1–9 |
| Shock and heat | 4–9 | 25–30 | 15–25 | 1–2 | $R_E$ 80–90 | 1.6 |
| Polycarbonate | 9–11 | 10–12 | 3–4 | 130 | $R_M$ 70 | 12–18 |
| Polyester | | | | | | |
| Cast rigid | 6–13 | 13–36 | 9–12F | 4 | $R_B$ 45–65 | 0.3 |
| Polyethylene | | | | | | |
| Low density | 1–3 | — | 0.3 | 50–800 | $S_D$ 73 | 20 |
| High density | 3–4 | 3–4 | 1–2F | 50/1000 | $S_D$ 63 | 1–5 |
| Polypropylene | 4–5 | 5–8 | 1.6 | 300 | $S_D$ 72 | 0.4–2 |
| Polystyrene | 5–8 | 11–16 | 4.6 | 1–3 | $R_M$ 72 | 0.6 |
| PVC | 1–4 | 1–2 | 0.03 | 350 | $S_A$ 50–100 | — |

*Source*: DatsKo, Joseph. *Materials Selection for Design and Manufacturing.* (Marcel Dekker Inc., 1997.)

# WEBSITES

*General engineering design websites*

| Name | Site | Specialty |
|---|---|---|
| Engineering Fundamentals | www.efunda.com | Superb resource for engineering knowledge and design information |
| ICrank | www.icrank.com | An excellent mechanical engineering oriented website |
| The Engineers Edge | www.engineersedge.com | A design, engineering and manufacturing database |
| The Engineering Tool Box | www.engineeringtool box.com | Resources, tools and basic information for engineering and design of technical applications |
| Machine Design | www.machinedesign.com | Excellent resource for new products, engineering knowledge and industry trends |
| HyperPhysics | hyperphysics.phy-astr. gsu.edu | Excellent website for engineering and physics based knowledge |
| Thomas Register of American Manufactures | www1.thomasregister.com | Exhaustive listing of American manufactures |
| United States Patent and Trademark Office | www.uspto.gov | Excellent resource to research patents |
| Digital Dutch | www.digitaldutch.com/ unitconverter | Free, easy to use unit converter |

*Ferrous materials websites*

| Name | Site | Specialty |
|---|---|---|
| MatWeb.com | www.matls.com | The best free material properties website |
| ASM International | www.asminternational.org | ASM International provides full online access to their comprehensive materials handbooks for members |
| Granta Design, Ltd. | www.grantadesign.com | Home to MatData.com, a source of material property data |
| Carpenter | www.cartech.com | Ferrous metals manufacture |
| Ryerson Tull | www.ryersontull.com | Materials distributor |

*Nonferrous materials websites*

| Name | Site | Specialty |
| --- | --- | --- |
| MatWeb.com | www.matls.com | The best free material properties website |
| ASM International | www.asminternational.org | ASM International provides full online access to their comprehensive materials handbooks for members |
| Granta Design, Ltd. | www.grantadesign.com | Home to MatData.com, a source of material property data |
| Alcoa, Inc. | www.alcoa.com | Aluminum manufacture |
| Ryerson Tull | www.ryersontull.com | Materials distributor |
| Copper and Brass Sales | www.copperandbrass.com | Materials distributor |

*Plastic materials websites*

| Name | Site | Specialty |
| --- | --- | --- |
| MatWeb.com | www.matls.com | The best free material properties website |
| ASM International | www.asminternational.org | ASM International provides full online access to their comprehensive materials handbooks for members |
| Granta Design, Ltd. | www.grantadesign.com | Home to MatData.com, a source of material property data |
| GE Plastics | www.geplastics.com | Materials manufacture |
| Bayer Plastics | www.bayer.com | Materials manufacture |
| Ticona Plastics | www.ticona.com | Materials manufacture |
| Port Plastics | www.portplastics.com | Materials distributor |

*Adhesive materials websites*

| Name | Site | Specialty |
| --- | --- | --- |
| Loctite | www.loctite.com | Adhesive manufacture |
| Masterbond | www.masterbond.com | Adhesive manufacture |
| 3M Company | www.3m.com/us/ mfg_industrial/adhesives | Adhesive manufacture |
| 3M Company | www.mmm.com/US/ mfg_industrial/indtape | Adhesive tape manufacture |
| Dow Corning | www.dowcorning.com | Adhesives manufacture. |
| Ellsworth Adhesives | www.ellsworth.com | Adhesives distributor |

*Engineering organizations websites*

| Name | Site | Specialty |
|------|------|-----------|
| National Society of Professional Engineers | www.nspe.org | Organization for Professional Engineers |
| American Society of Mechanical Engineers | www.asme.org | Mechanical Engineering |
| Institute of Electrical and Electronics Engineers | www.ieee.org | Electrical Engineering |

*Engineering specifications websites*

| Name | Site | Specialty |
|------|------|-----------|
| American National Standards Institute | www.ansi.org | Private standards organization |
| American Society of Mechanical Engineers | www.asme.org | Engineering standards organization |
| | www.bissc.org | Sanitation standards for the design and construction of bakery equipment and machinery |
| ASTM International | www.astm.org | Engineering standards organization |
| Acquisitions Streamlining and Standards Information System | assist.daps.dla.mil | United States government sponsored website providing free access to military (MIL) and handbooks (HDBK) specifications |

# Index